DUXBURY

Statistics
with STATA

Updated for Version 9

Lawrence C. Hamilton

University of New Hampshire

THOMSON

™

BROOKS/COLE

Australia • Brazil • Canada • Mexico • Singapore • Spain
United Kingdom • United States

Statistics with Stata: Updated for Version 9
Lawrence C. Hamilton

Publisher: Curt Hinrichs
Senior Assistant Editor: Ann Day
Editorial Assistant: Daniel Geller
Technology Project Manager: Fiona Chong
Marketing Manager: Joe Rogove
Marketing Assistant: Brian Smith
Executive Marketing Communications Manager: Darlene Amidon-Brent

Project Manager, Editorial Production: Kelsey McGee
Creative Director: Rob Hugel
Art Director: Lee Friedman
Print Buyer: Darlene Suruki
Permissions Editor: Kiely Sisk
Cover Designer: Denise Davidson/Simple Design
Cover Image: ©Imtek Imagineering/Masterfile
Cover Printing, Printing & Binding: Webcom

Printed in Canada
2 3 4 5 6 7 09 08 07

For more information about our products, contact us at:
Thomson Learning Academic Resource Center
1-800-423-0563
For permission to use material from this text or product, submit a request online at
http://www.thomsonrights.com.
Any additional questions about permissions can be submitted by email to
thomsonrights@thomson.com.

Library of Congress Control Number: 2005933358
ISBN- 13: 978-0-495-10972-3
ISBN- 10: 0-495-10972-X

Thomson Higher Education
10 Davis Drive
Belmont, CA 94002-3098
USA

Asia (including India)
Thomson Learning
5 Shenton Way
#01-01 UIC Building
Singapore 068808

Australia/New Zealand
Thomson Learning Australia
102 Dodds Street
Southbank, Victoria 3006
Australia

Canada
Thomson Nelson
1120 Birchmount Road
Toronto, Ontario M1K 5G4
Canada

UK/Europe/Middle East/Africa
Thomson Learning
High Holborn House
50–51 Bedford Row
London WC1R 4LR
United Kingdom

Latin America
Thomson Learning
Seneca, 53
Colonia Polanco
11560 Mexico
D.F. Mexico

Contents

Preface ix

1 Stata and Stata Resources **1**
A Typographical Note 1
An Example Stata Session 2
Stata's Documentation and Help Files 7
Searching for Information 8
Stata Corporation 9
Statalist 10
The *Stata Journal* 10
Books Using Stata 11

2 Data Management **12**
Example Commands 13
Creating a New Dataset 15
Specifying Subsets of the Data: in and if Qualifiers 19
Generating and Replacing Variables 23
Using Functions 27
Converting between Numeric and String Formats 32
Creating New Categorical and Ordinal Variables 35
Using Explicit Subscripts with Variables 38
Importing Data from Other Programs 39
Combining Two or More Stata Files 42
Transposing, Reshaping, or Collapsing Data 47
Weighting Observations 54
Creating Random Data and Random Samples 56
Writing Programs for Data Management 60
Managing Memory 61

3 Graphs **64**
Example Commands 65
Histograms 67
Scatterplots 72
Line Plots 77
Connected-Line Plots 83
Other Twoway Plot Types 84
Box Plots 90
Pie Charts 92
Bar Charts 94

Dot Plots 99
Symmetry and Quantile Plots 100
Quality Control Graphs 105
Adding Text to Graphs 109
Overlaying Multiple Twoway Plots 110
Graphing with Do-Files 115
Retrieving and Combining Graphs 116

4   Summary Statistics and Tables **120**
Example Commands 120
Summary Statistics for Measurement Variables 121
Exploratory Data Analysis 124
Normality Tests and Transformations 126
Frequency Tables and Two-Way Cross-Tabulations 130
Multiple Tables and Multi-Way Cross-Tabulations 133
Tables of Means, Medians, and Other Summary Statistics 136
Using Frequency Weights 138

5   ANOVA and Other Comparison Methods **141**
Example Commands 142
One-Sample Tests 143
Two-Sample Tests 146
One-Way Analysis of Variance (ANOVA) 149
Two- and N-Way Analysis of Variance 152
Analysis of Covariance (ANCOVA) 153
Predicted Values and Error-Bar Charts 155

6   Linear Regression Analysis **159**
Example Commands 159
The Regression Table 162
Multiple Regression 164
Predicted Values and Residuals 165
Basic Graphs for Regression 168
Correlations 171
Hypothesis Tests 175
Dummy Variables 176
Automatic Categorical-Variable Indicators and Interactions 183
Stepwise Regression 186
Polynomial Regression 188
Panel Data 191

7 Regression Diagnostics **196**
   Example Commands 196
   SAT Score Regression, Revisited 198
   Diagnostic Plots 200
   Diagnostic Case Statistics 205
   Multicollinearity 210

8 Fitting Curves **215**
   Example Commands 215
   Band Regression 217
   Lowess Smoothing 219
   Regression with Transformed Variables – 1 223
   Regression with Transformed Variables – 2 227
   Conditional Effect Plots 230
   Nonlinear Regression – 1 232
   Nonlinear Regression – 2 235

9 Robust Regression **239**
   Example Commands 240
   Regression with Ideal Data 240
   Y Outliers 243
   X Outliers (Leverage) 246
   Asymmetrical Error Distributions 248
   Robust Analysis of Variance 249
   Further `rreg` and `qreg` Applications 255
   Robust Estimates of Variance — 1 256
   Robust Estimates of Variance — 2 258

10 Logistic Regression **262**
   Example Commands 263
   Space Shuttle Data 265
   Using Logistic Regression 270
   Conditional Effect Plots 273
   Diagnostic Statistics and Plots 274
   Logistic Regression with Ordered-Category y 278
   Multinomial Logistic Regression 280

11 Survival and Event-Count Models **288**
   Example Commands 289
   Survival-Time Data 291
   Count-Time Data 293
   Kaplan–Meier Survivor Functions 295
   Cox Proportional Hazard Models 299
   Exponential and Weibull Regression 305
   Poisson Regression 309
   Generalized Linear Models 313

12 Principal Components, Factor, and Cluster Analysis **318**

    Example Commands 319
    Principal Components 320
    Rotation 322
    Factor Scores 323
    Principal Factoring 326
    Maximum-Likelihood Factoring 327
    Cluster Analysis — 1 329
    Cluster Analysis — 2 333

13 Time Series Analysis **339**

    Example Commands 339
    Smoothing 341
    Further Time Plot Examples 346
    Lags, Leads, and Differences 349
    Correlograms 351
    ARIMA Models 354

14 Introduction to Programming **361**

    Basic Concepts and Tools 361
    Example Program: Moving Autocorrelation 369
    Ado-File 373
    Help File 375
    Matrix Algebra 378
    Bootstrapping 382
    Monte Carlo Simulation 387

References **395**

Index **401**

Statistics with Stata is intended for students and practicing researchers, to bridge the gap between statistical textbooks and Stata's own documentation. In this intermediate role, it does not provide the detailed expositions of a proper textbook, nor does it come close to describing all of Stata's features. Instead, it demonstrates how to use Stata to accomplish a wide variety of statistical tasks. Chapter topics follow conceptual themes rather than focusing on particular Stata commands, which gives *Statistics with Stata* a different structure from the Stata reference manuals. The chapter on Data Management, for example, covers a variety of procedures for creating, updating, and restructuring data files. Chapters on Summary Statistics and Tables, ANOVA and Other Comparison Methods, and Fitting Curves, among others, have similarly broad themes that encompass a number of separate techniques.

The general topics of the first six chapters (through ordinary least squares regression) roughly parallel an introductory course in applied statistics, but with additional depth to cover practical issues often encountered by analysts — how to aggregate data, create dummy variables, draw publication-quality graphs, or translate ANOVA into regression, for instance. In Chapter 7 (Regression Diagnostics) and beyond, we move into the territory of advanced courses or original research. Here readers can find basic information and illustrations of how to obtain and interpret diagnostic statistics and graphs; perform robust, quantile, nonlinear, logit, ordered logit, multinomial logit, or Poisson regression; fit survival-time and event-count models; construct composite variables through factor analysis and principal components; divide observations into empirical types or clusters; and graph or model time-series data. Stata has worked hard in recent years to advance its state-of-the-art standing, and this effort is particularly apparent in the wide range of regression and model-fitting commands it now offers.

Finally, we conclude with a look at programming in Stata. Many readers will find that Stata does everything they need already, so they have no need to write original programs. For an active minority, however, programmability is one of Stata's principal attractions, and it certainly underlies Stata's currency and rapid advancement. This chapter opens the door for new users to explore Stata programming, whether for specialized data management tasks, to establish a new statistical capability, for Monte Carlo experiments, or for teaching.

Generally similar versions ("flavors") of Stata run on Windows, Macintosh, and Unix computers. Across all platforms, Stata uses the same commands, data files, and output. The versions differ in some details of screen appearance, menus, and file handling, where Stata follows the conventions native to each platform — such as \directory\filename file specifications under Windows, in contrast with the /directory/filename specifications under Unix. Rather than display all three, I employ Windows conventions, but users with other systems should find that only minor translations are needed.

Notes on the Fifth Edition

I began using Stata in 1985, the first year of its release. (Stata's 20th anniversary in 2005 was marked by a special issue of the *Stata Journal*, 2005:5(1), filled with historical articles and interviews including a brief history of *Statistics with Stata.*) Initially, Stata ran only on MS-DOS personal computers, but its PC orientation made it distinctly more modern than its main competitors — most of which had originated before the desktop revolution, in the 80-column punched-card Fortran environment of mainframes. Unlike mainframe statistical packages that believed each user was a stack of cards, Stata viewed the user as a conversation. Its interactive nature and integration of statistical procedures with data management and graphics supported the natural flow of analytical thought in ways that other programs did not. `graph` and `predict` soon became favorite commands. I was impressed enough by how it all fit together to start writing the original *Statistics with Stata*, published in 1989 for Stata version 2.

A great deal about Stata has changed since that book, in which I observed that "Stata is not a do-everything program . . . . The things it does, however, it does very well." The expansion of Stata's capabilities has been striking. This is very noticeable in the proliferation, and later in the steady rationalization, of model fitting procedures. William Gould's architecture for Stata, with its programming tools and unified syntax, has aged well and proven able to incorporate new statistical ideas as these were developed. The formidable list of modeling commands that begins Chapter 10, or of functions in Chapter 2, illustrate some of the ways that Stata became richer over the years. Suites of new techniques such as those for panel (`xt`), survey (`svy`), time series (`ts`), or survival time (`st`) data open worlds of possibility, as do programmable commands for nonlinear regression (`nl`) and general linear modeling (`glm`), or general procedures for maximum-likelihood estimation. Other critical expansions include the development of a matrix programming capability, and the wealth of new data-management features. Data management, with good reason, has been promoted from an incidental topic in the first *Statistics with Stata* to the second-longest chapter in this fifth edition.

Stata version 8 marked the most radical upgrade in Stata's history, led by the new menu system or GUI (graphical user interface), and completely redesigned graphing capabilities. A limited menu system, evolved from the student program StataQuest, had been available as an option since version 4, but Stata 8 for the first time incorporated an integrated menu interface offering a full range of alternatives to typed commands. These menus are more easily learned through exploration than by reading a book, so *Statistics with Stata* provides only general suggestions about menus at the beginning of each chapter. For the most part, this book employs commands to show what Stata can do; those commands' menu counterparts should be easy to discover.

The redesigned graphing capabilities of Stata 8 called for similarly sweeping changes in Chapter 3, turning it into the longest chapter in this edition. The topic itself is complex, as the thick *Graphics Reference Manual* (and other material scattered through the documentation) attests. Rather than try to condense the syntax-based reference manuals, I have taken a completely different, complementary approach based on examples. Chapter 3 thus provides an organized gallery of 49 diverse graphs, each with instructions for how it was drawn. Further examples appear throughout the book; even the last graphs in Chapter 14 demonstrate new variations. To an unexpected degree, *Statistics with Stata* became a showcase for the new graphics.

Less drastic but also noteworthy changes from the previous *Statistics with Stata* include new sections on panel data (Chapter 6), robust standard errors (Chapter 9), and cluster analysis (Chapter 12). The Example Commands sections have been revised and expanded in several chapters as well. Readers coming from the older *Statistics with Stata 5* will find more here that has changed, including new emphasis on Internet resources (Chapter 1), tools for data management (Chapter 2), table commands (Chapter 4), graphs for ANOVA (Chapter 5), a sharper look at multicollinearity (Chapter 8), robust and median-based analogues to ANOVA (Chapter 9), conditional effects plots for multinomial logit (Chapter 10), generalized linear models (Chapter 11), a new chapter on time series (Chapter 13), and a rewritten chapter on programming (Chapter 14). Other new Stata features, or improvements in old commands (including `graph` and `predict`), are scattered throughout the book. Because Stata now does so much, far beyond the scope of an introductory book, *Statistics with Stata* presents more procedures telegraphically in the "Example Commands" sections that begin most chapters, or in lists of options followed by advice that readers should "type `help whatever`" for details. Stata's online help and search features have advanced to keep pace with the program, so this is not idle advice.

Beyond the help files are Stata's web site, Internet and documentation search capabilities, user-community listserver, NetCourses, the *Stata Journal*, and of course Stata's formidable printed documentation — presently over 5,000 pages and growing. *Statistics with Stata* provides an accessible gateway to Stata; these other resources can help you go further.

Acknowledgments

Stata's architect, William Gould, deserves credit for originating the elegant program that *Statistics with Stata* describes. My editor, Curt Hinrichs, backed by some reviewers and persistent readers, persuaded me to begin the effort of writing a new book. Pat Branton at Stata Corporation, along with many others including Shannon Driver and Lisa Gilmore, gave invaluable feedback, often on short deadlines, once this effort got underway. It would not have been possible without their assistance. Alen Riley and Vince Wiggins responded quickly to my questions about graphics. James Hamilton contributed advice about time series. Leslie Hamilton proofread the final manuscript.

The book is built around data. To avoid endlessly recycling my own old examples, I drew on work by others including Carole Seyfrit, Sally Ward, Heather Turner, Rasmus Ole Rasmussen, Erich Buch, Paul Mayewski, Loren D. Meeker, and Dave Hamilton. Steve Selvin shared several of his examples from *Practical Biostatistical Methods* for Chapter 11. Data from agencies including Statistics Iceland, Statistics Greenland, Statistics Canada, the Northwest Atlantic Fisheries Organization, Greenland's Natural Resources Institute, and the Department of Fisheries and Oceans Canada can also be found among the examples. A presentation given by Brenda Topliss inspired the "gossip" programming example in Chapter 14. Other forms of encouragement or ideas came from Anna Kerttula, Richard Haedrich, Jeffrey Runge, Igor Belkin, James Morison, Oddmund Otterstad, James Tucker, and Cynthia M. Duncan.

To Leslie, Sarah, and Dave.

Stata and Stata Resources

Stata is a full-featured statistical program for Windows, Macintosh, and Unix computers. It combines ease of use with speed, a library of pre-programmed analytical and data-management capabilities, and programmability that allows users to invent and add further capabilities as needed. Most operations can be accomplished either via the pull-down menu system, or more directly via typed commands. Menus help newcomers to learn Stata, and help anyone to apply an unfamiliar procedure. The consistent, intuitive syntax of Stata commands frees experienced users to work more efficiently, and also makes it straightforward to develop programs for complex or repetitious tasks. Menu and command instructions can be mixed as needed during a Stata session. Extensive help, search, and link features make it easy to look up command syntax and other information instantly, on the fly.

After introductory information, we'll begin with an example Stata session to give you a sense of the "flow" of data analysis, and of how analytical results might be used. Later chapters explain in more detail. Even without explanations, however, you can see how straightforward the commands are — **use *filename*** to retrieve dataset *filename*, **summarize** when you want summary statistics, **correlate** to get a correlation matrix, and so forth. Alternatively, the same results can be obtained by making choices from the Data or Statistics menus.

Stata users have available a variety of resources to help them learn about Stata and solve problems at any level of difficulty. These resources come not just from Stata Corporation, but also from an active community of users. Sections of this chapter introduce some key resources — Stata's online help and printed documentation; where to phone, fax, write, or e-mail for technical help; Stata's web site (www.stata.com), which provides many services including updates and answers to frequently asked questions; the Statalist Internet forum; and the refereed *Stata Journal*.

A Typographical Note

This book employs several typographical conventions as a visual cue to how words are used:

- Commands typed by the user appear in a **bold Courier font**. When the whole command line is given, it starts with a period, as seen in a Stata Results window or log (output) file:

  ```
  . list year boats men penalty
  ```

- ***Variable*** or ***file*** names within these commands appear in italics to emphasize the fact that they are arbitrary and not a fixed part of the command.

- Names of *variables* or *files* also appear in italics within the main text to distinguish them from ordinary words.

- Items from Stata's menus are shown in an Arial font , with successive options separated by a dash. For example, we can open an existing dataset by selecting File – Open , and then finding and clicking on the name of the particular dataset. Note that some common menu actions can be accomplished either with text choices from Stata's top menu bar,

  File   Edit   Prefs   Data   Graphics   Statistics   User   Window   Help

  or with the row of icons below these. For example, selecting File – Open is equivalent to clicking the leftmost icon, an opening file folder: [icon] . One could also accomplish the same thing by typing a direct command of the form

  `. use filename`

- Stata output as seen in the Results window is shown in a `small Courier font`. The small font allows Stata's 80-column output to fit within the margins of this book.

Thus, we show the calculation of summary statistics for a variable named *penalty* as follows:

```
. summarize penalty

Variable |    Obs       Mean   Std. Dev.      Min        Max
---------+-----------------------------------------------------
 penalty |     10         63   59.59493       11        183
```

These typographic conventions exist only in this book, and not within the Stata program itself. Stata can display a variety of onscreen fonts, but it does not use italics in commands. Once Stata log files have been imported into a word processor, or a results table copied and pasted, you might want to format them in a Courier font, 10 point or smaller, so that columns will line up correctly.

In its commands and variable names, Stata is case sensitive. Thus, **summarize** is a command, but Summarize and SUMMARIZE are not. *Penalty* and *penalty* would be two different variables.

An Example Stata Session

As a preview showing Stata at work, this section retrieves and analyzes a previously-created dataset named *lofoten.dta*. Jentoft and Kristofferson (1989) originally published these data in an article about self-management among fishermen on Norway's arctic Lofoten Islands. There are 10 observations (years) and 5 variables, including *penalty*, a count of how many fishermen were cited each year for violating fisheries regulations.

If we might eventually want a record of our session, the best way to prepare for this is by opening a "log file" at the start. Log files contain commands and results tables, but not graphs. To begin a log file, click the scroll-shaped Begin Log icon, [icon] , and specify a name and folder for the resulting log file. Alternatively, a log file could be started by choosing File – Log – Begin from the top menu bar, or by typing a direct command such as

`. log using monday1`

Multiple ways of doing such things are common in Stata. Each has its own advantages, and each suits different situations or user tastes.

Log files can be created either in a special Stata format (.smcl), or in ordinary text or ASCII format (.log). A .smcl ("Stata markup and control language") file will be nicely formatted for viewing or printing within Stata. It could also contain hyperlinks that help to understand commands or error messages. .log (text) files lack such formatting, but are simpler to use if you plan later to insert or edit the output in a word processor. After selecting which type of log file you want, click Save . For this session, we will create a .smcl log file named *monday1.smcl*.

An existing Stata-format dataset named *lofoten.dta* will be analyzed here. To open or retrieve this dataset, we again have several options:

select File – Open – *lofoten.dta* using the top menu bar;

select 📂 – *lofoten.dta* ; or

type the command **use *lofoten*** .

Under its default Windows configuration, Stata looks for data files in folder C:\data. If the file we want is in a different folder, we could specify its location in the **use** command,

. **use *c:\books\sws8\chapter01\lofoten***

or change the session's default folder by issuing a **cd** (change directory) command:

. **cd *c:\books\sws8\chapter01***

. **use *lofoten***

Often, the simplest way to retrieve a file will be to choose File – Open and browse through folders in the usual way.

To see a brief description of the dataset now in memory, type

. **describe**

```
Contains data from C:\data\lofoten.dta
  obs:            10                          Jentoft & Kristoffersen '89
  vars:            5                          30 Jun 2005 10:36
  size:          130 (99.9% of memory free)
-------------------------------------------------------------------------------
              storage  display    value
variable name   type    format    label      variable label
-------------------------------------------------------------------------------
year            int     %9.0g                 Year
boats           int     %9.0g                 Number of fishing boats
men             int     %9.0g                 Number of fishermen
penalty         int     %9.0g                 Number of penalties
decade          byte    %9.0g     decade      Early 1970s or early 1980s
-------------------------------------------------------------------------------
Sorted by:  decade  year
```

Many Stata commands can be abbreviated to their first few letters. For example, we could shorten **describe** to just the letter **d** . Using menus, the same table could be obtained by choosing Data – Describe data – Describe variables in memory – OK.

This dataset has only 10 observations and 5 variables, so we can easily list its contents by typing the command **list** (or the letter **l** ; or Data – Describe data – List data – OK):

```
. list
```

```
        +----------------------------------------------+
        | year    boats    men    penalty    decade |
        |----------------------------------------------|
  1.    | 1971    1809    5281         71    1970s  |
  2.    | 1972    2017    6304        152    1970s  |
  3.    | 1973    2068    6794        183    1970s  |
  4.    | 1974    1693    5227         39    1970s  |
  5.    | 1975    1441    4077         36    1970s  |
        |----------------------------------------------|
  6.    | 1981    1540    4033         11    1980s  |
  7.    | 1982    1689    4267         15    1980s  |
  8.    | 1983    1842    4430         34    1980s  |
  9.    | 1984    1847    4622         74    1980s  |
 10.    | 1985    1365    3514         15    1980s  |
        +----------------------------------------------+
```

Analysis could begin with a table of means, standard deviations, minimum values, and maximum values (type **summarize** or **su**; or select Statistics – Summaries, tables, & tests – Summary statistics – Summary statistics – OK):

```
. summarize
```

Variable	Obs	Mean	Std. Dev.	Min	Max
year	10	1978	5.477226	1971	1985
boats	10	1731.1	232.1328	1365	2068
men	10	4854.9	1045.577	3514	6794
penalty	10	63	59.59493	11	183
decade	10	.5	.5270463	0	1

To print results from the session so far, bring the Results window to the front by clicking on this window or on ▦ (Bring Results Window to Front), and then click 🖶 (Print).

To copy a table, commands, or other information from the Results window into a word processor, again make sure that the Results window is in front by clicking on this window or on ▦. Drag the mouse to select the results you want, right-click the mouse, and then choose Copy Text from the mouse's menu. Finally, switch to your word processor and, at the desired insertion point either right-click and Paste or click a "clipboard" icon on the word processor's menu bar.

Did the number of penalties for fishing violations change over the two decades covered by these data? A table containing summary statistics for *penalty* at each value of *decade* shows that there were more penalties in the 1970s:

```
. tabulate decade, sum(penalty)
```

Early 1970s or early 1980s	Summary of Number of penalties Mean	Std. Dev.	Freq.
1970s	96.2	67.41439	5
1980s	29.8	26.281172	5
Total	63	59.594929	10

The same table could be obtained through menus: Statistics – Summaries, tables, & tests – Tables – One/two-way table of summary statistics, then fill in *decade* as variable 1, and *penalty* as the variable to be summarized. Although menu choices are often straightforward to use, you can see that they tend to be more complicated to describe than the simple text commands. From this point on, we will focus primarily on the commands, mentioning menu alternatives only occasionally. Fully exploring the menus, and working out how to use them to accomplish the same tasks, will be left to the reader. For similar reasons, the Stata reference manuals likewise take a command-based approach.

Perhaps the number of penalties declined because fewer people were fishing in the 1980s. The number of penalties correlates strongly ($r > .8$) with the number of boats and fishermen:

```
. correlate boats men penalty
(obs=10)

             |    boats      men  penalty
-------------+---------------------------
       boats |   1.0000
         men |   0.8748   1.0000
     penalty |   0.8259   0.9312   1.0000
```

A graph might help clarify these interrelationships. Figure 1.1 plots *men* and *penalty* against *year*, produced by the **graph twoway connected** command. In this example, we first ask for a twoway (two-variable) connected-line plot of *men* against *year*, using the left-hand *y* axis, **yaxis(1)**. After the separator **||** , we next ask for a connected-line plot of *penalty* against *year*, this time using the right-hand *y* axis, **yaxis(2)** . The resulting graph visualizes the correspondence between the number of fishermen and the number of penalties over time.

```
. graph twoway connected men year, yaxis(1)
      || connected penalty year, yaxis(2)
```

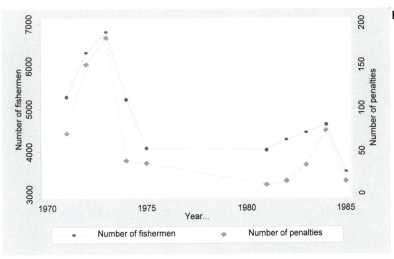

Figure 1.1

Because the years 1976 to 1980 are missing in these data, Figure 1.1 shows 1975 connected to 1981. For some purposes, we might hesitate to do this. Instead, we could either find the missing values or leave the gap unconnected by issuing a slightly more complicated set of commands.

To print this graph, click on the Graph window or on ▨ (Bring Graph Window to Front), and then click the Print icon 🖨 .

To copy the graph directly into a word processor or other document, bring the Graph window to the front, right-click on the graph, and select Copy . Switch to your word processor, go to the desired insertion point, and issue an appropriate "paste" command such as Edit – Paste, Edit – Paste Special (Metafile), or click a "clipboard" icon (different word processors will handle this differently).

To save the graph for future use, either right-click and Save, or select File – Save Graph from the top menu bar. The Save As Type submenu offers several different file formats to chose from. On a Windows system, the choices include

Stata graph (*.gph)   (A "live" graph, containing enough information for Stata to edit.)

As-is graph (*.gph)   (A more compact Stata graph format.)

Windows Metafile (*.wmf)

Enhanced Metafile (*.emf)

Portable Network Graphics (*.png)

TIFF (*.tif)

PostScript (*.ps)

Encapsulated PostScript with TIFF preview (*.eps)

Encapsulated PostScript (*.eps)

Regardless of which graphics format we want, it might be worthwhile also to save a copy of our graph in "live" .gph format. Live .gph graphs can later be retrieved, combined, recolored, or reformatted using the **graph use** or **graph combine** commands (Chapter 3).

Instead of using menus, graphs can be saved by adding a **saving(*filename*)** option to any **graph** command. To save a graph with the filename *figure1.gph*, add another separator **||**, a comma, and **saving(*figure1*)** . Chapter 3 explains more about the logic of **graph** commands. The complete command now contains the following (typed in the Stata Command window with as many spaces as you want, but no hard returns):

```
. graph twoway connected men year, yaxis(1)
     ||   connected penalty year, yaxis(2)
     ||   , saving(figure1)
```

Through all of the preceding analyses, the log file *monday1.smcl* has been storing our results. There are several possible ways to review this file to see what we have done:

File – Log – View – OK

▨ – View snapshot of log file – OK

typing the command **view *monday1.smcl***

We could print the log file by choosing 🖨 (Print) . Log files close automatically at the end of a Stata session, or earlier if instructed by one of the following:

    File – Log – Close

    ◇▨ – Close log file – OK

    typing the command **log close**

Once closed, the file *monday1.smcl* could be opened again through File – View during a subsequent Stata session. To make an output file that can be opened easily by your word processor, either translate the log file from .smcl (a Stata format) to .log (standard ASCII text format) by typing

. **translate** *monday1.smcl monday1.log*

or start out by creating the file in .log instead of .smcl format.

Stata's Documentation and Help Files

The complete Stata 9 Documentation Set includes over 6,000 pages in 15 volumes: a slim *Getting Started* manual (for example, *Getting Started with Stata for Windows*), the more extensive *User's Guide*, the encyclopedic three-volume *Base Reference Manual*, and separate reference manuals on data management, graphics, longitudinal and panel data, matrix programming (Mata), multivariate statistics, programming, survey data, survival analysis and epidemiological tables, and time series analysis. *Getting Started* helps you do just that, with the basics of installation, window management, data entry, printing, and so on. The *User's Guide* contains an extended discussion of general topics, including resources and troubleshooting. Of particular note for new users is the *User's Guide* section on "Commands everyone should know." The *Base Reference Manual* lists all Stata commands alphabetically. Entries for each command include the full command syntax, descriptions of all available options, examples, technical notes regarding formulas and rationale, and references for further reading. Data management, graphics, panel data, etc. are covered in the general references, but these complicated topics get more detailed treatment and examples in their own specialized manuals. A *Quick Reference and Index* volume rounds out the whole collection.

    When we are in the midst of a Stata session, it is often simpler to ask for onscreen help instead of consulting the manuals. Selecting Help from the top menu bar invokes a drop-down menu of further choices, including help on specific commands, general topics, online updates, the *Stata Journal*, or connections to Stata's web site (www.stata.com). Alternatively, we can bring the Viewer (👁) to front and use its Search or Contents features to find information. We can also use the **help** command. Typing **help correlate** , for example, causes help information to appear in a Viewer window. Like the reference manuals, onscreen help provides command syntax diagrams and complete lists of options. It also includes some examples, although often less detailed and without the technical discussions found in the manuals. The Viewer help has several advantages over the manuals, however. It can search for keywords in the documentation or on Stata's web site. Hypertext links take you directly to related entries. Onscreen help can also include material about recent updates, or the "unofficial" Stata programs that you have downloaded from Stata's web site or from other users.

Searching for Information

Selecting Help – Search – Search documentation and FAQs provides a direct way to search for information in Stata's documentation or in the web site's FAQs (frequently asked questions) and other pages. The equivalent Stata command is

```
. search keywords
```

Options available with **search** allow us to limit our search to the documentation and FAQs, to net resources including the *Stata Journal*, or to both. For example,

```
. search median regression
```

will search the documentation and FAQs for information indexed by both keywords, "median" and "regression." To search for these keywords across Stata's Internet resources in addition to the documentation and FAQs, type

```
. search median regression, all
```

Search results in the Viewer window contain clickable hyperlinks leading to further information or original citations.

One specialized use for the **search** command is to provide more information on those occasions when our command does not succeed as planned, but instead results in one of Stata's cryptic numerical error messages. For example, typing the one-word command **table** produces the error or "return code" r(100):

```
. table
varlist required
r(100);
```

The **table** command evidently requires a list of variables. Often, however, the meaning of an error message is less obvious. To learn more about what return code r(100) refers to, type

```
. search rc 100

Keyword search

        Keywords:   rc 100
          Search:   (1) Official help files, FAQs, Examples, SJs, and STBs

Search of official help files, FAQs, Examples, SJs, and STBs

[P]     error . . . . . . . . . . . . . . . . . . . . . . Return code 100
        varlist required;
        = exp required;
        using required;
        by() option required;
        Certain commands require a varlist or another element of the
        language.  The message specifies the required item that was
        missing from the command you gave.  See the command's syntax
        diagram.  For example, merge requires using be specified; perhaps,
        you meant to type append.  Or, ranksum requires a by() option;
        see [R] signrank.

(end of search)
```

Type **help search** for more about this command.

Stata Corporation

For orders, licensing, and upgrade information, you can contact Stata Corporation by e-mail at

stata@stata.com

or visit their web site at

http://www.stata.com

Stata's extensive web site contains a wealth of user-support information and links to resources. Stata Press also has its own web site, containing information about Stata publications including the datasets used for examples.

http://www.stata-press.com

Both web sites are well worth exploring.

The mailing or physical address is

Stata Corporation
4905 Lakeway Drive
College Station, TX 77845 USA

Telephone access includes an easy-to-remember 800 number.

telephone:	1-800-STATAPC	U.S.
	(1-800-782-8272)	
	1-800-248-8272	Canada
	1-979-696-4600	International
fax:	1-979-696-4601	

Online updates within major versions are free to licensed Stata users. These provide a fast and simple way to obtain the latest enhancements, bug fixes, etc. for your current version. To find out whether updates exist for your Stata, and initiate the simple online update process itself, type the command

```
. update query
```

Technical support can be obtained by sending e-mail messages *with your Stata serial number in the subject line* to

tech_support@stata.com

Before calling or writing for technical help, though, you might want to look at www.stata.com to see whether your question is a FAQ. The site also provides product, ordering, and help information; international notes; and assorted news and announcements. Much attention is given to user support, including the following:

FAQS — Frequently asked questions and their answers. If you are puzzled by something and can't find the answer in the manuals, check here next — it might be a FAQ. Example questions range from basic — "How can I convert other packages' files to Stata format data files?" — to more technical queries such as "How do I impose the restriction that rho is zero using the heckman command with full ml?"

UPDATES — Frequent minor updates or bug fixes, downloadable at no cost by licensed Stata users.

OTHER RESOURCES — Links and information including online Stata instruction (NetCourses); enhancements from the *Stata Journal*; an independent listserver (Statalist) for discussions among Stata users; a bookstore selling books about Stata and other up-to-date statistical references; downloadable datasets and programs for Stata-related books; and links to statistical web sites including Stata's competitors.

The following sections describe some of the most important user-support resources.

Statalist

Statalist provides a valuable online forum for communication among active Stata users. It is independent of Stata Corporation, although Stata programmers monitor it and often contribute to the discussion. To subscribe to Statalist, send an e-mail message to

```
majordomo@hsphsun2.harvard.edu
```

The body of this message should contain only the following words:

```
subscribe statalist
```

The list processor will acknowledge your message and send instructions for using the list, including how to post messages of your own. Any message sent to the following address goes out to all current subscribers:

```
statalist@hsphsun2.harvard.edu
```

Do *not* try to subscribe or unsubscribe by sending messages directly to the statalist address. This does not work, and your mistake goes to hundreds of subscribers. To unsubscribe from the list, write to the same majordomo address you used to subscribe:

```
majordomo@hsphsun2.harvard.edu
```

but send only the message

```
unsubscribe statalist
```

or send the equivalent message

```
signoff statalist
```

If you plan to be traveling or offline for a while, unsubscribing will keep your mailbox from filling up with Statalist messages. You can always re-subscribe.

Searchable Statalist archives are available at

http://www.stata.com/statalist/archive/

The material on Statalist includes requests for programs, solutions, or advice, as well as answers and general discussion. Along with the *Stata Journal* (discussed below), Statalist plays a major role in extending the capabilities both of Stata and of serious Stata users.

The *Stata Journal*

From 1991 through 2001, a bimonthly publication called the *Stata Technical Bulletin* (*STB*) served as a means of distributing new commands and Stata updates, both user-written and official. Accumulated *STB* articles were published in book form each year as *Stata Technical Bulletin Reprints*, which can be ordered directly from Stata Corporation.

With the growth of the Internet, instant communication among users became possible through vehicles such as Statalist. Program files could easily be downloaded from distant sources. A bimonthly printed journal and disk no longer provided the best avenues either for communicating among users, or for distributing updates and user-written programs. To adapt to a changing world, the *STB* had to evolve into something new.

The *Stata Journal* was launched to meet this challenge and the needs of Stata's broadening user base. Like the old *STB*, the *Stata Journal* contains articles describing new commands by users along with unofficial commands written by Stata Corporation employees. New commands are not its primary focus, however. The *Stata Journal* also contains refereed expository articles about statistics, book reviews, and a number of interesting columns, including "Speaking Stata" by Nicholas J. Cox, on effective use of the Stata programming language. The *Stata Journal* is intended for novice as well as experienced Stata users. For example, here are the contents from one recent issue:

"Exploratory analysis of single nucleotide polymorphism (SNP) for quantitative traits"	M.A. Cleves
"Value label utilities: labeldup and labelrename"	J. Weesie
"Multilingual datasets"	J. Weesie
"Multiple imputation of missing values: update"	P. Royston
"Estimation and testing of fixed-effect panel-data systems"	J.L. Blackwell, III
"Data inspection using biplots"	U. Kohler & M. Luniak
"Stata in space: Econometric analysis of spatially explicit raster data"	D. Müller
"Using the file command to produce formatted output for other applications"	E. Slaymaker
"Teaching statistics to physicians using Stata"	S.M. Hailpern
"Speaking Stata: Density probability plots"	N. J. Cox
"Review of *Regression Methods in Biostatistics: Linear, Logistic, Survival, and Repeated Measures Models*"	S. Lemeshow & M.L. Moeschberger

The *Stata Journal* is published quarterly. Subscriptions can be purchased directly from Stata Corporation by visiting www.stata.com.

Books Using Stata

In addition to Stata's own reference manuals, a growing library of books describe Stata, or use Stata to illustrate analytical techniques. These books include general introductions; disciplinary applications such as social science, biostatistics or econometrics; and focused texts concerning survey analysis, experimental data, categorical dependent variables, and other subjects. The Bookstore pages on Stata's web site have up-to-date lists, with descriptions of content:

http://www.stata.com/bookstore/

This online bookstore provides a central place to learn about and order Stata-relevant books from many different publishers.

2

Data Management

The first steps in data analysis involve organizing the raw data into a format usable by Stata. We can bring new data into Stata in several ways: type the data from the keyboard; read a text or ASCII file containing the raw data; paste data from a spreadsheet into the Editor; or, using a third-party data transfer program, translate the dataset directly from a system file created by another spreadsheet, database, or statistical program. Once Stata has the data in memory, we can save the data in Stata format for easy retrieval and updating in the future.

Data management encompasses the initial tasks of creating a dataset, editing to correct errors, and adding internal documentation such as variable and value labels. It also encompasses many other jobs required by ongoing projects, such as adding new observations or variables; reorganizing, simplifying, or sampling from the data; separating, combing, or collapsing datasets; converting variable types; and creating new variables through algebraic or logical expressions. When data-management tasks become complex or repetitive, Stata users can write their own programs to automate the work. Although Stata is best known for its analytical capabilities, it possesses a broad range of data-management features as well. This chapter introduces some of the basics.

The *User's Guide* provides an overview of the different methods for inputting data, followed by eight rules for determining which input method to use. Input, editing, and many other operations discussed in this chapter can be accomplished through the Data menus. Data menu subheadings refer to the general category of task:

Describe data

Data editor

Data browser (read-only editor)

Create or change variables

Sort

Combine datasets

Labels

Notes

Variable utilities

Matrices

Other utilities

Example Commands

. **append using** *olddata*
Reads previously-saved dataset *olddata.dta* and adds all its observations to the data currently in memory. Subsequently typing **save** *newdata,* **replace** will save the combined dataset as *newdata.dta*.

. **browse**
Opens the spreadsheet-like Data Browser for viewing the data. The Browser looks similar to the Data Editor, but it has no editing capability, so there is no risk of inadvertently changing your data. Alternatively, click [icon].

. **browse** *boats men if year* **> 1980**
Opens the Data Browser showing only the variables *boats* and *men* for observations in which *year* is greater than 1980. This example illustrates the **if** qualifier, which can be used to focus the operation of many Stata commands.

. **compress**
Automatically converts all variables to their most efficient storage types to conserve memory and disk space. Subsequently typing the command **save** *filename,* **replace** will make these changes permanent.

. **drawnorm** *z1 z2 z3,* **n(5000)**
Creates an artificial dataset with 5,000 observations and three random variables, *z1, z2,* and *z3*, sampled from uncorrelated standard normal distributions. Options could specify other means, standard deviations, and correlation or covariance matrices.

. **edit**
Opens the spreadsheet-like Data Editor where data can be entered or edited. Alternatively, choose Window – Data Editor or click [icon].

. **edit** *boats year men*
Opens the Data Editor with only the variables *boats*, *year*, and *men* (in that order) visible and available for editing.

. **encode** *stringvar,* **gen(***numvar***)**
Creates a new variable named *numvar*, with labeled numerical values based on the string (non-numeric) variable *stringvar*.

. **format** *rainfall* **%8.2f**
Establishes a fixed (**f**) display format for numeric variable *rainfall*: 8 columns wide, with two digits always shown after the decimal.

. **generate** *newvar* **= (***x* **+** *y***)/100**
Creates a new variable named *newvar*, equal to *x* plus *y* divided by 100.

. **generate** *newvar* **= uniform()**
Creates a new variable with values sampled from a uniform random distribution over the interval ranging from 0 to nearly 1, written [0,1).

. **infile** *x y z* **using** *data.raw*
Reads an ASCII file named *data.raw* containing data on three variables: *x, y,* and *z*. The values of these variables are separated by one or more white-space characters — blanks, tabs, and newlines (carriage return, linefeed, or both) — or by commas. With white-space

delimiters, missing values are represented by periods, not blanks. With comma-delimited data, missing values are represented by a period or by two consecutive commas. Stata also provides for extended missing values, which we will discuss later. Other commands are better suited for reading tab-delimited, comma-delimited, or fixed-column raw data; type **help infiling** for more infomation.

. **list**

Lists the data in default or "table" format. If the dataset contains many variables, table format becomes hard to read, and **list, display** produces better results. See **help list** for other options controlling the format of data lists.

. **list** *x y z* **in 5/20**

Lists the *x*, *y*, and *z* values of the 5th through 20th observations, as the data are presently sorted. The **in** qualifier works in similar fashion with most other Stata commands as well.

. **merge** *id* **using** *olddata*

Reads the previously-saved dataset *olddata.dta* and matches observations from *olddata* with observations in memory that have identical *id* values. Both *olddata* (the "using" data) and the data currently in memory (the "master" data) must already be sorted by *id*.

. **replace** *oldvar* = 100 * *oldvar*

Replaces the values of *oldvar* with 100 times their previous values.

. **sample 10**

Drops all the observations in memory except for a 10% random sample. Instead of selecting a certain percentage, we could select a certain number of cases. For example, **sample 55, count** would drop all but a random sample of size $n = 55$.

. **save** *newfile*

Saves the data currently in memory, as a file named *newfile.dta*. If *newfile.dta* already exists, and you want to write over the previous version, type **save** *newfile*, **replace**. Alternatively, use the menus: File – Save or File – Save As . To save *newfile.dta* in the format of Stata version 7, type **saveold** *newfile* .

. **set memory 24m**

(Windows or Unix systems only) Allocates 24 megabytes of memory for Stata data. The amount set could be greater or less than the current allocation. Virtual memory (disk space) is used if the request exceeds physical memory. Type **clear** to drop the current data from memory before using **set memory** .

. **sort** *x*

Sorts the data from lowest to highest values of *x*. Observations with missing *x* values appear last after sorting because Stata views missing values as very high numbers. Type **help gsort** for a more general sorting command that can arrange values in either ascending or descending order and can optionally place the missing values first.

. **tabulate** *x* **if** *y* > 65

Produces a frequency table for *x* using only those observations that have *y* values above 65. The **if** qualifier works similarly with most other Stata commands.

. **use** *oldfile*

Retrieves previously-saved Stata-format dataset *oldfile.dta* from disk, and places it in memory. If other data are currently in memory, and you want to discard those data without saving them, type **use** *oldfile***, clear** . Alternatively, these tasks can be accomplished through File – Open or by clicking ⬚ .

Creating a New Dataset

Data that were previously saved in Stata format can be retrieved into memory either by typing a command of the form **use** *filename*, or by menu selections. This section describes basic methods for creating a Stata-format dataset in the first place, using as our example the 1995 data on Canadian provinces and territories listed in Table 2.1. (From the Federal, Provincial and Territorial Advisory Committee on Population Health, 1996. Canada's newest territory, Nunavut, is not listed here because it was part of the Northwest Territories until 1999.)

Table 2.1: Data on Canada and Its Provinces

Place	1995 Pop. (1000's)	Unemployment Rate (percent)	Male Life Expectancy	Female Life Expectancy
Canada	29606.1	10.6	75.1	81.1
Newfoundland	575.4	19.6	73.9	79.8
Prince Edward Island	136.1	19.1	74.8	81.3
Nova Scotia	937.8	13.9	74.2	80.4
New Brunswick	760.1	13.8	74.8	80.6
Quebec	7334.2	13.2	74.5	81.2
Ontario	11100.3	9.3	75.5	81.1
Manitoba	1137.5	8.5	75.0	80.8
Saskatchewan	1015.6	7.0	75.2	81.8
Alberta	2747.0	8.4	75.5	81.4
British Columbia	3766.0	9.8	75.8	81.4
Yukon	30.1	—	71.3	80.4
Northwest Territories	65.8	—	70.2	78.0

The simplest way to create a dataset from Table 2.1 is through Stata's spreadsheet-like Data Editor, which is invoked either by clicking ▦ , selecting Window – Data Editor from the top menu bar, or by typing the command **edit** . Then begin typing values for each variable, in columns that Stata automatically calls *var1*, *var2*, etc. Thus, *var1* contains place names (Canada, Newfoundland, etc.); *var2*, populations; and so forth.

We can assign more descriptive variable names by double-clicking on the column headings (such as *var1*) and then typing a new name in the resulting dialog box — eight characters or fewer works best, although names with up to 32 characters are allowed. We can also create variable labels that contain a brief description. For example, *var2* (population) might be renamed *pop*, and given the variable label "Population in 1000s, 1995".

Renaming and labeling variables can also be done outside of the Data Editor through the **rename** and **label variable** commands:

. `rename` *var2 pop*

. `label variable` *pop* `"Population in 1000s, 1995"`

Cells left empty, such as employment rates for the Yukon and Northwest Territories, will automatically be assigned Stata's system (default) missing value code, a period. At any time, we can close the Data Editor and then save the dataset to disk. Clicking 🗔 or Window – Data Editor brings the Editor back.

If the first value entered for a variable is a number, as with population, unemployment, and life expectancy, then Stata assumes that this column is a "numerical variable" and it will thereafter permit only numerical values. Numerical values can also begin with a plus or minus sign, include decimal points, or be expressed in scientific notation. For example, we could represent Canada's population as $2.96061e+7$, which means 2.96061×10^7 or about 29.6 million people. Numerical values *should not include any commas*, such as 29,606,100. If we did happen to put commas within the first value typed in a column, Stata would interpret this as a "string variable" (next paragraph) rather than as a number.

If the first value entered for a variable includes non-numerical characters, as did the place names above (or "1,000" with the comma), then Stata thereafter considers this column to be a string variable. String variable values can be almost any combination of letters, numbers, symbols, or spaces up to 80 characters long in Intercooled or Small Stata, and up to 244 characters in Stata/SE. We can thus store names, quotations, or other descriptive information. String variable values can be tabulated and counted, but do not allow the calculation of means, correlations, or most other statistics. In the Data Editor or Data Browser, string variable values appear in red, so we can visually distinguish the two variable types.

After typing in the information from Table 2.1 in this fashion, we close the Data Editor and save our data, perhaps with the name *canada0.dta*:

. `save` *canada0*

Stata automatically adds the extension .dta to any dataset name, unless we tell it to do otherwise. If we already had saved and named an earlier version of this file, it is possible to write over that with the newest version by typing

. `save, replace`

At this point, our new dataset looks like this:

```
. describe

Contains data from C:\data\canada0.dta
  obs:             13
  vars:             5                              3 Jul 2005 10:30
  size:           533 (99.9% of memory free)
-------------------------------------------------------------------------------
               storage   display    value
variable name    type    format     label      variable label
-------------------------------------------------------------------------------
var1            str21    %21s
pop             float    %9.0g                  Population in 1000s, 1995
var3            float    %9.0g
var4            float    %9.0g
var5            float    %9.0g
-------------------------------------------------------------------------------
Sorted by:
```

```
. list

     +---------------------------------------------------------------+
     |                       var1      pop   var3    var4    var5 |
     |---------------------------------------------------------------|
  1. |                     Canada  29606.1   10.6    75.1    81.1 |
  2. |               Newfoundland    575.4   19.6    73.9    79.8 |
  3. |       Prince Edward Island    136.1   19.1    74.8    81.3 |
  4. |               Nova Scotia    937.8   13.9    74.2    80.4 |
  5. |             New Brunswick    760.1   13.8    74.8    80.6 |
     |---------------------------------------------------------------|
  6. |                    Quebec   7334.2   13.2    74.5    81.2 |
  7. |                   Ontario  11100.3    9.3    75.5    81.1 |
  8. |                  Manitoba   1137.5    8.5      75    80.8 |
  9. |              Saskatchewan   1015.6      7    75.2    81.8 |
 10. |                   Alberta     2747    8.4    75.5    81.4 |
     |---------------------------------------------------------------|
 11. |          British Columbia     3766    9.8    75.8    81.4 |
 12. |                    Yukon     30.1      .    71.3    80.4 |
 13. |     Northwest Territories     65.8      .    70.2      78 |
     +---------------------------------------------------------------+
```

```
. summarize

    Variable |      Obs        Mean    Std. Dev.       Min        Max
-------------+--------------------------------------------------------
        var1 |        0
         pop |       13    4554.769    8214.304       30.1    29606.1
        var3 |       11    12.10909    4.250048          7       19.6
        var4 |       13    74.29231    1.673052       70.2       75.8
        var5 |       13    80.71539    .9754027         78       81.8
```

Examining such output tables gives us a chance to look for errors that should be corrected. The **summarize** table, for instance, provides several numbers useful in proofreading, including the count of nonmissing observations (always 0 for string variables) and the minimum and maximum for each variable. Substantive interpretation of the summary statistics would be premature at this point, because our dataset contains one observation (Canada) that represents a combination of the other 12 provinces and territories.

The next step is to make our dataset more self-documenting. The variables could be given more descriptive names, such as the following:

```
. rename var1 place

. rename var3 unemp
```

```
. rename var4 mlife

. rename var5 flife
```

Stata also permits us to add several kinds of labels to the data. **label data** describes the dataset as a whole. For example,

```
. label data "Canadian dataset 0"
```

label variable describes an individual variable. For example,

```
. label variable place "Place name"

. label variable unemp "% 15+ population unemployed, 1995"

. label variable mlife "Male life expectancy years"

. label variable flife "Female life expectancy years"
```

By labeling data and variables, we obtain a dataset that is more self-explanatory:

```
. describe

Contains data from C:\data\canada0.dta
  obs:            13                        Canadian dataset 0
  vars:            5                        3 Jul 2005 10:45
  size:          533 (99.9% of memory free)
-------------------------------------------------------------------------
              storage  display     value
variable name  type    format      label      variable label
-------------------------------------------------------------------------
place          str21   %21s                    Place name
pop            float   %9.0g                    Population in 1000s, 1995
unemp          float   %9.0g                    % 15+ population unemployed,
                                                 1995
mlife          float   %9.0g                    Male life expectancy years
flife          float   %9.0g                    Female life expectancy years
-------------------------------------------------------------------------
Sorted by:
```

Once labeling is completed, we should save the data to disk by using File – Save or typing

```
. save, replace
```

We can later retrieve these data any time through ⌗, File – Open, or by typing

```
. use c:\data\canada0
(Canadian dataset 0)
```

We can then proceed with a new analysis. We might notice, for instance, that male and female life expectancies correlate positively with each other and also negatively with the unemployment rate. The life expectancy–unemployment rate correlation is slightly stronger for males.

```
. correlate unemp mlife flife
(obs=11)

         |    unemp     mlife     flife
---------+---------------------------
   unemp |   1.0000
   mlife |  -0.7440    1.0000
   flife |  -0.6173    0.7631    1.0000
```

The order of observations within a dataset can be changed through the **sort** command. For example, to rearrange observations from smallest to largest in population, type

. `sort` *`pop`*

String variables are sorted alphabetically instead of numerically. Typing **sort** *place* will rearrange observations putting Alberta first, British Columbia second, and so on.

We can control the order of variables in the data, using the **order** command. For example, we could make unemployment rate the second variable, and population last:

. `order` *`place unemp mlife flife pop`*

The Data Editor also has buttons that perform these functions. The Sort button applies to the column currently highlighted by the cursor. The << and >> buttons move the current variable to the beginning or end of the variable list, respectively. As with any other editing, these changes only become permanent if we subsequently save our data.

The Data Editor's Hide button does not rearrange the data, but rather makes a column temporarily invisible on the spreadsheet. This feature is convenient if, for example, we need to type in more variables and want to keep the province names or some other case identification column in view, adjacent to the "active" column where we are entering data.

We can also restrict the Data Editor beforehand to work only with certain variables, in a specified order, or with a specified range of values. For example,

. `edit` *`place mlife flife`*

or

. `edit` *`place unemp if pop > 100`*

The last example employs an **if** qualifier, an important tool described in the next section.

Specifying Subsets of the Data: `in` and `if` Qualifiers

Many Stata commands can be restricted to a subset of the data by adding an **in** or **if** qualifier. (Qualifiers are also available for many menu selections: look for an if/in or by/if/in tab along the top of the menu.) **in** specifies the observation numbers to which the command applies. For example, **list in 5** tells Stata to list only the 5th observation. To list the 1st through 20th observations, type

. `list in 1/20`

The letter **l** denotes the last case, and **-4**, for example, the fourth-from-last. Thus, we could list the four most populous Canadian places (which will include Canada itself) as follows:

. `sort` *`pop`*

. `list` *`place pop`* `in -4/l`

Note the important, although typographically subtle, distinction between **1** (number one, or first observation) and **l** (letter "el," or last observation). The **in** qualifier works in a similar way with most other analytical or data-editing commands. It always refers to the data *as presently sorted*.

The **if** qualifier also has broad applications, but it selects observations based on specific variable values. As noted, the observations in *canada0.dta* include not only 12 Canadian provinces or territories, but also Canada as a whole. For many purposes, we might want to exclude Canada from analyses involving the 12 territories and provinces. One way to do so is to restrict the analysis to only those places with populations below 20 million (20,000 thousand); that is, every place except Canada:

```
. summarize if pop < 20000
Variable |     Obs        Mean    Std. Dev.      Min        Max
---------+-----------------------------------------------------
   place |       0
     pop |      12    2467.158    3435.521       30.1    11100.3
   unemp |      10       12.26     4.44877          7       19.6
   mlife |      12      74.225    1.728965       70.2       75.8
   flife |      12    80.68333      1.0116         78       81.8
```

Compare this with the earlier **summarize** output to see how much has changed. The previous mean of population, for example, was grossly misleading because it counted every person twice.

The " **<** " (is less than) sign is one of six *relational operators*:

==	is equal to
!=	is not equal to (**~=** also works)
>	is greater than
<	is less than
>=	is greater than or equal to
<=	is less than or equal to

A double equals sign, " **==** ", denotes the logical test, "*Is* the value on the left side the same as the value on the right?" To Stata, a single equals sign means something different: "*Make* the value on the left side be the same as the value on the right." The single equals sign is not a relational operator and cannot be used within **if** qualifiers. Single equals signs have other meanings. They are used with commands that generate new variables, or replace the values of old ones, according to algebraic expressions. Single equals signs also appear in certain specialized applications such as weighting and hypothesis tests.

Any of these relational operators can be used to select observations based on their values for numerical variables. Only two operators, **==** and **!=**, make sense with string variables. To use string variables in an **if** qualifier, enclose the target value in double quotes. For example, we could get a summary excluding Canada (leaving in the 12 provinces and territories):

```
. summarize if place != "Canada"
```

Two or more relational operators can be combined within a single **if** expression by the use of *logical operators*. Stata's logical operators are the following:

&	and
|	or (symbol is a vertical bar, not the number one or letter "el")
!	not (**~** also works)

The Canadian territories (Yukon and Northwest) both have fewer than 100,000 people. To find the mean unemployment and life expectancies for the 10 Canadian provinces only, excluding both the smaller places (territories) and the largest (Canada), we could use this command:

```
. summarize unemp mlife flife if pop > 100 & pop < 20000

Variable |     Obs        Mean    Std. Dev.        Min         Max
---------+-----------------------------------------------------------
   unemp |      10       12.26     4.44877          7        19.6
   mlife |      10       74.92    .6051633       73.9        75.8
   flife |      10       80.98     .586515       79.8        81.8
```

Parentheses allow us to specify the precedence among multiple operators. For example, we might list all the places that either have unemployment below 9, *or* have life expectancies of at least 75.4 for men *and* 81.4 for women:

```
. list if unemp < 9 | (mlife >= 75.4 & flife >= 81.4)

     +-----------------------------------------------------+
     |              place       pop   unemp   mlife   flife |
     |-----------------------------------------------------|
  8. |           Manitoba    1137.5     8.5      75    80.8 |
  9. |       Saskatchewan    1015.6       7    75.2    81.8 |
 10. |            Alberta      2747     8.4    75.5    81.4 |
 11. |   British Columbia      3766     9.8    75.8    81.4 |
     +-----------------------------------------------------+
```

A note of caution regarding missing values: Stata ordinarily shows missing values as a period, but in some operations (notably **sort** and **if**, although not in statistical calculations such as means or correlations), these same missing values are treated as if they were large positive numbers. Watch what happens if we sort places from lowest to highest unemployment rate, and then ask to see places with unemployment rates above 15%:

```
. sort unemp

. list if unemp > 15

     +-----------------------------------------------------------+
     |                   place      pop   unemp   mlife   flife |
     |-----------------------------------------------------------|
 10. |   Prince Edward Island    136.1    19.1    74.8    81.3 |
 11. |          Newfoundland     575.4    19.6    73.9    79.8 |
 12. |                 Yukon      30.1       .    71.3    80.4 |
 13. |   Northwest Territories    65.8       .    70.2      78 |
     +-----------------------------------------------------------+
```

The two places with missing unemployment rates were included among those "greater than 15." In this instance the result is obvious, but with a larger dataset we might not notice. Suppose that we were analyzing a political opinion poll. A command such as the following would tabulate the variable *vote* not only for people with ages older than 65, as intended, but also for any people whose *age* values were missing:

```
. tabulate vote if age > 65
```

Where missing values exist, we might have to deal with them explicitly as part of the **if** expression.

```
. tabulate vote if age > 65 & age < .
```

A less-than inequality such as **age < .** is a general way to select observations with nonmissing values. Stata permits up to 27 different missing values codes, although we are using only the default " . " here. The other 26 codes are represented internally as numbers even larger than " . ", so **<** **.** avoids them all. Type **help missing** for more details.

The **in** and **if** qualifiers set observations aside temporarily so that a particular command does not apply to them. These qualifiers have no effect on the data in memory, and the next command will apply to all observations, unless it too has an **in** or **if** qualifier. To drop variables from the data in memory, use the **drop** command. For example, to drop *mlife* and *flife* from memory, type

. **drop** *mlife flife*

We can drop observations from memory by using either the **in** qualifier or the **if** qualifier. Because we earlier sorted on *unemp*, the two territories occupy the 12th and 13th positions in the data. Canada itself is 6th. One way to drop these three nonprovinces employs the **in** qualifier. **drop in 12/13** means "drop the 12th through the 13th observations."

. **list**

```
     +------------------------------------------+
     |                    place     pop   unemp |
     |------------------------------------------|
 1.  |            Saskatchewan   1015.6       7 |
 2.  |                Alberta      2747     8.4 |
 3.  |                Manitoba   1137.5     8.5 |
 4.  |                 Ontario  11100.3     9.3 |
 5.  |        British Columbia     3766     9.8 |
     |------------------------------------------|
 6.  |                 Canada   29606.1    10.6 |
 7.  |                 Quebec    7334.2    13.2 |
 8.  |          New Brunswick     760.1    13.8 |
 9.  |            Nova Scotia     937.8    13.9 |
10.  |   Prince Edward Island     136.1    19.1 |
     |------------------------------------------|
11.  |            Newfoundland    575.4    19.6 |
12.  |                  Yukon      30.1       . |
13.  | Northwest Territories      65.8       . |
     +------------------------------------------+
```

. **drop in 12/13**
(2 observations deleted)

. **drop in 6**
(1 observation deleted)

The same change could have been accomplished through an **if** qualifier, with a command that says "drop if *place* equals Canada or population is less than 100."

. **drop if** *place* **==** **"Canada" |** *pop* **< 100**
(3 observations deleted)

After dropping Canada, the territories, and the variables *mlife* and *flife*, we have the following reduced dataset:

. **list**

```
     +--------------------------------------+
     |                  place    pop  unemp |
     |--------------------------------------|
 1.  |          Saskatchewan  1015.6      7 |
 2.  |               Alberta    2747    8.4 |
 3.  |              Manitoba  1137.5    8.5 |
 4.  |               Ontario 11100.3    9.3 |
 5.  |      British Columbia    3766    9.8 |
```

```
      |-----------------------------------------|
  6.  |               Quebec   7334.2   13.2 |
  7.  |        New Brunswick    760.1   13.8 |
  8.  |          Nova Scotia    937.8   13.9 |
  9.  | Prince Edward Island    136.1   19.1 |
 10.  |         Newfoundland    575.4   19.6 |
      +-----------------------------------------+
```

We can also drop selected variables or observations through the Delete button in the Data Editor.

Instead of telling Stata which variables or observations to drop, it sometimes is simpler to specify which to keep. The same reduced dataset could have been obtained as follows:

```
. keep place pop unemp
```

```
. keep if place != "Canada" & pop >= 100
(3 observations deleted)
```

Like any other changes to the data in memory, none of these reductions affect disk files until we save the data. At that point, we will have the option of writing over the old dataset (**save, replace**) and thus destroying it, or just saving the newly modified dataset with a new name (by choosing File – Save As , or by typing a command with the form **save** **newname**) so that both versions exist on disk.

Generating and Replacing Variables

The **generate** and **replace** commands allow us to create new variables or change the values of existing variables. For example, in Canada, as in most industrial societies, women tend to live longer than men. To analyze regional variations in this gender gap, we might retrieve dataset *canada1.dta* and generate a new variable equal to female life expectancy (*flife*) minus male life expectancy (*mlife*). In the main part of a **generate** or **replace** statement (unlike **if** qualifiers) we use a single equals sign.

```
. use canada1, clear
(Canadian dataset 1)
```

```
. generate gap = flife - mlife
```

```
. label variable gap "Female-male gap life expectancy"
```

```
. describe
Contains data from C:\data\canada1.dta
  obs:           13                        Canadian dataset 1
  vars:           6                        3 Jul 2005 10:48
  size:          585 (99.9% of memory free)
-------------------------------------------------------------------
              storage   display    value
variable name   type    format     label     variable label
-------------------------------------------------------------------
place          str21    %21s                 Place name
pop            float    %9.0g                Population in 1000s, 1995
unemp          float    %9.0g                % 15+ population unemployed,
                                               1995
mlife          float    %9.0g                Male life expectancy years
flife          float    %9.0g                Female life expectancy years
gap            float    %9.0g                Female-male gap life expectancy
-------------------------------------------------------------------
Sorted by:
```

```
. list place flife mlife gap
```

```
     +-----------------------------------------------------+
     |                   place    flife    mlife       gap |
     |-----------------------------------------------------|
 1.  |                  Canada     81.1     75.1         6 |
 2.  |            Newfoundland     79.8     73.9  5.900002 |
 3.  |   Prince Edward Island     81.3     74.8       6.5 |
 4.  |             Nova Scotia     80.4     74.2  6.200005 |
 5.  |           New Brunswick     80.6     74.8  5.799995 |
     |-----------------------------------------------------|
 6.  |                  Quebec     81.2     74.5  6.699997 |
 7.  |                 Ontario     81.1     75.5  5.599998 |
 8.  |                Manitoba     80.8       75  5.800003 |
 9.  |            Saskatchewan     81.8     75.2  6.600006 |
10.  |                 Alberta     81.4     75.5  5.900002 |
     |-----------------------------------------------------|
11.  |        British Columbia     81.4     75.8  5.599998 |
12.  |                   Yukon     80.4     71.3  9.099998 |
13.  |   Northwest Territories       78     70.2  7.800003 |
     +-----------------------------------------------------+
```

For the province of Newfoundland, the true value of *gap* should be $79.8 - 73.9 = 5.9$ years, but the output shows this value as 5.900002 instead. Like all computer programs, Stata stores numbers in binary form, and 5.9 has no exact binary representation. The small inaccuracies that arise from approximating decimal fractions in binary are unlikely to affect statistical calculations much because calculations are done in double precision (8 bytes per number). They appear disconcerting in data lists, however. We can change the display format so that Stata shows only a rounded-off version. The following command specifies a fixed display format four numerals wide, with one digit to the right of the decimal:

```
. format gap %4.1f
```

Even when the display shows 5.9, however, a command such as the following will return no observations:

```
. list if gap == 5.9
```

This occurs because Stata believes the value does not exactly equal 5.9. (More technically, Stata stores *gap* values in single precision but does all calculations in double, and the single- and double-precision approximations of 5.9 are not identical.)

Display formats, as well as variables names and labels, can also be changed by double-clicking on a column in the Data Editor. Fixed numeric formats such as **%4.1f** are one of the three most common numeric display format types. These are

%*w*.*d*g**   General numeric format, where *w* specifies the total width or number of columns displayed and *d* the minimum number of digits that must follow the decimal point. Exponential notation (such as 1.00e+07, meaning 1.00×10^7 or 10 million) and shifts in the decimal-point position will be used automatically as needed, to display values in an optimal (but varying) fashion.

%*w*.*d*f**   Fixed numeric format, where *w* specifies the total width or number of columns displayed and *d* the fixed number of digits that must follow the decimal point.

%*w*.*d*e**   Exponential numeric format, where *w* specifes the total width or number of columns displayed and *d* the fixed number of digits that must follow the decimal point.

For example, as we saw in Table 2.1, the 1995 population of Canada was approximately 29,606,100 people, and the Yukon Territory population was 30,100. Below we see how these two numbers appear under several different display formats:

format	Canada	Yukon
%9.0g	2.96e+07	30100
%9.1f	29606100.0	30100.0
%12.5e	2.96061e+07	3.01000e+04

Although the displayed values look different, their internal values are identical. Statistical calculations remain unaffected by display formats. Other numerical display formatting options include the use of commas, left- and right-justification, or leading zeroes. There also exist special formats for dates, time series variables, and string variables. Type **help format** for more information.

    replace can make the same sorts of calculations as **generate**, but it changes values of an existing variable instead of creating a new variable. For example, the variable *pop* in our dataset gives population in thousands. To convert this to simple population, we just multiply ("*** " means multiply) all values by 1,000:

```
. replace pop = pop * 1000
```

    replace can make such wholesale changes, or it can be used with **in** or **if** qualifiers to selectively edit the data. To illustrate, suppose that we had questionnaire data with variables including *age* and year born (*born*). A command such as the following would correct one or more typos where a subject's age had been incorrectly typed as 229 instead of 29:

```
. replace age = 29 if age == 229
```

Alternatively, the following command could correct an error in the value of *age* for observation number 1453:

```
. replace age = 29 in 1453
```

For a more complicated example,

```
. replace age = 2005-born if age >= .  |  age < 2005-born
```

This replaces values of variable *age* with 2005 minus the year of birth if *age* is missing or if the reported age is less than 2005 minus the year of birth.

    generate and **replace** provide tools to create categorical variables as well. We noted earlier that our Canadian dataset includes several types of observations: 2 territories, 10 provinces, and 1 country combining them all. Although **in** and **if** qualifiers allow us to separate these, and **drop** can eliminate observations from the data, it might be most convenient to have a categorical variable that indicates the observation's "type." The following example shows one way to create such a variable. We start by generating *type* as a constant, equal to 1 for each observation. Next, we replace this with the value 2 for the Yukon and Northwest Territories, and with 3 for Canada. The final steps involve labeling new variable *type* and defining labels for values 1, 2, and 3.

```
. use canada1, clear
(Canadian dataset 1)
. generate type = 1
```

```
. replace type = 2 if place == "Yukon" | place == "Northwest
    Territories"
(2 real changes made)
. replace type = 3 if place == "Canada"
(1 real change made)
. label variable type "Province, territory or nation"
. label values type typelbl
. label define typelbl 1 "Province" 2 "Territory" 3 "Nation"
. list place flife mlife gap type
```

```
      +---------------------------------------------------------------+
      |                  place   flife   mlife        gap        type |
      |---------------------------------------------------------------|
  1.  |                 Canada    81.1    75.1          6      Nation |
  2.  |            Newfoundland    79.8    73.9   5.900002    Province |
  3.  |   Prince Edward Island    81.3    74.8        6.5    Province |
  4.  |            Nova Scotia    80.4    74.2   6.200005    Province |
  5.  |          New Brunswick    80.6    74.8   5.799995    Province |
      |---------------------------------------------------------------|
  6.  |                 Quebec    81.2    74.5   6.699997    Province |
  7.  |                Ontario    81.1    75.5   5.599998    Province |
  8.  |               Manitoba    80.8      75   5.800003    Province |
  9.  |           Saskatchewan    81.8    75.2   6.600006    Province |
 10.  |                Alberta    81.4    75.5   5.900002    Province |
      |---------------------------------------------------------------|
 11.  |       British Columbia    81.4    75.8   5.599998    Province |
 12.  |                  Yukon    80.4    71.3   9.099998   Territory |
 13.  | Northwest Territories      78    70.2   7.800003   Territory |
      +---------------------------------------------------------------+
```

As illustrated, labeling the values of a categorical variable requires two commands. The **label define** command specifies what labels go with what numbers. The **label values** command specifies to which variable these labels apply. One set of labels (created through one **label define** command) can apply to any number of variables (that is, be referenced in any number of **label values** commands). Value labels can have up to 32,000 characters, but work best for most purposes if they are not too long.

generate can create new variables, and **replace** can produce new values, using any mixture of old variables, constants, random values, and expressions. For numeric variables, the following *arithmetic operators* apply:

> + add
>
> – subtract
>
> * multiply
>
> / divide
>
> ^ raise to power

Parentheses will control the order of calculation. Without them, the ordinary rules of precedence apply. Of the arithmetic operators, only addition, "+", works with string variables, where it connects two string values into one.

Although their purposes differ, **generate** and **replace** have similar syntax. Either can use any mathematically or logically feasible combination of Stata operators and **in** or **if** qualifiers. These commands can also employ Stata's broad array of special functions, introduced in the following section.

Using Functions

This section lists many of the functions available for use with **generate** or **replace**. For example, we could create a new variable named *loginc*, equal to the natural logarithm of *income*, by using the natural log function **ln** within a **generate** command:

. **generate** *loginc* = **ln**(*income*)

ln is one of Stata's *mathematical functions*. These functions are as follows:

abs(x)	Absolute value of x.
acos(x)	Arc-cosine returning radians. Because 360 degrees = 2π radians, **acos**(x)***180/_pi** gives the arc-cosine returning degrees (_pi denotes the mathematical constant π).
asin(x)	Arc-sine returning radians.
atan(x)	Arc-tangent returning radians.
atan2(y,x)	Two-argument arc-tangent returning radians.
atanh(x)	Arc-hyperbolic tangent returning radians.
ceil(x)	Integer n such that $n-1 < x \le n$
cloglog(x)	Complementary log-log of x: $\ln(-\ln(1-x))$
comb(n,k)	Combinatorial function (number of possible combinations of n things taken k at a time).
cos(x)	Cosine of radians. To find the cosine of y degrees, type **generate y = cos**(y ***_pi/180**)
digamma(x)	$d\ln\Gamma(x)/dx$
exp(x)	Exponential (e to power).
floor(x)	Integer n such that $n \le x < n+1$
trunc(x)	Integer obtained by truncating x towards zero.
invcloglog(x)	Inverse of the complementary log-log: $1 - \exp(-\exp(x))$
invlogit(x)	Inverse of logit of x: $\exp(x)/(1 + \exp(x))$
ln(x)	Natural (base e) logarithm. For any other base number B, to find the base B logarithm of x, type **generate y = ln**(x)**/ln**(B)
lnfactorial(x)	Natural log of factorial. To find x factorial, type **generate y = round**(**exp**(**lnfact**(x)**,1**)
lngamma(x)	Natural log of $\Gamma(x)$. To find $\Gamma(x)$, type **generate y = exp**(**lngamma**(x))
log(x)	Natural logarithm; same as **ln**(x)
log10(x)	Base 10 logarithm.
logit(x)	Log of odds ratio of x: $\ln(x/(1-x))$
max(**x1,x2,..,xn**)	Maximum of $x1, x2, ..., xn$.
min(**x1,x2,..,xn**)	Minimum of $x1, x2, ..., xn$

`mod(x,y)`	Modulus of x with respect to y.
`reldif(x,y)`	Relative difference: $|x-y|/(|y|+1)$
`round(x)`	Round x to nearest whole number.
`round(x,y)`	Round x in units of y.
`sign(x)`	-1 if $x<0$, 0 if $x=0$, $+1$ if $x>0$
`sin(x)`	Sine of radians.
`sqrt(x)`	Square root.
`total(x)`	Running sum of x (also see **help egen**)
`tan(x)`	Tangent of radians.
`tanh(x)`	Hyperbolic tangent of x.
`trigamma(x)`	$d^2 \ln\Gamma(x)/dx^2$

Many *probability functions* exit as well, and are listed below. Consult **help probfun** and the reference manuals for important details, including definitions, constraints on parameters, and the treatment of missing values.

`betaden(a,b,x)`	Probability density of the beta distribution.
`Binomial(n,k,p)`	Probability of k or more successes in n trials when the probability of a success on a single trial is p.
`binormal(h,k,r)`	Joint cumulative distribution of bivariate normal with correlation r.
`chi2(n,x)`	Cumulative chi-squared distribution with n degrees of freedom.
`chi2tail(n,x)`	Reverse cumulative (upper-tail, survival) chi-squared distribution with n degrees of freedom.. $\quad$ chi2tail(n,x) = $1 -$ chi2(n,x)
`dgammapda(a,x)`	Partial derivative of the cumulative gamma distribution gammap(a,x) with respect to a.
`dgammapdx(a,x)`	Partial derivative of the cumulative gamma distribution gammap(a,x) with respect to x.
`dgammapdada(a,x)`	2nd partial derivative of the cumulative gamma distribution gammap(a,x) with respect to a.
`dgammapdadx(a,x)`	2nd partial derivative of the cumulative gamma distribution gammap(a,x) with respect to a and x.
`dgammapdxdx(a,x)`	2nd partial derivative of the cumulative gamma distribution gammap(a,x) with respect to x.
`F(n1,n2,f)`	Cumulative F distribution with $n1$ numerator and $n2$ denominator degrees of freedom.
`Fden(n1,n2,f)`	Probability density function for the F distribution with $n1$ numerator and $n2$ denominator degrees of freedom.
`Ftail(n1,n2,f)`	Reverse cumulative (upper-tail, survival) F distribution with $n1$ numerator and $n2$ denominator degrees of freedom. Ftail($n1,n2,f$) = $1 -$ F($n1,n2,f$)

`gammaden(a,b,g,x)` Probability density function for the gamma family, where gammaden(a,1,0,x) = the probability density function for the cumulative gamma distribution gammap(a,x).

`gammap(a,x)` Cumulative gamma distribution for a; also known as the incomplete gamma function.

`ibeta(a,b,x)` Cumulative beta distribution for a, b; also known as the incomplete beta function.

`invbinomial(n,k,P)` Inverse binomial. For $P \leq 0.5$, probability p such that the probability of observing k or more successes in n trials is P; for $P > 0.5$, probability p such that the probability of observing k or fewer successes in n trials is $1 - P$.

`invchi2(n,p)` Inverse of `chi2()`. If chi2(n,x) = p, then invchi2(n,p) = x

`invchi2tail(n,p)` Inverse of `chi2tail()`
If chi2tail(n,x) = p, then invchi2tail(n,p) = x

`invF(n1,n2,p)` Inverse cumulative F distribution.
If F($n1$,$n2$,f) = p, then invF($n1$,$n2$,p) = f

`invFtail(n1,n2,p)` Inverse reverse cumulative F distribution.
If Ftail($n1$,$n2$,f) = p, then invFtail($n1$,$n2$,p) = f

`invgammap(a,p)` Inverse cumulative gamma distribution.
If gammap(a,x) = p, then invgammap(a,p) = x

`invibeta(a,b,p)` Inverse cumulative beta distribution.
If ibeta(a,b,x) = p, then invibeta(a,b,p) = x

`invnchi2(n,L,p)` Inverse cumulative noncentral chi-squared distribution.
If nchi2(n,L,x) = p, then invnchi2(n,L,p) = x

`invnFtail(n1,n2,L,p)` Inverse reverse cumulative noncentral F distribution.
If nFtail($n1$,$n2$,L,f) = p, then invnFtail($n1$,$n2$,L, p) = f

`invnibeta(a,b,L,p)` Inverse cumulative noncentral beta distribution.
If nibeta(a,b,L,x) = p, then invnibeta(a,b,L,p) = x

`invnormal(p)` Inverse cumulative standard normal distribution.
If normal(z) = p, then invnormal(p) = z

`invttail(n,p)` Inverse reverse cumulative Student's t distribution.
If ttail(n,t) = p, then invttail(n,p) = t

`nbetaden(a,b,L,x)` Noncentral beta density with shape parameters a, b, noncentrality parameter L.

`nchi2(n,L,x)` Cumulative noncentral chi-squared distribution with n degrees of freedom and noncentrality parameter L.

`nFden(n1,n2,L,x)` Noncentral F density with $n1$ numerator and $n2$ denominator degrees of freedom, noncentrality parameter L.

`nFtail(n1,n2,L,x)` Reverse cumulative (upper-tail, survival) noncentral F distribution with $n1$ numerator and $n2$ denominator degrees of freedom, noncentrality parameter L.

`nibeta(a,b,L,x)`	Cumulative noncentral beta distribution with shape parameters a and b, and noncentrality parameter L.
`normal(z)`	Cumulative standard normal distribution.
`normalden(z)`	Standard normal density, mean 0 and standard deviation 1.
`normalden(z,s)`	Normal density, mean 0 and standard deviation s.
`normalden(x,m,s)`	Normal density, mean m and standard deviation s.
`npnchi2(n,x,p)`	Noncentrality parameter L for the noncentral cumulative chi-squared distribution. If $nchi2(n,L,x) = p$, then $npnchi2(n,x,p) = L$
`tden(n,t)`	Probability density function of Student's t distribution with n degrees of freedom.
`ttail(n,t)`	Reverse cumulative (upper-tail) Student's t distribution with n degrees of freedom. This function returns probability $T > t$.
`uniform()`	Pseudo-random number generator, returning values from a uniform distribution theoretically ranging from 0 to nearly 1, written [0,1).

Nothing goes inside the parentheses with **`uniform()`**. Optionally, we can control the pseudo-random generator's starting seed, and hence the stream of "random" numbers, by first issuing a **`set seed #`** command — where # could be any integer from 0 to $2^{31} - 1$ inclusive. Omitting the **`set seed`** command corresponds to **`set seed 123456789`**, which will always produce the same stream of numbers.

Stata provides more than 40 *date functions* and date-related *time series functions*. A listing can be found in Chapter 27 of the *User's Guide*, or by typing **`help datefun`**. Below are some examples of date functions. "Elapsed date" in these functions refers to the number of days since January 1, 1960.

`date(s₁,s₂[,y])`	Elapsed date corresponding to s_1. s_1 is a string variable indicating the date in virtually any format. Months can be spelled out, abbreviated to three characters, or given as numbers; years can include or exclude the century; blanks and punctuation are allowed. s_2 is any permutation of m, d, and [##]y with their order defining the order that month, day and year occur in s_1. ## gives the century for two-digit years in s_1; the default is 19y.
`d(l)`	A date literal convenience function. For example, typing d(2jan1960) is equivalent to typing 1.
`mdy(m,d,y)`	Elapsed date corresponding to m, d, and y.
`day(e)`	Numeric day of the month corresponding to e, the elapsed date.
`month(e)`	Numeric month corresponding to e, the elapsed date.
`year(e)`	Numeric year corresponding to e, the elapsed date.
`dow(e)`	Numeric day of the week corresponding to e, the elapsed date.
`doy(e)`	Numeric day of the year corresponding to e, the elapsed date.
`week(e)`	Numeric week of the year corresponding to e, the elapsed date.
`quarter(e)`	Numeric quarter of the year corresponding to e, the elapsed date.
`halfyear(e)`	Numeric half of the corresponding to e, the elapsed date.

Some useful special functions include the following:

autocode(*x,n,xmin,xmax*) Forms categories from *x* by partitioning the interval from *xmin* to *xmax* into *n* equal-length intervals and returning the upper bound of the interval that contains *x*.

cond(*x,a,b*)   Returns *a* if *x* evaluates to "true" and *b* if *x* evaluates to "false."

                . **generate y = cond(*inc1* > *inc2*, *inc1*, *inc2*)**
creates the variable *y* as the maximum of *inc1* and *inc2* (assuming neither is missing).

group(*x*)   Creates a categorical variable that divides the data *as presently sorted* into *x* subsamples that are as nearly equal-sized as possible.

trunc(*x*)   Returns the integer obtained by truncating (dropping fractional parts of) *x*.

max($x_1, x_2, \ldots, x_n$)   Returns the maximum of $x_1, x_2, \ldots, x_n$. Missing values are ignored. For example, **max(3+2,1)** evaluates to 5.

min($x_1, x_2, \ldots, x_n$)   Returns the minimum of $x_1, x_2, \ldots, x_n$.

recode($x, x_1, x_2, \ldots, x_n$)   Returns missing if *x* is missing, x_1 if $x < x_1$, or x_2 if $x < x_2$, and so on.

round(*x,y*)   Returns *x* rounded to the nearest *y*.

sign(*x*)   Returns – 1 if *x* < 0, 0 if *x* = 0, and +1 if *x* > 0 (missing if *x* is missing).

total(*x*)   Returns the running sum of *x*, treating missing values as zero.

String functions, not described here, help to manipulate and evaluate string variables. Type **help strfun** for a complete list of string functions. The reference manuals and *User's Guide* give examples and details of these and other functions.

Multiple functions, operators, and qualifiers can be combined in one command as needed. The functions and algebraic operators just described can also be used in another way that does not create or change any dataset variables. The **display** command performs a single calculation and shows the results onscreen. For example:

```
. display 2+3
5

. display log10(10^83)
83

. display invttail(120,.025) * 34.1/sqrt(975)
2.1622305
```

Thus, **display** works as an onscreen statistical calculator.

Unlike a calculator, **display**, **generate**, and **replace** have direct access to Stata's statistical results. For example, suppose that we summarized the unemployment rates from dataset *canada1.dta*:

```
. summarize unemp
```

Variable	Obs	Mean	Std. Dev.	Min	Max
unemp	11	12.10909	4.250048	7	19.6

After **summarize**, Stata temporarily stores the mean as a macro named r(mean).

```
. display r(mean)
12.109091
```

We could use this result to create variable *unempDEV*, defined as deviations from the mean:

```
. gen unempDEV = unemp - r(mean)
(2 missing values generated)
```

```
. summ unemp unempDEV
```

Variable	Obs	Mean	Std. Dev.	Min	Max
unemp	11	12.10909	4.250048	7	19.6
unempDEV	11	4.33e-08	4.250048	-5.109091	7.49091

Stata also provides another variable-creation command, **egen** ("extensions to **generate**"), which has its own set of functions to accomplish tasks not easily done by **generate**. These include such things as creating new variables from the sums, maxima, minima, medians, interquartile ranges, standardized values, or moving averages of existing variables or expressions. For example, the following command creates a new variable named *zscore*, equal to the standardized (mean 0, variance 1) values of *x*:

```
. egen zscore = std(x)
```

Or, the following command creates new variable *avg*, equal to the row mean of each observation's values on *x*, *y*, *z*, and *w*, ignoring any missing values.

```
. egen avg = rowmean(x,y,z,w)
```

To create a new variable named *sum*, equal to the row sum of each observation's values on *x*, *y*, *z*, and *w*, treating missing values as zeroes, type

```
. egen sum = rowsum(x,y,z,w)
```

The following command creates new variable *xrank*, holding ranks corresponding to values of *x*: *xrank* = 1 for the observation with highest *x*. *xrank* = 2 for the second highest, and so forth.

```
. egen xrank = rank(x)
```

Consult **help egen** for a complete list of **egen** functions, or the reference manuals for further examples.

Converting between Numeric and String Formats

Dataset *canada2.dta* contains one string variable, *place*. It also has a labeled categorical variable, *type*. Both seem to have nonnumerical values.

```
. use canada2, clear
(Canadian dataset 2)
```

```
. list place type
```

```
      +-----------------------------------+
      |                   place     type |
      |-----------------------------------|
  1.  |                  Canada    Nation |
  2.  |            Newfoundland  Province |
  3.  |   Prince Edward Island  Province |
  4.  |             Nova Scotia  Province |
```

```
  5. |          New Brunswick     Province |
     |-----------------------------------|
  6. |                 Quebec     Province |
  7. |                Ontario     Province |
  8. |               Manitoba     Province |
  9. |           Saskatchewan     Province |
 10. |                Alberta     Province |
     |-----------------------------------|
 11. |       British Columbia     Province |
 12. |                  Yukon    Territory |
 13. | Northwest Territories    Territory |
     +-----------------------------------+
```

Beneath the labels, however, *type* remains a numeric variable, as we can see if we ask for the **nolabel** option:

```
. list place type, nolabel

          +-----------------------------+
          |                 place   type |
          |-----------------------------|
       1. |                Canada     3 |
       2. |          Newfoundland     1 |
       3. |   Prince Edward Island    1 |
       4. |           Nova Scotia     1 |
       5. |         New Brunswick     1 |
          |-----------------------------|
       6. |                Quebec     1 |
       7. |               Ontario     1 |
       8. |              Manitoba     1 |
       9. |          Saskatchewan     1 |
      10. |               Alberta     1 |
          |-----------------------------|
      11. |       British Columbia    1 |
      12. |                 Yukon     2 |
      13. | Northwest Territories     2 |
          +-----------------------------+
```

String and labeled numeric variables look similar when listed, but they behave differently when analyzed. Most statistical operations and algebraic relations are not defined for string variables, so we might want to have both string and labeled-numeric versions of the same information in our data. The **encode** command generates a labeled-numeric variable from a string variable. The number 1 is given to the alphabetically first value of the string variable, 2 to the second, and so on. In the following example, we create a labeled numeric variable named *placenum* from the string variable *place*:

```
. encode place, gen(placenum)
```

The opposite conversion is possible, too: The **decode** command generates a string variable using the values of a labeled numeric variable. Here we create string variable *typestr* from numeric variable *type*:

```
. decode type, gen(typestr)
```

When listed, the new numeric variable *placenum*, and the new string variable *typestr*, look similar to the originals:

```
. list place placenum type typestr

     +---------------------------------------------------------------------+
     |                   place               placenum       type   typestr |
     |---------------------------------------------------------------------|
  1. |                  Canada                 Canada      Nation    Nation |
  2. |             Newfoundland           Newfoundland   Province  Province |
  3. |     Prince Edward Island   Prince Edward Island   Province  Province |
  4. |             Nova Scotia             Nova Scotia   Province  Province |
  5. |            New Brunswick           New Brunswick   Province  Province |
     |---------------------------------------------------------------------|
  6. |                  Quebec                 Quebec    Province  Province |
  7. |                 Ontario                Ontario    Province  Province |
  8. |                Manitoba               Manitoba    Province  Province |
  9. |            Saskatchewan           Saskatchewan    Province  Province |
 10. |                 Alberta                Alberta    Province  Province |
     |---------------------------------------------------------------------|
 11. |        British Columbia       British Columbia    Province  Province |
 12. |                   Yukon                  Yukon   Territory Territory |
 13. |   Northwest Territories   Northwest Territories   Territory Territory |
     +---------------------------------------------------------------------+
```

But with the **nolabel** option, the differences become visible. Stata views *placenum* and *type* basically as numbers.

```
. list place placenum type typestr, nolabel

     +---------------------------------------------------------------------+
     |                   place              placenum  type    typestr |
     |---------------------------------------------------------------------|
  1. |                  Canada   3.00000000000000e+00     3     Nation |
  2. |             Newfoundland   6.00000000000000e+00     1   Province |
  3. |     Prince Edward Island   1.00000000000000e+01     1   Province |
  4. |             Nova Scotia   8.00000000000000e+00     1   Province |
  5. |            New Brunswick   5.00000000000000e+00     1   Province |
     |---------------------------------------------------------------------|
  6. |                  Qucbec   1.10000000000000e+01     1   Province |
  7. |                 Ontario   9.00000000000000e+00     1   Province |
  8. |                Manitoba   4.00000000000000e+00     1   Province |
  9. |            Saskatchewan   1.20000000000000e+01     1   Province |
 10. |                 Alberta   1.00000000000000e+00     1   Province |
     |---------------------------------------------------------------------|
 11. |        British Columbia   2.00000000000000e+00     1   Province |
 12. |                   Yukon   1.30000000000000e+01     2  Territory |
 13. |   Northwest Territories   7.00000000000000e+00     2  Territory |
     +---------------------------------------------------------------------+
```

Statistical analyses, such as finding means and standard deviations, work only with basically numeric variables. For calculation purposes, numeric variables' labels do not matter.

```
. summarize place placenum type typestr

Variable |       Obs        Mean    Std. Dev.       Min        Max
---------+---------------------------------------------------------
   place |         0
placenum |        13           7     3.89444         1         13
    type |        13    1.307692    .6304252         1          3
 typestr |         0
```

Occasionally we encounter a string variable where the values are all or mostly numbers. To convert these string values into their numerical counterparts, use the **real** function. For example, the variable *siblings* below is a string variable, although it only has one value, "4 or more," that could not be represented just as easily by a number.

```
. describe siblings
    1. siblings    str9   %9s                    Number of siblings (string)
. list
        +----------+
        | siblings |
        |----------|
     1. |        0 |
     2. |        1 |
     3. |        2 |
     4. |        3 |
     5. | 4 or more |
        +----------+
. generate sibnum = real(siblings)
(1 missing value generated)
```

The new variable *sibnum* is numeric, with a missing value where *siblings* had "4 or more."

```
. list
        +--------------------+
        | siblings   sibnum |
        |--------------------|
     1. |        0        0 |
     2. |        1        1 |
     3. |        2        2 |
     4. |        3        3 |
     5. | 4 or more        . |
        +--------------------+
```

The **destring** command provides a more flexible method for converting string variables to numeric. In the example above, we could have accomplished the same thing by typing

```
. destring siblings, generate(sibnum) force
```

See **help destring** for information about syntax and options.

Creating New Categorical and Ordinal Variables

A previous section illustrated how to construct a categorical variable called *type* to distinguish among territories, provinces, and nation in our Canadian dataset. You can create categorical or ordinal variables in many other ways. This section gives a few examples.

type has three categories:

```
. tabulate type
Province,   |
territory or|
nation      |      Freq.      Percent        Cum.
------------+-----------------------------------
   Province |         10        76.92       76.92
  Territory |          2        15.38       92.31
     Nation |          1         7.69      100.00
------------+-----------------------------------
      Total |         13       100.00
```

For some purposes, we might want to re-express a multicategory variable as a set of dichotomies or "dummy variables," each coded 0 or 1. **tabulate** will create dummy variables automatically if we add the **generate** option. In the following example, this results in a set of variables called *type1*, *type2*, and *type3*, each representing one of the three categories of *type*:

```
. tabulate type, generate(type)
Province,    |
territory or|
nation       |       Freq.      Percent        Cum.
------------+-----------------------------------
   Province |          10        76.92        76.92
  Territory |           2        15.38        92.31
     Nation |           1         7.69       100.00
------------+-----------------------------------
      Total |          13       100.00
```

```
. describe
Contains data from C:\data\canada2.dta
  obs:              13                          Canadian dataset 2
  vars:             10                          3 Jul 2005 10:48
  size:            637 (99.9% of memory free)
-------------------------------------------------------------------------
              storage  display    value
variable name  type    format     label      variable label
-------------------------------------------------------------------------
place         str21    %21s                   Place name
pop           float    %9.0g                  Population in 1000s, 1995
unemp         float    %9.0g                  % 15+ population unemployed,
                                                1995
mlife         float    %9.0g                  Male life expectancy years
flife         float    %9.0g                  Female life expectancy years
gap           float    %9.0g                  Female-male gap life expectancy
type          byte     %9.0g      typelbl     Province, territory or nation
type1         byte     %8.0g                  type==Province
type2         byte     %8.0g                  type==Territory
type3         byte     %8.0g                  type==Nation
-------------------------------------------------------------------------
Sorted by:
    Note:  dataset has changed since last saved
```

```
. list place type type1-type3
     +--------------------------------------------------------------+
     |                    place       type    type1   type2   type3 |
     |--------------------------------------------------------------|
  1. |                   Canada     Nation        0       0       1 |
  2. |             Newfoundland   Province        1       0       0 |
  3. |    Prince Edward Island   Province        1       0       0 |
  4. |              Nova Scotia   Province        1       0       0 |
  5. |            New Brunswick   Province        1       0       0 |
     |--------------------------------------------------------------|
  6. |                   Quebec   Province        1       0       0 |
  7. |                  Ontario   Province        1       0       0 |
  8. |                 Manitoba   Province        1       0       0 |
  9. |             Saskatchewan   Province        1       0       0 |
 10. |                  Alberta   Province        1       0       0 |
     |--------------------------------------------------------------|
 11. |         British Columbia   Province        1       0       0 |
 12. |                    Yukon  Territory        0       1       0 |
 13. |    Northwest Territories  Territory        0       1       0 |
     +--------------------------------------------------------------+
```

Re-expressing categorical information as a set of dummy variables involves no loss of information; in this example, *type1* through *type3* together tell us exactly as much as *type* itself does. Occasionally, however, analysts choose to re-express a measurement variable in categorical or ordinal form, even though this *does* result in a substantial loss of information. For example, *unemp* in *canada2.dta* gives a measure of the unemployment rate. Excluding Canada itself from the data, we see that *unemp* ranges from 7% to 19.6%, with a mean of 12.26:

```
. summarize unemp if type != 3
Variable |      Obs        Mean    Std. Dev.       Min        Max
---------+-----------------------------------------------------------
   unemp |       10       12.26     4.44877          7       19.6
```

Having Canada in the data becomes a nuisance at this point, so we drop it:

```
. drop if type == 3
(1 observation deleted)
```

Two commands create a dummy variable named *unemp2* with values of 0 when unemployment is below average (12.26), 1 when unemployment is equal to or above average, and missing when *unemp* is missing. In reading the second command, recall that Stata's sorting and relational operators treat missing values as very large numbers.

```
. generate unemp2 = 0 if unemp < 12.26
(7 missing values generated)
. replace unemp2 = 1 if unemp >= 12.26 & unemp < .
(5 real changes made)
```

We might want to group the values of a measurement variable, thereby creating an ordered-category or ordinal variable. The **autocode** function (see "Using Functions" earlier in this chapter) provides automatic grouping of measurement variables. To create new ordinal variable *unemp3*, which groups values of *unemp* into three equal-width groups over the interval from 5 to 20, type

```
. generate unemp3 = autocode(unemp,3,5,20)
(2 missing values generated)
```

A list of the data shows how the new dummy (*unemp2*) and ordinal (*unemp3*) variables correspond to values of the original measurement variable *unemp*.

```
. list place unemp unemp2 unemp3

     +--------------------------------------------------+
     |                 place   unemp   unemp2   unemp3 |
     |--------------------------------------------------|
  1. |          Newfoundland    19.6        1       20 |
  2. | Prince Edward Island     19.1        1       20 |
  3. |           Nova Scotia    13.9        1       15 |
  4. |         New Brunswick    13.8        1       15 |
  5. |               Quebec     13.2        1       15 |
     |--------------------------------------------------|
  6. |              Ontario      9.3        0       10 |
  7. |             Manitoba      8.5        0       10 |
  8. |         Saskatchewan        7        0       10 |
  9. |              Alberta      8.4        0       10 |
 10. |     British Columbia      9.8        0       10 |
     |--------------------------------------------------|
 11. |               Yukon         .        .        . |
 12. | Northwest Territories       .        .        . |
     +--------------------------------------------------+
```

Both strategies just described dealt appropriately with missing values, so that Canadian places with missing values on *unemp* likewise receive missing values on the variables derived from *unemp*. Another possible approach works best if our data contain no missing values. To illustrate, we begin by dropping the Yukon and Northwest Territories:

```
. drop if unemp >= .
(2 observations deleted)
```

A greater-than-or-equal-to inequality such as **unemp >= .** will select any user-specified missing value codes, in addition to the default code ".". Type **help missing** for details.

Having dropped observations with missing values, we now can use the **group** function to create an ordinal variable not with approximately equal-width groupings, as **autocode** did, but instead with groupings of approximately equal size . We do this in two steps. First, sort the data (assuming no missing values) on the variable of interest. Second, generate a new variable using the **group(#)** function, where # indicates the number of groups desired. The example below divides our 10 Canadian provinces into 5 groups.

```
. sort unemp

. generate unemp5 = group(5)

. list place unemp unemp2 unemp3 unemp5
```

```
      +-----------------------------------------------------------+
      |                  place   unemp   unemp2   unemp3   unemp5 |
      |-----------------------------------------------------------|
  1.  |           Saskatchewan       7        0       10        1 |
  2.  |               Alberta      8.4        0       10        1 |
  3.  |               Manitoba     8.5        0       10        2 |
  4.  |                Ontario     9.3        0       10        2 |
  5.  |       British Columbia     9.8        0       10        3 |
      |-----------------------------------------------------------|
  6.  |                 Quebec    13.2        1       15        3 |
  7.  |          New Brunswick    13.8        1       15        4 |
  8.  |            Nova Scotia    13.9        1       15        4 |
  9.  | Prince Edward Island      19.1        1       20        5 |
 10.  |            Newfoundland    19.6        1       20        5 |
      +-----------------------------------------------------------+
```

Another difference is that **autocode** assigns values equal to the upper bound of each interval, whereas **group** simply assigns 1 to the first group, 2 to the second, and so forth.

Using Explicit Subscripts with Variables

When Stata has data in memory, it also defines certain system variables that describe those data. For example, $\_N$ represents the total number of observations. $\_n$ represents the observation number: $\_n = 1$ for the first observation, $\_n = 2$ for the second, and so on to the last observation ($\_n = \_N$). If we issue a command such as the following, it creates a new variable, *caseID*, equal to the number of each observation as presently sorted:

```
. generate caseID = _n
```

Sorting the data another way will change each observation's value of $\_n$, but its *caseID* value will remain unchanged. Thus, if we do sort the data another way, we can later return to the earlier order by typing

```
. sort caseID
```

Creating and saving unique case identification numbers that store the order of observations at an early stage of dataset development can greatly facilitate later data management.

We can use explicit subscripts with variable names, to specify particular observation numbers. For example, the 6th observation in dataset *canada1.dta* (if we have not dropped or re-sorted anything) is Quebec. Consequently, *pop[6]* refers to Quebec's population, 7334 thousand.

```
. display pop[6]
7334.2002
```

Similarly, *pop[12]* is the Yukon's population:

```
. display pop[12]
30.1
```

Explicit subscripting and the _n system variable have additional relevance when our data form a series. If we had the daily stock market price of a particular stock as a variable named *price*, for instance, then either *price* or, equivalently, *price*[_n] denotes the value of the _nth observation or day. *price*[_n–1] denotes the previous day's price, and *price*[_n+1] denotes the next. Thus, we might define a new variable *difprice*, which is equal to the change in *price* since the previous day:

```
. generate difprice = price - price[_n-1]
```

Chapter 13, on time series analysis, returns to this topic.

Importing Data from Other Programs

Previous sections illustrated how to enter and edit data by typing into the Data Editor. If our original data reside in an appropriately formatted spreadsheet, a shortcut can speed up this work: we might be able to copy and paste multi-column blocks of data (not including column labels) directly from the spreadsheet into Stata's Data Editor. This requires some care and perhaps experimentation, because Stata will interpret any column containing non-numeric values as representing a string variable. Single columns (variables) of data could also be pasted into the Data Editor from a text or word processor document. Once data have been successfully pasted into Editor columns, we assign variable names, labels, and so on in the usual manner.

These Data Editor methods are quick and easy, but for larger projects it is important to have tools that work directly with computer files created by other programs. Such files fall into two general categories: raw-data ASCII (text) files, which can be read into Stata with the appropriate Stata commands; and system files, which must be translated to Stata format by a special third-party program before Stata can read them.

To illustrate ASCII file methods, we return to the Canadian data of Table 2.1. Suppose that, instead of typing these data into Stata's Data Editor, we typed them into our word processor, with at least one space between each value. String values must be in double quotes if they contain internal spaces, as does "Prince Edward Island". For other string values, quotes are optional. Word processors allow the option of saving documents as ASCII (text) files, a simpler and more universal type than the word processor's usual saved-file format. We can thus create an ASCII file named *canada.raw* that looks something like this:

```
"Canada"  29606.1  10.6   75.1   81.1
"Newfoundland"  575.4   19.6   73.9   79.8
"Prince Edward Island"  136.1   19.1   74.8   81.3
"Nova Scotia"  937.8   13.9   74.2   80.4
"New Brunswick"  760.1   13.8   74.8   80.6
"Quebec"  7334.2   13.2   74.5   81.2
"Ontario"  11100.3   9.3   75.5   81.1
"Manitoba"  1137.5   8.5   75   80.8
"Saskatchewan"  1015.6   7   75.2   81.8
"Alberta"  2747   8.4   75.5   81.4
"British Columbia"  3766   9.8   75.8   81.4
"Yukon"  30.1   .   71.3   80.4
"Northwest Territories"  65.8   .   70.2   78
```

Note the use of periods, not blanks, to indicate missing values for the Yukon and Northwest Territories. If the dataset should have five variables, then for every observation, exactly five values (including periods for missing values) must exist.

infile reads into memory an ASCII file, such as *canada.raw*, in which the values are separated by one or more whitespace characters — blanks, tabs, and newlines (carriage return, line feed, or both) — or by commas. Its basic form is

. **infile** *variable-list* **using** *filename.raw*

With purely numeric data, the variable list could be omitted, in which case Stata assigns the names *var1*, *var2*, *var3*, and so forth. On the other hand, we might want to give each variable a distinctive name. We also need to identify string variables individually. For *canada.raw*, the **infile** command might be

. **infile str30** *place pop unemp mlife flife* **using** *canada.raw*, **clear**
(13 observations read)

The **infile** variable list specifies variables in the order that they appear in the data file. The **clear** option drops any current data from memory before reading in the new file.

If any string variables exist, their names must each be preceded by a **str**# statement. **str30** , for example, informs Stata that the next-named variable (*place*) is a string variable with as many as 30 characters. Actually, none of the Canadian place names involve more than 21 characters, but we do not need to know that in advance. It is often easier to overestimate string variable lengths. Then, once data are in memory, use **compress** to ensure that no variable takes up more space than it needs. The **compress** command automatically changes all variables to their most memory-efficient storage type.

. **compress**
place was str30 now str21

. **describe**

```
Contains data
  obs:            13
  vars:            5
  size:          533 (99.9% of memory free)
-------------------------------------------------------------------------------
                storage   display     value
variable name   type      format      label         variable label
-------------------------------------------------------------------------------
place           str21     %21s
pop             float     %9.0g
unemp           float     %9.0g
```

```
mlife           float   %9.0g
flife           float   %9.0g
-----------------------------------------------------------------------
Sorted by:
```

We can now proceed to label variables and data as described earlier. At any point, the commands **save** *canada0* (or **save** *canada0,* **replace**) would save the new dataset in Stata format, as file *canada0.dta*. The original raw-data file, *canada.raw*, remains unchanged on disk.

If our variables have non-numeric values (for example, "male" and "female") that we want to store as labeled numeric variables, then adding the option **automatic** will accomplish this. For example, we might read in raw survey data through this **infile** command:

. **infile** *gender age income vote* **using** *survey.raw,* **automatic**

Spreadsheet and database programs commonly write ASCII files that have only one observation per line, with values separated by tabs or commas. To read these files into Stata, use **insheet** . Its general syntax resembles that of **infile** , with options telling Stata whether the data delimited by tabs, commas, or other characters. For example, assuming tab-delimited data,

. **insheet** *variable-list* **using** *filename.raw,* **tab**

Or, assuming comma-delimited data with the first row of the file containing variable names (also comma-delimited),

. **insheet** *variable-list* **using** *filename.raw,* **comma names**

With **insheet** we do not need to separately identify string variables. If we include no variable list, and do not have variable names in the file's first row, Stata automatically assigns the variable names *var1*, *var2*, *var3*, ... Errors will occur if some values in our ASCII file are not separated by tabs, commas, or some other delimiter as specified in the **insheet** command.

Raw data files created by other statistical packages can be in "fixed-column" format, where the values are not necessarily delimited at all, but do occupy predefined column positions. Both **infile** and the more specialized command **infix** permit Stata to read such files. In the command syntax itself, or in a "data dictionary" existing in a separate file or as the first part of the data file, we have to specify exactly how the columns should be read.

Here is a simple example. Data exist in an ASCII file named *nfresour.raw*:

```
198624087641691000
198725247430001044
198825138637481086
198925358964371140
1990    8615731195
1991    7930001262
```

These data concern natural resource production in Newfoundland. The four variables occupy fixed column positions: columns 1–4 are the years (1986...1991); columns 5–8 measure forestry production in thousands of cubic meters (2408...missing); columns 9–14 measure mine production in thousands of dollars (764,169...793,000); and columns 15–18 are the consumer price index relative to 1986 (1000...1262). Notice that in fixed-column format, unlike space or tab-delimited files, blanks indicate missing values, and the raw data contain no decimal points. To read *nfresour.raw* into Stata, we specify each variable's column position:

```
. infix year 1-4 wood 5-8 mines 9-14 CPI 15-18
      using nfresour.raw, clear
(6 observations read)
. list
```

```
    +----------------------------+
    | year    wood    mines    CPI |
    |----------------------------|
 1. | 1986    2408   764169   1000 |
 2. | 1987    2524   743000   1044 |
 3. | 1988    2513   863748   1086 |
 4. | 1989    2535   896437   1140 |
 5. | 1990       .   861573   1195 |
    |----------------------------|
 6. | 1991       .   793000   1262 |
    +----------------------------+
```

More complicated fixed-column formats might require a data "dictionary." Data dictionaries can be straightforward, but they offer many possible choices. Typing **help infix** or **help infile2** obtains brief outlines of these commands. For more examples and explanation, consult the *User's Guide* and reference manuals. Stata also can load, write, or view data from OBDC (Open Database Connectivity) sources; see **help obdc**.

What if we need to export data from Stata to some other, non-OBDC program? The **outfile** command writes ASCII files to disk. A command such as the following will create a space-delimited ASCII file named *canada6.raw*, containing whatever data were in memory:

```
. outfile using canada6
```

The **infile**, **insheet**, **infix**, and **outfile** commands just described all manipulate raw data in ASCII files. A second, very quick, possibility is to copy your data from Stata's Browser and paste this directly into a spreadsheet such as Excell. Often the best option, however, is to transfer data directly between the specialized system files saved by various spreadsheet, database, or statistical programs. Several third-party programs perform such translations. Stat/Transfer, for example, will transfer data across many different formats including dBASE, Excel, FoxPro, Gauss, JMP, Lotus, MATLAB, Minitab, OSIRIS, Paradox, S-Plus, SAS, SPSS, SYSTAT, and Stata. It is available through Stata Corporation (www.stata.com) or from its maker, Circle Systems (www.stattransfer.com). Transfer programs prove indispensable for analysts working in multi-program environments or exchanging data with colleagues.

Combining Two or More Stata Files

We can combine Stata datasets in two general ways: **append** a second dataset that contains additional observations; or **merge** with other datasets that contain new variables or values. In keeping with this chapter's Canadian theme, we will illustrate these procedures using data on Newfoundland. File *newf1.dta* records the province's population for 1985 to 1989.

```
. use newf1, clear
(Newfoundland 1985-89)

. describe

Contains data from C:\data\newf1.dta
  obs:             5                          Newfoundland 1985-89
  vars:            2                          3 Jul 2005 10:49
  size:           50 (99.9% of memory free)
-------------------------------------------------------------------------------
              storage  display   value
variable name   type   format    label      variable label
-------------------------------------------------------------------------------
year            int    %9.0g                 Year
pop             float  %9.0g                 Population
-------------------------------------------------------------------------------
Sorted by:

. list

     +---------------+
     | year     pop |
     |---------------|
  1. | 1985   580700 |
  2. | 1986   580200 |
  3. | 1987   568200 |
  4. | 1988   568000 |
  5. | 1989   570000 |
     +---------------+
```

File *newf2.dta* has population and unemployment counts for some later years:

```
. use newf2
(Newfoundland 1990-95)

. describe

Contains data from C:\data\newf2.dta
  obs:             6                          Newfoundland 1990-95
  vars:            3                          3 Jul 2005 10:49
  size:           84 (99.9% of memory free)
-------------------------------------------------------------------------------
              storage  display   value
variable name   type   format    label      variable label
-------------------------------------------------------------------------------
year            int    %9.0g                 Year
pop             float  %9.0g                 Population
jobless         float  %9.0g                 Number of people unemployed
-------------------------------------------------------------------------------
Sorted by:

. list

     +------------------------+
     | year     pop   jobless |
     |------------------------|
  1. | 1990   573400    42000 |
  2. | 1991   573500    45000 |
  3. | 1992   575600    49000 |
  4. | 1993   584400    49000 |
  5. | 1994   582400    50000 |
     |------------------------|
  6. | 1995   575449        . |
     +------------------------+
```

To combine these datasets, with *newf2.dta* already in memory, we use the **append** command:

```
. append using newf1
```

```
. list

       +------------------------------+
       | year      pop    jobless |
       |------------------------------|
  1.   | 1990    573400     42000 |
  2.   | 1991    573500     45000 |
  3.   | 1992    575600     49000 |
  4.   | 1993    584400     49000 |
  5.   | 1994    582400     50000 |
       |------------------------------|
  6.   | 1995    575449        . |
  7.   | 1985    580700        . |
  8.   | 1986    580200        . |
  9.   | 1987    568200        . |
 10.   | 1988    568000        . |
       |------------------------------|
 11.   | 1989    570000        . |
       +------------------------------+
```

Because variable *jobless* occurs in *newf2* (1990 to 1995) but not in *newf1*, its 1985 to 1989 values are missing in the combined dataset. We can now put the observations in order from earliest to latest and save these combined data as a new file, *newf3.dta*:

```
. sort year

. list

       +------------------------------+
       | year      pop    jobless |
       |------------------------------|
  1.   | 1985    580700        . |
  2.   | 1986    580200        . |
  3.   | 1987    568200        . |
  4.   | 1988    568000        . |
  5.   | 1989    570000        . |
       |------------------------------|
  6.   | 1990    573400     42000 |
  7.   | 1991    573500     45000 |
  8.   | 1992    575600     49000 |
  9.   | 1993    584400     49000 |
 10.   | 1994    582400     50000 |
       |------------------------------|
 11.   | 1995    575449        . |
       +------------------------------+
```

```
. save newf3
```

append might be compared to lengthening a sheet of paper (that is, the dataset in memory) by taping a second sheet with new observations (rows) to its bottom. **merge**, in its simplest form, corresponds to "widening" our sheet of paper by taping a second sheet to its right side, thereby adding new variables (columns). For example, dataset *newf4.dta* contains further Newfoundland time series: the numbers of births and divorces over the years 1980 to 1994. Thus it has some observations in common with our earlier dataset *newf3.dta*, as well as one variable (*year*) in common, but it also has two new variables not present in *newf3.dta*.

```
. use newf4
(Newfoundland 1980-94)

. describe

Contains data from C:\data\newf4.dta
  obs:           15                        Newfoundland 1980-94
  vars:           3                        3 Jul 2005 10:49
  size:         150  (99.9% of memory free)
```

```
-------------------------------------------------------------------------
                storage  display    value
variable name   type     format     label     variable label
-------------------------------------------------------------------------
year            int      %9.0g                 Year
births          int      %9.0g                 Number of births
divorces        int      %9.0g                 Number of divorces
-------------------------------------------------------------------------
Sorted by:
```

. **list**

```
      +--------------------------+
      | year    births   divorces |
      |--------------------------|
  1.  | 1980    10332       555 |
  2.  | 1981    11310       569 |
  3.  | 1982     9173       625 |
  4.  | 1983     9630       711 |
  5.  | 1984     8560       590 |
      |--------------------------|
  6.  | 1985     8080       561 |
  7.  | 1986     8320       610 |
  8.  | 1987     7656      1002 |
  9.  | 1988     7396       884 |
 10.  | 1989     7996       981 |
      |--------------------------|
 11.  | 1990     7354       973 |
 12.  | 1991     6929       912 |
 13.  | 1992     6689       867 |
 14.  | 1993     6360       930 |
 15.  | 1994     6295       933 |
      +--------------------------+
```

We want to merge *newf3* with *newf4*, matching observations according to *year* wherever possible. To accomplish this, both datasets must be sorted by the index variable (which in this example is *year*). We earlier issued a **sort year** command before saving *newf3.dta*, so we now do the same with *newf4.dta*. Then we merge the two, specifying *year* as the index variable to match.

. **sort** *year*

. **merge** *year* **using** *newf3*

. **describe**

```
Contains data from newf4.dta
  obs:             16                    Newfoundland 1980-94
  vars:             6                    3 Jul 2005 10:49
  size:           304 (99.9% of memory free)
-------------------------------------------------------------------------
                storage  display    value
variable name   type     format     label     variable label
-------------------------------------------------------------------------
year            int      %9.0g                 Year
births          int      %9.0g                 Number of births
divorces        int      %9.0g                 Number of divorces
pop             float    %9.0g                 Population
jobless         float    %9.0g                 Number of people unemployed
_merge          byte     %8.0g
-------------------------------------------------------------------------
Sorted by:
     Note:  dataset has changed since last saved
```

. **list**

```
+----------------------------------------------------------+
      | year    births   divorces      pop   jobless   _merge |
      |----------------------------------------------------------|
  1.  | 1980    10332       555         .         .        1 |
  2.  | 1981    11310       569         .         .        1 |
  3.  | 1982     9173       625         .         .        1 |
  4.  | 1983     9630       711         .         .        1 |
  5.  | 1984     8560       590         .         .        1 |
      |----------------------------------------------------------|
  6.  | 1985     8080       561     580700         .        3 |
  7.  | 1986     8320       610     580200         .        3 |
  8.  | 1987     7656      1002     568200         .        3 |
  9.  | 1988     7396       884     568000         .        3 |
 10.  | 1989     7996       981     570000         .        3 |
      |----------------------------------------------------------|
 11.  | 1990     7354       973     573400     42000        3 |
 12.  | 1991     6929       912     573500     45000        3 |
 13.  | 1992     6689       867     575600     49000        3 |
 14.  | 1993     6360       930     584400     49000        3 |
 15.  | 1994     6295       933     582400     50000        3 |
      |----------------------------------------------------------|
 16.  | 1995        .         .     575449         .        2 |
      +----------------------------------------------------------+
```

In this example, we simply used **merge** to add new variables to our data, matching observations. By default, whenever the same variables are found in both datasets, those of the "master" data (the file already in memory) are retained and those of the "using" data are ignored. The **merge** command has several options, however, that override this default. A command of the following form would allow any *missing values* in the master data to be replaced by corresponding nonmissing values found in the using data (here, *newf5.dta*):

. **merge** *year* using *newf5*, **update**

Or, a command such as the following causes *any values* from the master data to be replaced by nonmissing values from the using data, if the latter are different:

. **merge** *year* using *newf5*, **update replace**

Suppose that the values of an index variable occur more than once in the master data; for example, suppose that the year 1990 occurs twice. Then values from the using data with *year* = 1990 are matched with each occurrence of *year* = 1990 in the master data. You can use this capability for many purposes, such as combining background data on individual patients with data on any number of separate doctor visits they made. Although **merge** makes this and many other data-management tasks straightforward, analysts should look closely at the results to be certain that the command is accomplishing what they intend.

As a diagnostic aid, **merge** automatically creates a new variable called *_merge*. Unless **update** was specified, *_merge* codes have the following meanings:

    1   Observation from the master dataset only.

    2   Observation from the using dataset only.

    3   Observation from both master and using data (using values ignored if different).

If the **update** option was specified, *_merge* codes convey what happened:

    1   Observation from the master dataset only.

    2   Observation from the using dataset only.

    3   Observation from both, master data agrees with using.

    4   Observation from both, master data updated if missing.

    5   Observation from both, master data replaced if different.

Before performing another **merge** operation, it will be necessary to discard or rename this variable. For example,

```
. drop _merge
```

Or,

```
. rename _merge _merge1
```

We can merge multiple datasets with a single **merge** command. For example, if *newf5.dta* through *newf8.dta* are four datasets, each sorted by the variable *year*, then merging all four with the master dataset could be accomplished as follows.

```
. merge year using newf5 newf6 newf7 newf8, update replace
```

Other **merge** options include checks on whether the merging-variable values are unique, and the ability to specify which variables to keep for the final dataset. Type **help merge** for details.

Transposing, Reshaping, or Collapsing Data

Long after a dataset has been created, we might discover that for some analytical purposes it has the wrong organization. Fortunately, several commands facilitate drastic restructuring of datasets. We will illustrate these using data (*growth1.dta*) on recent population growth in five eastern provinces of Canada. In these data, unlike our previous examples, province names are represented by a numerical variable with eight-character labels.

```
. use growth1, clear
(Eastern Canada growth)

. describe
Contains data from C:\data\growth1.dta
  obs:            5                          Eastern Canada growth
  vars:           5                          3 Jul 2005 10:48
  size:         105 (99.9% of memory free)
-------------------------------------------------------------------------
              storage  display     value
variable name  type    format      label    variable label
-------------------------------------------------------------------------
provinc2       byte    %8.0g       provinc2 Eastern Canadian province
grow92         float   %9.0g                Pop. gain in 1000s, 1991-92
grow93         float   %9.0g                Pop. gain in 1000s, 1992-93
grow94         float   %9.0g                Pop. gain in 1000s, 1993-94
grow95         float   %9.0g                Pop. gain in 1000s, 1994-95
-------------------------------------------------------------------------
Sorted by:
```

```
. list

     +-------------------------------------------------+
     | provinc2   grow92   grow93   grow94   grow95 |
     |-------------------------------------------------|
1. | New Brun       10      2.5      2.2      2.4 |
2. | Newfound       4.5      .8       -3     -5.8 |
3. | Nova Sco       12.1     5.8      3.5      3.9 |
4. |   Ontario      174.9   169.1    120.9    163.9 |
5. |    Quebec      80.6    77.4     48.5     47.1 |
     +-------------------------------------------------+
```

In this organization, population growth for each year is stored as a separate variable. We could analyze changes in the mean or variation of population growth from year to year. On the other hand, given this organization, Stata could not readily draw a simple time plot of population growth against year, nor can Stata find the correlation between population growth in New Brunswick and Newfoundland. All the necessary information is here, but such analyses require different organizations of the data.

One simple reorganization involves transposing variables and observations. In effect, the dataset rows become its columns, and vice versa. This is accomplished by the **xpose** command. The option **clear** is required with this command, because it always clears the present data from memory. Including the **varname** option creates an additional variable (named _varname) in the transposed dataset, containing original variable names as strings.

```
. xpose, clear varname

. describe

Contains data
  obs:              5
  vars:             6
  size:           160 (99.9% of memory free)
-----------------------------------------------------------------------------
                 storage  display     value
variable name     type    format      label        variable label
-----------------------------------------------------------------------------
v1                float   %9.0g
v2                float   %9.0g
v3                float   %9.0g
v4                float   %9.0g
v5                float   %9.0g
_varname          str8    %9s
-----------------------------------------------------------------------------
Sorted by:
     Note:   dataset has changed since last saved

. list

     +-------------------------------------------------+
     |   v1      v2      v3      v4      v5   _varname |
     |-------------------------------------------------|
1. |    1       2       3       4       5   provinc2 |
2. |   10      4.5     12.1    174.9    80.6    grow92 |
3. |  2.5      .8      5.8     169.1    77.4    grow93 |
4. |  2.2      -3      3.5     120.9    48.5    grow94 |
5. |  2.4     -5.8     3.9     163.9    47.1    grow95 |
     +-------------------------------------------------+
```

Value labels are lost along the way, so provinces in the transposed dataset are indicated only by their numbers (1 = New Brunswick, 2 = Newfoundland, and so on). The second through last values in each column are the population gains for that province, in thousands.

Thus, variable *v1* has a province identification number (1, meaning New Brunswick) in its first row, and New Brunswick's population growth values for 1992 to 1995 in its second through fifth rows. We can now find correlations between population growth in different provinces, for instance, by typing a **correlate** command with **in 2/5** (second through fifth observations only) qualifier:

```
. correlate v1-v5 in 2/5
(obs=4)
          |       v1       v2       v3       v4       v5
----------+---------------------------------------------
       v1 |   1.0000
       v2 |   0.8058   1.0000
       v3 |   0.9742   0.8978   1.0000
       v4 |   0.5070   0.4803   0.6204   1.0000
       v5 |   0.6526   0.9362   0.8049   0.6765   1.0000
```

The strongest correlation appears between the growth of neighboring maritime provinces New Brunswick (*v1*) and Nova Scotia (*v3*): *r* = .9742. Newfoundland's (*v2*) growth has a much weaker correlation with that of Ontario (*v4*): *r* = .4803.

More sophisticated restructuring is possible through the **reshape** command. This command switches datasets between two basic configurations termed "wide" and "long." Dataset *growth1.dta* is initially in wide format.

```
. use growth1, clear
(Eastern Canada growth)

. list

     +------------------------------------------------+
     | provinc2   grow92   grow93   grow94   grow95 |
     |------------------------------------------------|
  1. | New Brun        10      2.5      2.2      2.4 |
  2. | Newfound       4.5       .8       -3     -5.8 |
  3. | Nova Sco      12.1      5.8      3.5      3.9 |
  4. |  Ontario     174.9    169.1    120.9    163.9 |
  5. |   Quebec      80.6     77.4     48.5     47.1 |
     +------------------------------------------------+
```

A **reshape** command switches this to long format.

```
. reshape long grow, i(provinc2) j(year)
(note: j = 92 93 94 95)

Data                               wide   ->   long
-----------------------------------------------------------------
Number of obs.                        5   ->        20
Number of variables                   5   ->         3
j variable (4 values)                     ->      year
xij variables:
              grow92 grow93 ... grow95   ->      grow
-----------------------------------------------------------------
```

Listing the data shows how they were reshaped. A **sepby()** option with the **list** command produces a table with horizontal lines visually separating the provinces, instead of every five observations (the default).

. **list, sepby**(*provinc2*)

```
     +-------------------------+
     | provinc2   year   grow |
     |-------------------------|
  1. | New Brun     92     10  |
  2. | New Brun     93    2.5  |
  3. | New Brun     94    2.2  |
  4. | New Brun     95    2.4  |
     |-------------------------|
  5. | Newfound     92    4.5  |
  6. | Newfound     93     .8  |
  7. | Newfound     94     -3  |
  8. | Newfound     95   -5.8  |
     |-------------------------|
  9. | Nova Sco     92   12.1  |
 10. | Nova Sco     93    5.8  |
 11. | Nova Sco     94    3.5  |
 12. | Nova Sco     95    3.9  |
     |-------------------------|
 13. |  Ontario     92  174.9  |
 14. |  Ontario     93  169.1  |
 15. |  Ontario     94  120.9  |
 16. |  Ontario     95  163.9  |
     |-------------------------|
 17. |   Quebec     92   80.6  |
 18. |   Quebec     93   77.4  |
 19. |   Quebec     94   48.5  |
 20. |   Quebec     95   47.1  |
     +-------------------------+
```

. **label data** "Eastern Canadian growth--long"

. **label variable** *grow* "Population growth in 1000s"

. **save** *growth2*
file C:\data\growth2.dta saved

The **reshape** command above began by stating that we want to put the dataset in **long** form. Next, it named the new variable to be created, *grow*. The **i**(*provinc2*) option specified the observation identifier, or the variable whose unique values denote logical observations. In this example, each province forms a logical observation. The **j**(*year*) option specifies the sub-observation identifier, or the variable whose unique values (within each logical observation) denote sub-observations. Here, the sub-observations are years within each province.

Figure 2.1 shows a possible use for the long-format dataset. With one **graph** command, we can now produce time plots comparing the population gains in New Brunswick, Newfoundland, and Nova Scotia (observations for which *provinc2* < 4). The **graph** command on the following page calls for connected-line plots of *grow* (as *y*-axis variable) against *year* (*x* axis) if *province2* < 4, with horizontal lines at *y* = 0 (zero population growth), and separate plots for each value of *provinc2*.

```
. graph twoway connected grow year if provinc2 < 4, yline(0)
     by(provinc2)
```

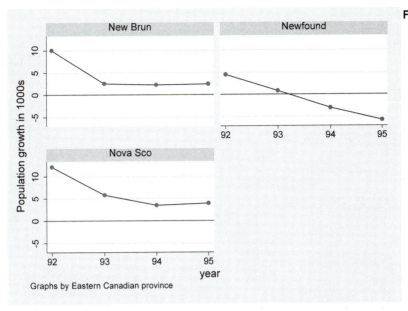

Figure 2.1

Declines in their fisheries during the early 1990s contributed to economic hardships in these three provinces. Growth slowed dramatically in New Brunswick and Nova Scotia, while Newfoundland (the most fisheries-dependent province) actually lost population.

reshape works equally well in reverse, to switch data from "long" to "wide" format. Dataset *growth3.dta* serves as an example of long format.

```
. use growth3, clear
(Eastern Canadian growth--long)

. list, sepby(provinc2)

     +---------------------------+
     | provinc2      grow   year |
     |---------------------------|
  1. | New Brun        10     92 |
  2. | New Brun       2.5     93 |
  3. | New Brun       2.2     94 |
  4. | New Brun       2.4     95 |
     |---------------------------|
  5. | Newfound       4.5     92 |
  6. | Newfound        .8     93 |
  7. | Newfound        -3     94 |
  8. | Newfound      -5.8     95 |
     |---------------------------|
  9. | Nova Sco      12.1     92 |
 10. | Nova Sco       5.8     93 |
 11. | Nova Sco       3.5     94 |
 12. | Nova Sco       3.9     95 |
     |---------------------------|
 13. |  Ontario     174.9     92 |
 14. |  Ontario     169.1     93 |
 15. |  Ontario     120.9     94 |
 16. |  Ontario     163.9     95 |
     |---------------------------|
```

```
17. |    Quebec     80.6      92 |
18. |    Quebec     77.4      93 |
19. |    Quebec     48.5      94 |
20. |    Quebec     47.1      95 |
     +-----------------------+
```

To convert this to wide format, we use **reshape wide** :

. **reshape wide** *grow*, **i**(*provinc2*) **j**(*year*)

(note: j = 92 93 94 95)

```
Data                                 long   ->   wide
-------------------------------------------------------------------
Number of obs.                          20   ->       5
Number of variables                      3   ->       5
j variable (4 values)                 year   ->   (dropped)
xij variables:
                                      grow   ->   grow92 grow93 ... grow95
-------------------------------------------------------------------
```

. **list**

```
    +---------------------------------------------+
    | provinc2   grow92   grow93   grow94   grow95 |
    |---------------------------------------------|
 1. | New Brun       10      2.5      2.2      2.4 |
 2. | Newfound      4.5       .8       -3     -5.8 |
 3. | Nova Sco     12.1      5.8      3.5      3.9 |
 4. |  Ontario    174.9    169.1    120.9    163.9 |
 5. |   Quebec     80.6     77.4     48.5     47.1 |
    +---------------------------------------------+
```

Notice that we have recreated the organization of dataset *growth1.dta*.

Another important tool for restructuring datasets is the **collapse** command, which creates an aggregated dataset of statistics (for example, means, medians, or sums). The long *growth3* dataset has four observations for each province:

. **use** *growth3*, **clear**
(Eastern Canadian growth--long)

. **list**, **sepby**(*provinc2*)

```
    +------------------------+
    | provinc2    grow   year |
    |------------------------|
 1. | New Brun      10     92 |
 2. | New Brun     2.5     93 |
 3. | New Brun     2.2     94 |
 4. | New Brun     2.4     95 |
    |------------------------|
 5. | Newfound     4.5     92 |
 6. | Newfound      .8     93 |
 7. | Newfound      -3     94 |
 8. | Newfound    -5.8     95 |
    |------------------------|
 9. | Nova Sco    12.1     92 |
10. | Nova Sco     5.8     93 |
11. | Nova Sco     3.5     94 |
12. | Nova Sco     3.9     95 |
    |------------------------|
13. |  Ontario   174.9     92 |
14. |  Ontario   169.1     93 |
15. |  Ontario   120.9     94 |
16. |  Ontario   163.9     95 |
```

```
     |-------------------------|
17.  |   Quebec    80.6    92  |
18.  |   Quebec    77.4    93  |
19.  |   Quebec    48.5    94  |
20.  |   Quebec    47.1    95  |
     +-------------------------+
```

We might want to aggregate the different years into a mean growth rate for each province. In the collapsed dataset, each observation will correspond to one value of the **by( )** variable, that is, one province.

```
. collapse (mean) grow, by(provinc2)
. list
      +----------------------+
      | provinc2        grow |
      |----------------------|
1.    | New Brun       4.275 |
2.    | Newfound  -.8750001  |
3.    | Nova Sco       6.325 |
4.    |  Ontario       157.2 |
5.    |   Quebec        63.4 |
      +----------------------+
```

For a slightly more complicated example, suppose we had a dataset similar to *growth3.dta* but also containing the variables *births*, *deaths*, and *income*. We want an aggregate dataset with each province's total numbers of births and deaths over these years, the mean income (to be named *meaninc*), and the median income (to be named *medinc*). If we do not specify a new variable name, as with *grow* in the previous example, or *births* and *deaths*, the collapsed variable takes on the same name as the old variable.

```
. collapse (sum) births deaths (mean) meaninc = income
      (median) medinc = income, by(provinc2)
```

collapse can create variables based on the following summary statistics:

mean	Means (the default; used if the type of statistic is not specified)
sd	Standard deviations
sum	Sums
rawsum	Sums ignoring optionally specified weight
count	Number of nonmissing observations
max	Maximums
min	Minimums
median	Medians
p1	1st percentiles
p2	2nd percentiles (and so forth to **p99**)
iqr	Interquartile ranges

Weighting Observations

Stata understands four types of weighting:

aweight Analytical weights, used in weighted least squares (WLS) regression and similar procedures.

fweight Frequency weights, counting the number of duplicated observations. Frequency weights must be integers.

iweight Importance weights, however you define "importance."

pweight Probability or sampling weights, equal to the inverse of the probability that an observation is included due to sampling strategy.

Researchers sometimes speak of "weighted data." This might mean that the original sampling scheme selected observations in a deliberately disproportionate way, as reflected by weights equal to 1/(*probability of selection*). Appropriate use of **pweight** can compensate for disproportionate sampling in certain analyses. On the other hand, "weighted data" might mean something different — an aggregate dataset, perhaps constructed from a frequency table or cross-tabulation, with one or more variables indicating how many times a particular value or combination of values occurred. In that case, we need **fweight**.

Not all types of weighting have been defined for all types of analyses. We cannot, for example, use **pweight** with the **tabulate** command. Using weights in any analysis requires a clear understanding of what we want weighting to accomplish in that particular analysis. The weights themselves can be any variable in the dataset.

The following small dataset (*nfschool.dta*), containing results from a survey of 1,381 rural Newfoundland high school students, illustrates a simple application of frequency weighting.

```
. describe
Contains data from C:\data\nfschool.dta
  obs:            6                        Newf.school/univer.(Seyfrit 93)
 vars:            3                        3 Jul 2005 10:50
 size:           48 (99.9% of memory free)
--------------------------------------------------------------------------------
              storage  display    value
variable name  type    format     label      variable label
--------------------------------------------------------------------------------
univers        byte    %8.0g      yes        Expect to attend university?
year           byte    %8.0g                 What year of school now?
count          int     %8.0g                 observed frequency
--------------------------------------------------------------------------------
Sorted by:

. list, sep(3)
     +----------------------+
     | univers  year  count |
     |----------------------|
  1. |      no    10    210 |
  2. |      no    11    260 |
  3. |      no    12    274 |
     |----------------------|
  4. |     yes    10    224 |
  5. |     yes    11    235 |
  6. |     yes    12    178 |
     +----------------------+
```

At first glance, the dataset seems to contain only 6 observations, and when we cross-tabulate whether students expect to attend a university (*univers*) by their current year in high school (*year*), we get a table with one observation per cell.

```
. tabulate univers year
```

Expect to attend university ?	What year of school now? 10	11	12	Total
no	1	1	1	3
yes	1	1	1	3
Total	2	2	2	6

To understand these data, we need to apply frequency weights. The variable *count* gives frequencies: 210 of these students are tenth graders who said they did not expect to attend a university, 260 are eleventh graders who said no, and so on. Specifying **[fweight = count]** obtains a cross-tabulation showing responses of all 1,381 students.

```
. tabulate univers year [fweight = count]
```

Expect to attend university ?	What year of school now? 10	11	12	Total
no	210	260	274	744
yes	224	235	178	637
Total	434	495	452	1,381

Carrying the analysis further, we might add options asking for a table with column percentages (**col**), no cell frequencies (**nof**), and a χ^2 test of independence (**chi2**). This reveals a statistically significant relationship ($P = .001$). The percentage of students expecting to go to college declines with each year of high school.

```
. tabulate univers year [fw = count], col nof chi2
```

Expect to attend university ?	What year of school now? 10	11	12	Total
no	48.39	52.53	60.62	53.87
yes	51.61	47.47	39.38	46.13
Total	100.00	100.00	100.00	100.00

Pearson chi2(2) = 13.8967   Pr = 0.001

Survey data often reflect complex sampling designs, based on one or more of the following:

disproportionate sampling — for example, oversampling particular subpopulations, in order to get enough cases to draw conclusions about them.

clustering — for example, selecting voting precincts at random, and then sampling individuals within the selected precincts.

stratification — for example, dividing precincts into "urban" and "rural" strata, and then sampling precincts and/or individuals within each stratum.

Complex sampling designs require specialized analytical tools. **pweights** and Stata's ordinary analytical commands do not suffice.

Stata's procedures for complex survey data include special tabulation, means, regression, logit, probit, tobit, and Poisson regression commands. Before applying these commands, users must first set up their data by identifying variables that indicate the PSUs (primary sampling units) or clusters, strata, finite population correction, and probability weights. This is accomplished through the **svyset** command. For example:

```
. svyset precinct [pweight=invPsel], strata(urb_rur) fpc(finite)
```

For each observation in this example, the value of variable *princinct* identifies PSU or cluster. Values of *urb_rur* identify the strata, *finite* gives the finite population correction, and *invPsel* gives the probability weight or inverse of the probability of selection. After the data have been **svyset** and saved, the survey analytical procedures are relatively straightforward. Commands are typically prefixed by **svy:** , as in

```
svy:   mean income
```

or

```
svy:   regress income education experience gender
```

The *Survey Data Reference Manual* contains full details and examples of Stata's extensive survey-analysis capabilities. For online guidance, type **help svy** and follow the links to particular commands.

Creating Random Data and Random Samples

The pseudo-random number function **uniform()** lies at the heart of Stata's ability to generate random data or to sample randomly from the data at hand. The *Base Reference Manual* (Functions) provides a technical description of this 32-bit pseudo-random generator. If we presently have data in memory, then a command such as the following creates a new variable named *randnum*, having apparently random 16-digit values over the interval [0,1) for each case in the data.

```
. generate randnum = uniform()
```

Alternatively, we might create a random dataset from scratch. Suppose we want to start a new dataset containing 10 random values. We first clear any other data from memory (if they were valuable, **save** them first). Next, set the number of observations desired for the new dataset. Explicitly setting the seed number makes it possible to later reproduce the same "random" results. Finally, we generate our random variable.

```
. clear
. set obs 10
obs was 0, now 10
. set seed 12345
. generate randnum = uniform()
```

```
. list
     +----------+
     | randnum  |
     |----------|
  1. |  .309106 |
  2. | .6852276 |
  3. | .1277815 |
  4. | .5617244 |
  5. | .3134516 |
     |----------|
  6. | .5047374 |
  7. | .7232868 |
  8. | .4176817 |
  9. | .6768828 |
 10. | .3657581 |
     +----------+
```

In combination with Stata's algebraic, statistical, and special functions, **uniform()** can simulate values sampled from a variety of theoretical distributions. If we want *newvar* sampled from a uniform distribution over [0,428) instead of the usual [0,1), we type

```
. generate newvar = 428 * uniform()
```

These will still be 16-digit values. Perhaps we want only integers from 1 to 428 (inclusive):

```
. generate newvar = 1 + trunc(428 * uniform())
```

To simulate 1,000 rolls of a six-sided die, type

```
. clear

. set obs 1000
obs was 0, now 1000

. generate roll = 1 + trunc(6 * uniform())

. tabulate roll
```

```
        die |      Freq.     Percent        Cum.
------------+-----------------------------------
          1 |        171       17.10       17.10
          2 |        164       16.40       33.50
          3 |        150       15.00       48.50
          4 |        170       17.00       65.50
          5 |        169       16.90       82.40
          6 |        176       17.60      100.00
------------+-----------------------------------
      Total |       1000      100.00
```

We might theoretically expect 16.67% ones, 16.67% twos, and so on, but in any one sample like these 1,000 "rolls," the observed percentages will vary randomly around their expected values.

To simulate 1,000 rolls of a pair of six-sided dice, type

```
. generate dice = 2 + trunc(6 * uniform()) + trunc(6 * uniform())

. tabulate dice
```

```
       dice |      Freq.     Percent        Cum.
------------+-----------------------------------
          2 |         26        2.60        2.60
          3 |         62        6.20        8.80
          4 |         78        7.80       16.60
          5 |        120       12.00       28.60
          6 |        153       15.30       43.90
          7 |        149       14.90       58.80
          8 |        146       14.60       73.40
```

```
    9 |           96           9.60        83.00
   10 |           88           8.80        91.80
   11 |           53           5.30        97.10
   12 |           29           2.90       100.00
------------+-------------------------------------
 Total |         1000         100.00
```

We can use _n to begin an artificial dataset as well. The following commands create a new 5,000-observation dataset with one variable named *index*, containing values from 1 to 5,000.

```
. set obs 5000
obs was 0, now 5000

. generate index = _n

. summarize
   Variable |        Obs        Mean    Std. Dev.        Min        Max
------------+--------------------------------------------------------------
      index |       5000      2500.5     1443.52          1       5000
```

It is possible to generate variables from a normal (Gaussian) distribution using **uniform()**. The following example creates a dataset with 2,000 observations and 2 variables, z from an N(0,1) population, and x from N(500,75).

```
. clear

. set obs 2000
obs was 0, now 2000

. generate z = invnormal(uniform())

. generate x = 500 + 75*invnormal(uniform())
```

The actual sample means and standard deviations differ slightly from their theoretical values:

```
. summarize
   Variable |        Obs        Mean    Std. Dev.        Min        Max
------------+--------------------------------------------------------------
          z |       2000    .0375032    1.026784  -3.536209   4.038878
          x |       2000     503.322    75.68551   244.3384   743.1377
```

If z follows a normal distribution, $v = e^z$ follows a lognormal distribution. To form a lognormal variable v based upon a standard normal z,

```
. generate v = exp(invnormal(uniform()))
```

To form a lognormal variable w based on an N(100,15) distribution,

```
. generate w = exp(100 + 15*invnormal(uniform()))
```

Taking logarithms, of course, normalizes a lognormal variable.

To simulate y values drawn randomly from an exponential distribution with mean and standard deviation $\mu = \sigma = 3$,

```
. generate y = -3 * ln(uniform())
```

For other means and standard deviations, substitute other values for 3.

X1 follows a χ^2 distribution with one degree of freedom, which is the same as a squared standard normal:

```
. generate X1 = (invnormal(uniform()))^2
```

By similar logic, *X2* follows a χ^2 with two degrees of freedom:

```
. generate X2 = (invnormal(uniform()))^2 + (invnormal(uniform()))^2
```

Other statistical distributions, including t and F, can be simulated along the same lines. In addition, programs have been written for Stata to generate random samples following distributions such as binomial, Poisson, gamma, and inverse Gaussian.

Although **invnormal(uniform())** can be adjusted to yield normal variates with particular correlations, a much easier way to do this is through the **drawnorm** command. To generate 5,000 observations from N(0,1), type

```
. clear

. drawnorm z, n(5000)

. summ

    Variable |      Obs        Mean    Std. Dev.      Min         Max
-------------+--------------------------------------------------------
           z |     5000   -.0005951    1.019788   -4.518918    3.923464
```

Below, we will create three further variables. Variable $x1$ is from an N(0,1) population, variable $x2$ is from N(100,15), and $x3$ is from N(500,75). Furthermore, we define these variables to have the following population correlations:

	x1	x2	x3
x1	1.0	0.4	−0.8
x2	0.4	1.0	0.0
x3	−0.8	0.0	1.0

The procedure for creating such data requires first defining the correlation matrix C, and then using C in the **drawnorm** command:

```
. mat C = (1, .4, -.8 \ .4, 1, 0 \ -.8, 0, 1)

. drawnorm x1 x2 x3, means(0,100,500) sds(1,15,75) corr(C)

. summarize x1-x3

    Variable |      Obs        Mean    Std. Dev.      Min         Max
-------------+--------------------------------------------------------
          x1 |     5000    .0024364     1.01648   -3.478467    3.598916
          x2 |     5000    100.1826    14.91325    46.13897    150.7634
          x3 |     5000    500.7747    76.93925    211.5596    769.6074

. correlate x1-x3
(obs=5000)
             |      x1        x2        x3
-------------+---------------------------
          x1 |   1.0000
          x2 |   0.3951    1.0000
          x3 |  -0.8134   -0.0072    1.0000
```

Compare the sample variables' correlations and means with the theoretical values given earlier. Random data generated in this fashion can be viewed as samples drawn from theoretical populations. We should not expect the samples to have exactly the theoretical population parameters (in this example, an $x3$ mean of 500, $x1$–$x2$ correlation of 0.4, $x1$–$x3$ correlation of −.8, and so forth).

The command **sample** makes unobtrusive use of **uniform**'s random generator to obtain random samples of the data in memory. For example, to discard all but a 10% random sample of the original data, type

```
. sample 10
```

When we add an **in** or **if** qualifier, **sample** applies only to those observations meeting our criteria. For example,

```
. sample 10 if age < 26
```

would leave us with a 10% sample of those observations with *age* less than 26, plus 100% of the original observations with *age* $\geq$ 26.

We could also select random samples of a particular size. To discard all but 90 randomly-selected observations from the dataset in memory, type

```
. sample 90, count
```

The sections in Chapter 14 on bootstrapping and Monte Carlo simulations provide further examples of random sampling and random variable generation.

Writing Programs for Data Management

Data management on larger projects often involves repetitive or error-prone tasks that are best handled by writing specialized Stata programs. Advanced programming can become very technical, but we can also begin by writing simple programs that consist of nothing more than a sequence of Stata commands, typed and saved as an ASCII file. ASCII files can be created using your favorite word processor or text editor, which should offer "ASCII text file" among its options under File – Save As. An even easier way to create such text files is through Stata's Do-file Editor, which is brought up by clicking Window – Do-file Editor or the icon. Alternatively, bring up the Do-file Editor by typing the command **doedit**, or **doedit** *filename* if *filename* exists.

For example, using the Do-file Editor we might create a file named *canada.do* (which contains the commands to read in a raw data file named *canada.raw*), then label the dataset and its variables, compress it, and save it in Stata format. The commands in this file are identical to those seen earlier when we went through the example step by step.

```
infile str30 place pop unemp mlife flife using canada.raw
label data "Canadian dataset 1"
label variable pop "Population in 1000s, 1995"
label variable unemp "% 15+ population unemployed, 1995"
label variable mlife "Male life expectancy years"
label variable flife "Female life expectancy years"
compress
save canada1, replace
```

Once this *canada.do* file has been written and saved, simply typing the following command causes Stata to read the file and run each command in turn:

```
. do canada
```

Such batch-mode programs, termed "do-files," are usually saved with a .do extension. More elaborate programs (defined by do-files or "automatic do" files) can be stored in memory, and can call other programs in turn — creating new Stata commands and opening worlds of possibility for adventurous analysts. The Do-file Editor has several other features that you might find useful. Chapter 3 describes a simple way to use do-files in building graphs. For further information, see the *Getting Started* manual on Using the Do-file Editor.

Stata ordinarily interprets the end of a command line as the end of that command. This is reasonable onscreen, where the line can be arbitrarily long, but does not work as well when we are typing commands in a text file. One way to avoid line-length problems is through the #delimit command, which can set some other character as the end-of-command delimiter. In the following example, we make a semicolon the delimiter; then type two long commands that do not end until a semicolon appears; and then finally reset the delimiter to its usual value, a carriage return (cr):

```
#delimit ;
infile str30 place pop unemp mlife flife births deaths
    marriage medinc mededuc using newcan.raw;
order place pop births deaths marriage medinc mededuc
    unemp mlife flife;
#delimit cr
```

Stata normally pauses each time the Results window becomes full of information, and waits to proceed until we press any key (or 🔘). Instead of pausing, we can ask Stata to continue scrolling until the output is complete. Typed in the Command window or as part of a program, the command

. **set more off**

calls for continuous scrolling. This is convenient if our program produces much screen output that we don't want to see, or if it is writing to a log file that we will examine later. Typing

. **set more on**

returns to the usual mode of waiting for keyboard input before scrolling.

Managing Memory

When we **use** or File – Open a dataset, Stata reads the disk file and loads it into memory. Loading the data into memory permits rapid analysis, but it is only possible if the dataset can fit within the amount of memory currently allocated to Stata. If we try to open a dataset that is too large, we get an elaborate error message saying "no room to add more observations," and advising what to do next.

. **use C:\data*gbank2.dta***

```
(Scientific surveys off S. Newfoundland)
no room to add more observations
    An attempt was made to increase the number of observations beyond what is
    currently possible.  You have the following alternatives:
    1.  Store your variables more efficiently; see help compress.  (Think of
        Stata's data area as the area of a rectangle; Stata can trade off width
        and length.)
    2.  Drop some variables or observations; see help drop.
```

```
    3.  Increase the amount of memory allocated to the data area using the set
        memory command; see help memory.
r(901);
```

Small Stata allocates a fixed amount of memory to data, and this limit cannot be changed. Intercooled Stata and Stata/SE versions are flexible, however. Default allocations equal 1 megabyte for Intercooled, and 10 megabytes for Stata/SE. If we have Intercooled or Stata/SE, running on a computer with enough physical memory, we can set Stata's memory allocation higher with the **set memory** command. To allocate 20 megabytes to data, type

```
. set memory 20m
Current memory allocation

                        current                                    memory usage
        settable         value    description                      (1M = 1024k)
        -------------------------------------------------------------------------
        set maxvar        5000    max. variables allowed               1.733M
        set memory         20M    max. data space                     20.000M
        set matsize        400    max. RHS vars in models              1.254M
                                                                    -----------
                                                                       22.987M
```

If there are data already in memory, first type the command **clear** to remove them. To reset the memory allocation "permanently," so it will be the same next time we start up, type

```
. set memory 20m, permanently
```

In the example given earlier, *gbank2.dta* is a 11.3-megabyte dataset that would not fit into the default allocation. Asking for a 20-megabyte allocation has now given us more than enough room for these data.

```
Contains data from C:\data\gbank2.dta
  obs:        74,078                       Spring scientific surveys NAFO
                                           3KLNOPQ, 1971-93
  vars:           44                       2 Mar 2000 21:28
  size:    11,333,934 (46.0% of memory free)
-------------------------------------------------------------------------------
              storage  display   value
variable name   type   format    label    variable label
-------------------------------------------------------------------------------
id            float    %9.0g               original case number
rec_type      byte     %4.0g
vessel        byte     %4.0g               Vessel
trip          int      %8.0g               Trip number
set           int      %8.0g               Set number
rank          int      %8.0g
assembla      str7     %7s
year          byte     %4.0g               Year
month         byte     %4.0g               Month
day           byte     %4.0g               Day
set_type      byte     %8.0g    set_type   Set type
stratum       int      %8.0g               Stratum or line fished
division      str2     %2s                 NAFO division
unit_are      str3     %3s                 Nfld. area grid map square
light         int      %8.0g               Light conditions
wind_dir      byte     %4.0g               Wind direction
wind_for      byte     %4.0g               Wind force
sea           byte     %4.0g
bottom        byte     %4.0g               Type of bottom
time_mid      int      %8.0g               Time (midpoint)
duration      byte     %8.0g               Duration of set
tow_dist      int      %8.0g               Distance towed
```

```
gear_op          byte     %4.0g              Operation of gear
depthcat         byte     %4.0g              Category of depth
min_dept         int      %8.0g              Depth (minumum)
max_dept         int      %8.0g              Depth (maximum)
bot_dept         int      %8.0g              Depth (bottom if MWT)
temp_sur         int      %8.0g              Temperature (surface)
tempcat          byte     %8.0g              Category of temperature
temp_fs_         int      %8.0g              Temperature (fishing depth)
lat              float    %9.0g              Latitude (decimal)
long             float    %9.0g              Longitude (decimal)
pos_meth         byte     %4.0g
gear             int      %8.0g              Gear
total            byte     %9.0g
species          int      %8.0g              Species
number           long     %9.0g              Number of individual fish
weight           double   %9.0g              Catch weight in kilograms
latin            str31    %31s               Species -- Latin name
common           str27    %27s               Species -- common name
surtemp          float    %9.0g              Surface temperature degrees C
fishtemp         float    %9.0g              Fishing depth temperature C
depth            int      %9.0g              Mean trawl depth in meters
ispecies         byte     %9.0g              Indicator species
-------------------------------------------------------------------------
Sorted by:  id
```

Dataset *gbank2.dta* contains 74,078 observations from scientific surveys of fish populations on Newfoundland's Grand Banks, conducted over the years 1971 to 1993. When we **describe** the data (above), Stata reports "46.09% of memory free," meaning not 46% of the computer's total resources, but 46% of the 20 megabytes we allocated for Stata data. It is usually advisable to ask for more memory than our data actually require. Many statistical and data-management operations consume additional memory, in part because they temporarily create new variables as they work.

It is possible to **set memory** to values higher than the computer's available physical memory. In that case, Stata uses "virtual memory," which is really disk storage. Although virtual memory allows bypassing hardware limitations, it can be terribly slow. If you regularly work with datasets that push the limits of your computer, you might soon conclude that it is time to buy more memory.

Type **help limits** to see a list of limitations in Stata, not only on dataset size but also other dimensions including matrix size, command lengths, lengths of names, and numbers of variables in commands. Some of these limitations can be adjusted by the user.

3

Graphs

Graphs appear in every chapter of this book — one indication of their value and integration with other analyses in Stata. Indeed, graphics have always been one of Stata's strong suits, and reason enough for many users to choose Stata over other packages. The **graph** command evolved incrementally from Stata versions 1 through 7. Stata version 8 marked a major step forward, however. **graph** underwent a fundamental redesign, expanding its capabilities for sophisticated, publication-quality analytical graphics. Output appearance and choices were much improved as well. With the new **graph** command syntax and defaults, or alternatively through the new menus, attractive (and publishable) basic graphs are quite easy to draw. Graphically ambitious users who visualize non-basic graphs will find their efforts supported by a truly impressive array of tools and options, described in the 500-page *Graphics Reference Manual*.

In the much shorter space of this chapter, the spectrum from elementary to creative graphing will be covered taking an example- rather than syntax-oriented approach (see the *Graphics Reference Manual* or **help graph** for thorough coverage of syntax). We begin by illustrating seven basic types of graphs.

histogram	histograms
graph twoway	two-variable scatterplots, line plots, and many others
graph matrix	scatterplot matrices
graph box	box plots
graph pie	pie charts
graph bar	bar charts
graph dot	dot plots

For each of these basic types, there exist many options. That is especially true for the versatile **twoway** type.

More specialized graphs such as symmetry plots, quantile plots, and quantile–normal plots exist as well, for examining details of variable distributions. A few examples of these, and also of graphs for industrial quality control, appear in this chapter. Type **help graph_other** for more details.

Finally, the chapter concludes with techniques particularly useful in building data-rich, self-contained graphics for publication. Such techniques include adding text to graphs, overlaying multiple twoway plots, retrieving and reformatting saved graphs, and combining multiple graphs into one. As our graphing commands grow more complicated, simple batch programs (do-files) can help to write and re-use them. The full range of graphical choices goes far beyond what this book can cover, but the concluding examples point out a few of the possibilities. Later chapters supply further examples.

The Graphics menu provides point-and-click access to most of these graphing procedures.

A note to long-time Stata users: The graphical capabilities of Stata 8 and 9 outshine those of earlier versions. For analysts comfortable with old Stata, there is much new material to learn. Menus allow a quick entry, and the new graphics commands, like the old ones, follow a consistent logic that becomes clear with practice. Fortunately, the changeover need not be sudden. Version 7-style graphics remain available if needed. They have been moved to the command **graph7**. For example, an old-version scatterplot would formerly have been drawn by the command

```
. graph income education
```

which does not work in the newer Stata. Instead, the command

```
. graph7 income education
```

will reproduce the familiar old type of graph. The options of **graph7** are similar to those of the old-style **graph**. To see an updated version of this same scatterplot, type the new graphics command

```
. graph twoway scatter income education
```

Further examples of new commands appear in the next section, which should give a sense of what has changed (and what is familiar) with the redesigned graphical capabilities.

Example Commands

```
. histogram y, frequency
```
Draws histogram of variable y, showing frequencies on the vertical axis.

```
. histogram y, start(0) width(10) norm fraction
```
Draws histogram of y with bins 10 units wide, starting at 0. Adds a normal curve based on the sample mean and standard deviation, and shows fraction of the data on the vertical axis.

```
. histogram y, by(x, total) fraction
```
In one figure, draws separate histograms of y for each value of x, and also a "total" histogram for the sample as a whole.

```
. kdensity x, generate(xpoints xdensity) width(20) biweight
```
Produces and graphs kernel density estimate of the distribution of x. Two new variables are created: *xpoints* containing the x values at which the density is estimated, and *xdensity* with the density estimates themselves. **width(20)** specifies the halfwidth of the kernel, in units of the variable x. (If **width()** is not specified, the default follows a simple formula for "optimal.") The **biweight** option in this example calls for a biweight kernel, instead of the default **epanechnikov**.

```
. graph twoway scatter y x
```
Displays a basic two-variable scatterplot of y against x.

. `graph twoway lfit` *y* *x*   `||`   `scatter` *y* *x*
Visualizes the linear regression of *y* on *x* by overlaying two **twoway** graphs: the regression (linear fit or `lfit`) line, and the *y* vs. *x* scatterplot To include a 95% confidence band for the regression line, replace `lfit` with `lfitci` .

. `graph twoway scatter` *y* *x*, `xlabel(0(10)100) ylabel(-3(1)6, horizontal)`
Constructs scatterplot of *y* vs. *x*, with *x* axis labeled at 0, 10, ..., 100. *y* axis is labeled at –3, –2, ..., 6, with labels written horizontally instead of vertically (the default).

. `graph twoway scatter` *y* *x*, `mlabel(`*country*`)`
Constructs scatterplot of *y* vs. *x*, with data points (markers) labeled by the values of variable *country*.

. `graph twoway scatter` *y* *x1*, `by(`*x2*`)`
In one figure, draws separate *y* vs. *x1* scatterplots for each value of *x2*.

. `graph twoway scatter` *y* *x1* `[fweight = `*population*`]`, `msymbol(Oh)`
Draws a scatterplot of *y* vs. *x1*. Marker symbols are hollow circles **(Oh)**, with their size (area) proportional to frequency-weight variable *population*.

. `graph twoway connected` *y* *time*
A basic time plot of *y* against *time*. Data points are shown connected by line segments. To include line segments but no data-point markers, use `line` instead of `connected` :
  . `graph twoway line` *y* *time*

. `graph twoway line` *y1* *y2* *time*
Draws a time plot (in this example, a line plot) with two *y* variables that both have the same scale, and are graphed against an *x* variable named *time*.

. `graph twoway line` *y1* *time*, `yaxis(1)` `||` `line` *y2* *time*, `yaxis(2)`
Draws a time plot with two *y* variables that have different scales, by overlaying two individual line plots. The left-hand *y* axis, **yaxis(1)**, gives the scale for *y1*, while the right-hand *y* axis, **yaxis(2)**, gives the scale for *y2*.

. `graph matrix` *x1* *x2* *x3* *x4* *y*
Constructs a scatterplot matrix, showing all possible scatterplot pairs among the variables listed.

. `graph box` *y1* *y2* *y3*
Constructs box plots of variables *y1*, *y2*, and *y3*.

. `graph box` *y*, `over(`*x*`) yline(.22)`
Constructs box plots of *y* for each value of *x*, and draws a horizontal line at *y* = .22.

. `graph pie` *a* *b* *c*, `pie`
Draws one pie chart with slices indicating the relative amounts of variables *a*, *b*, and *c*. The variables must have similar units.

. `graph bar (sum)` *a* *b* *c*
Shows the sums of variables *a*, *b*, and *c* as side-by-side bars in a bar chart. To obtain means instead of sums, type `graph bar (mean)` *a* *b* *c* . Other options include bars representing medians, percentiles, or counts of each variable.

. `graph bar (mean)` *a*, `over(`*x*`)`
Draws a bar chart showing the mean of variable *a* at each value of variable *x*.

. `graph bar (asis)` *a b c*`, over(`*x*`) stack`

> Draws a bar chart in which the values ("as is") of variables *a*, *b*, and *c* are stacked on top of one another, at each value of variable *x*.

. `graph dot (median)` *y*`, over(`*x*`)`

> Draws a dot plot, in which dots along a horizontal scale mark the median value of *y* at each level of *x*. Other options include means, percentiles, or counts of each variable.

. `qnorm` *y*

> Draws a quantile–normal plot (normal probability plot) showing quantiles of *y* versus corresponding quantiles of a normal distribution.

. `rchart` *x1 x2 x3 x4 x5*`, connect(1)`

> Constructs a quality-control R chart graphing the range of values represented by variables *x1 – x5*.

Graph options, such as those controlling titles, labels, and tick marks on the axes are common across graph types wherever this makes sense. Moreover, the underlying logic of Stata's graph commands is consistent from one type to the next. These common elements are the key to gaining graph-building fluency, as the basics begin to fall into place.

Histograms

Histograms, displaying the distribution of measurement variables, are most easily produced with their own command **histogram**. For examples, we turn to *states.dta*, which contains selected environment and education measures on the 50 U.S. states plus the District of Columbia (data from the League of Conservation Voters 1991; National Center for Education Statistics 1992, 1993; World Resources Institute 1993).

. `use states`
(U.S. states data 1990-91)

. `describe`

```
Contains data from c:\data\states.dta
  obs:            51                          U.S. states data 1990-91
  vars:           21                          4 Jul 2005 12:07
  size:        4,080 (99.9% of memory free)
-------------------------------------------------------------------------------
              storage  display    value
variable name   type   format     label      variable label
-------------------------------------------------------------------------------
state          str20   %20s                   State
region         byte    %9.0g      region      Geographical region
pop            float   %9.0g                  1990 population
area           float   %9.0g                  Land area, square miles
density        float   %7.2f                  People per square mile
metro          float   %5.1f                  Metropolitan area population, %
waste          float   %5.2f                  Per capita solid waste, tons
energy         int     %8.0g                  Per capita energy consumed, Btu
miles          int     %8.0g                  Per capita miles/year, 1,000
toxic          float   %5.2f                  Per capita toxics released, lbs
green          float   %5.2f                  Per capita greenhouse gas, tons
house          byte    %8.0g                  House '91 environ. voting, %
senate         byte    %8.0g                  Senate '91 environ. voting, %
csat           int     %9.0g                  Mean composite SAT score
vsat           int     %8.0g                  Mean verbal SAT score
```

```
msat            int     %8.0g           Mean math SAT score
percent         byte    %9.0g           % HS graduates taking SAT
expense         int     %9.0g           Per pupil expenditures prim&sec
income          long    %10.0g          Median household income, $1,000
high            float   %9.0g           % adults HS diploma
college         float   %9.0g           % adults college degree
--------------------------------------------------------------------------
Sorted by:  state
```

Figure 3.1 shows a simple histogram of *college*, the percentage of a state's over-25 population with a bachelor's degree or higher. It was produced by the following command:

```
. histogram college, frequency title("Figure 3.1")
```

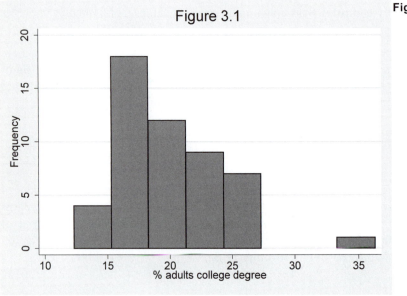

Under the Prefs – Graph Preferences menus, we have the choice of several pre-designed "schemes" for the default colors and shading of our graphs. Custom schemes can be defined as well. The examples in this book employ the s2 mono (monochrome) scheme, which among other things calls for shaded margins around each graph. The s1 mono scheme does not have such margins. Experimenting with the different monochrome and color schemes helps to determine which works best for a particular purpose. A graph drawn and saved under one scheme can subsequently be retrieved and re-saved under a different one, as described later in this chapter.

Options can be listed in any order following the comma in a graph command. Figure 3.1 illustrates two options: frequency (instead of density, the default) is shown on the vertical axis; and the title "Figure 3.1" appears over the graph. Once a graph is onscreen, menu choices provide the easiest way to print it, save it to disk, or cut and paste it into another program such as a word processor.

Figure 3.1 reveals the positive skew of this distribution, with a mode above 15 and an outlier around 35. It is hard to describe the graph more specifically because the bars do not line up with *x*-axis tick marks. Figure 3.2 contains a version with several improvements (based on some quick experiments to find the right values):

1. The *x* axis is labeled from 12 to 34, in increments of 2.

2. The *y* axis is labeled from 0 to 12, in increments of 2.

3. Tick marks are drawn on the *y* axis from 1 to 13, in increments of 2.

4. The histogram's first bar (bin) starts at 12.

5. The width of each bar (bin) is 2.

```
. histogram college, frequency title("Figure 3.2") xlabel(12(2)34)
      ylabel(0(2)12) ytick(1(2)13) start(12) width(2)
```

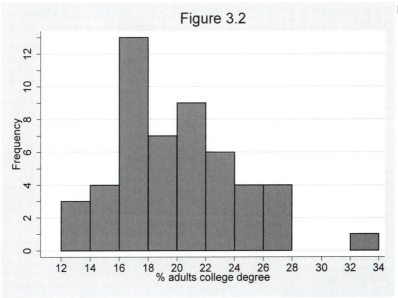

Figure 3.2

Figure 3.2

Figure 3.2 helps us to describe the distribution more specifically. For example, we now see that in 13 states, the percent with college degrees is between approximately 16 and 18.

Other useful **histogram** options include:

bin(#)   Draw a histogram with # bins (bars). We can specify either **bin(#)** or, as in Figure 3.2, **start(#)** and **width(#)** — but not both.

percent   Show percentages on the vertical axis. **ylabel** and **ytick** then refer to percentage values. Another possibility, **frequency**, is illustrated in Figure 3.2. We could also ask for **fraction** of the data. The default histogram shows **density**, meaning that bars are scaled so that the sum of their areas equals 1.

gap(#)   Leave a gap between bars. # is relative, $0 \le \# < 100$; experiment to find a suitable value.

addlabels   Label the heights of histogram bars. A separate option, **addlabopts**, controls the how the labels look.

discrete   Specify discrete data, requiring one bar for each value of *x*.

norm   Overlay a normal curve on the histogram, based on sample mean and standard deviation.

kdensity Overlay a kernal-density estimate on the histogram. The option **kdenopts** controls density computation; see **help kdensity** for details.

With histograms or most other graphs, we can also override the defaults and specify our own titles for the horizontal and vertical axes. The option **ytitle** controls *y*-axis titles, and **xtitle** controls *x*-axis titles. Figure 3.3 illustrates such titles, together with some other histogram options. Note the incremental buildup from basic (Figure 3.1) to more elaborate (Figure 3.3) graphs. This is the usual pattern of graph construction in Stata: we start simply, then experimentally add options to earlier commands retrieved from the Review window, as we work toward an image that most clearly presents our findings. Figure 3.3 actually is over-elaborate, but drawn here to show off multiple options.

```
. histogram college, frequency title("Figure 3.3") ylabel(0(2)12)
     ytick(1(2)13) xlabel(12(2)34) start(12) width(2) addlabel
     norm gap(15)
```

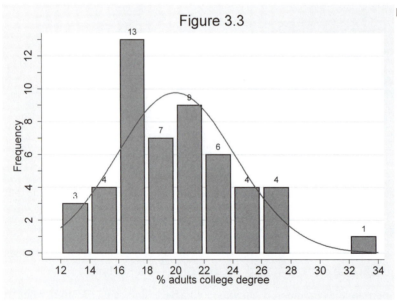

Figure 3.3

Figure 3.3

Suppose we want to see how the distribution of *college* varies by *region*. The **by** option obtains a separate histogram for each value of *region*. Other options work as they do for single histograms. Figure 3.4 shows an example in which we ask for percentages on the vertical axis, and the data grouped into 8 bins.

. histogram *college*, by(*region*) percent bin(8)

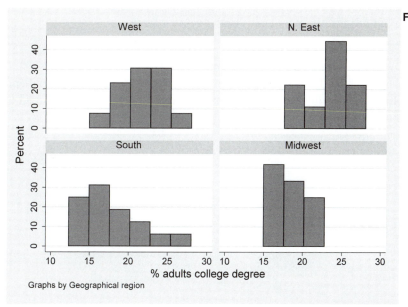

Figure 3.4

Figure 3.5, below, contains a similar set of four regional graphs, but includes a fifth that shows the distribution for all regions combined.

. histogram *college*, percent bin(8) by(*region*, total)

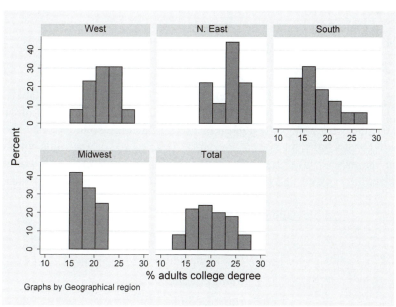

Figure 3.5

Axis labeling, tick marks, titles, and the **by**(*varname*) or **by**(*varname*, **total**) options work in a similar fashion with other Stata graphing commands, as seen in the following sections.

Scatterplots

Basic scatterplots are obtained through commands of the general form

. `graph twoway scatter y x`

where *y* is the vertical or *y*-axis variable, and *x* the horizontal or *x*-axis one. For example, again using the *states.dta* dataset, we could plot *waste* (per capita solid wastes) against *metro* (percent population in metropolitan areas), with the result shown in Figure 3.6. Each point in Figure 3.6 represents one of the 50 U.S. states (or Washington DC).

. `graph twoway scatter waste metro`

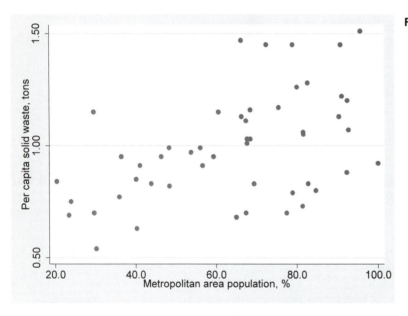

Figure 3.6

As with histograms, we can use **xlabel**, **xtick**, **xtitle**, etc. to control axis labels, tick marks, or titles. Scatterplots also allow control of the shape, color, size, and other attributes of markers. Figure 3.6 employs the default markers, which are solid circles. The same effect would result if we included the option **msymbol(circle)**, or wrote this option in abbreviated form as **msymbol(O)**. **msymbol(diamond)** or **msymbol(D)** would produce a graph with diamond markers, and so forth. The following table lists possible shapes.

msymbol( )	Abbreviation	Description
circle	O	circle, solid
diamond	D	diamond, solid
triangle	T	triangle, solid
square	S	square, solid
plus	+	plus sign
x	X	letter x
smcircle	o	small circle, solid

`smdiamond`	`d`	small diamond, solid
`smsquare`	`s`	small square, solid
`smtriangle`	`t`	small triangle, solid
`smplus`	`smplus`	small plus sign
`smx`	`x`	small letter x
`circle_hollow`	`Oh`	circle, hollow
`diamond_hollow`	`Dh`	diamond, hollow
`triangle_hollow`	`Th`	triangle, hollow
`square_hollow`	`Sh`	square, hollow
`smcircle_hollow`	`oh`	small circle, hollow
`smdiamond_hollow`	`dh`	small diamond, hollow
`smtriangle_hollow`	`th`	small triangle, hollow
`smsquare_hollow`	`sh`	small square, hollow
`point`	`p`	very small dot
`none`	`i`	invisible

The **mcolor** option controls marker colors. For example, the command

`. graph twoway scatter waste metro, msymbol(S) mcolor(purple)`

would produce a scatterplot in which the symbols were large purple squares. Type **help colorstyle** for a list of available colors.

One interesting possibility with scatterplots is to make symbol size (area) proportional to a third variable, thereby giving the data points different visual "weight." For example, we might redraw the scatterplot of *waste* against *metro*, but make the symbols size reflect each state's population (*pop*). This can be done as shown in Figure 3.7, using the **fweight[ ]** (frequency weight) feature. Hollow circles, **msymbol(Oh)**, provide a suitable shape.

Frequency weights are useful with some other graph types as well. Weighting can be a deceptively complex topic, because "weights" come in several types, and have different meanings in different contexts. For an overview of weighting in Stata, type **help weight**.

`. graph twoway scatter *waste* *metro* [fweight = *pop*], msymbol(Oh)`

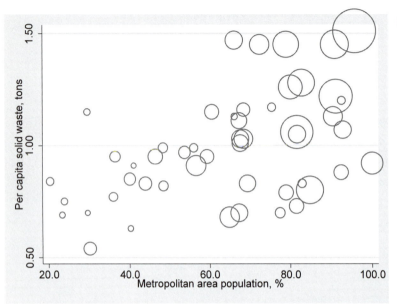

Figure 3.7

New in Stata 9, density-distribution sunflower plots provide an alternative to scatterplots with high-density data. Basically, they resemble scatterplots in which some of the individual data points are replaced with sunflower-like symbols to indicate more than one observation at that location. Figure 3.8 shows a sunflower-plot version of Figure 3.6, in which some of the flower symbols (those with four "petals") represent up to four individual data points, or states. A table printed after the **sunflower** command provides a key regarding how many observations each flower represents. The number of petals and the darkness of the flower correspond to the density of data.

`. sunflower *waste* *metro*, addplot(lfit *waste* *metro*)`

```
Bin width           =    11.3714
Bin height          =    .286522
Bin aspect ratio    =   .0218209
Max obs in a bin    =          4
Light               =          3
Dark                =         13
X-center            =      67.55
Y-center            =        .96
Petal weight        =          1
```

flower type	petal weight	No. of petals	No. of flowers	estimated obs.	actual obs.
none				23	23
light	1	3	5	15	15
light	1	4	3	12	12
				50	50

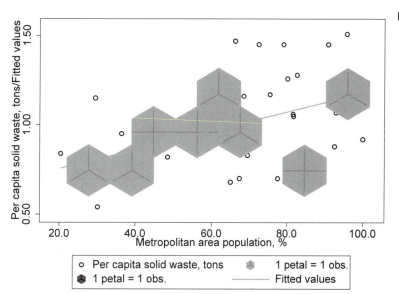

Figure 3.8

Sunflower plots are particularly helpful with large datasets, or when many observations plot at similar (or identical) coordinates. The example in Figure 3.8 includes a regression line, essentially a `twoway lfit` plot that has been overlaid or added to the sunflower plot by specifying the option **addplot(lfit *waste metro*)**.

Markers in an ordinary scatterplot can be identified by labels. For example, we might want to name the states in a scatterplot such as Figure 3.6. Fifty state names, however, would turn the graph into a visual jumble. Concentrating on one region such as the West seems more promising. An **if** qualifier accomplishes this, producing the results seen in Figure 3.9 on the following page.

. `graph twoway scatter` *waste metro* `if` *region*`==1, mlabel(`*state*`)`

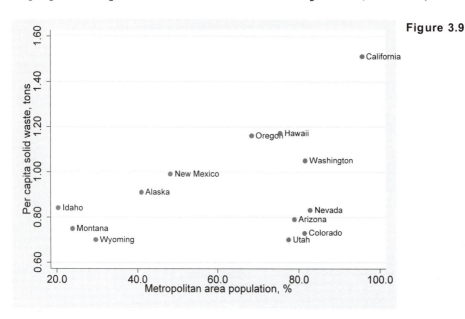

Figure 3.9

Figure 3.10 (below) shows separate *waste – metro* scatterplots for each region. The relationship between these two variables appears noticeably steeper in the South and Midwest than it does in the West and Northeast, an impression we will later confirm. The **ylabel** and **xlabel** options in this example give the *y*- and *x*-axis labels three-digit (maximum) fixed display formats with no decimals, making them easier to read in the small subplots.

. `graph twoway scatter` *waste metro*`, by(`*region*`)`
    `ylabel(, format(%3.0f)) xlabel(, format(%3.0f))`

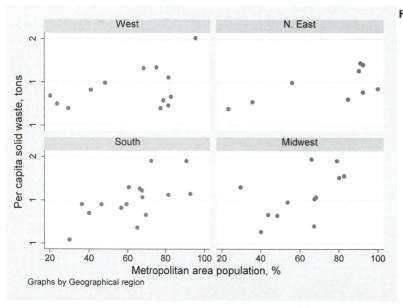

Figure 3.10

Scatterplot matrices, produced by **graph matrix**, prove useful in multivariate analysis. They provide a compact display of the relationships between a number of variable pairs, allowing the analyst to scan for signs of nonlinearity, outliers, or clustering that might affect statistical modeling. Figure 3.11 shows a scatterplot matrix involving three variables from *states.dta*.

```
. graph matrix miles metro income waste, half msymbol(oh)
```

Figure 3.11

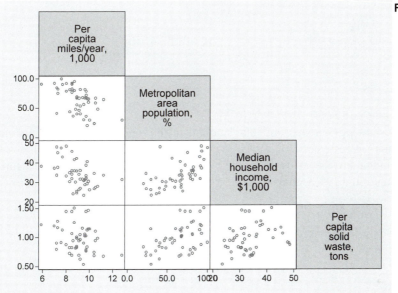

The **half** option specified that Figure 3.11 should include only the lower triangular part of the matrix. The upper triangular part is symmetrical and, for many purposes, redundant. **msymbol(oh)** called for small hollow circles as markers, just as we might with a scatterplot. Control of the axes is more complicated, because there are as many axes as variables; type **help graph_matrix** for details.

When the variables of interest include one dependent or "effect" variable, and several independent or "cause" variables, it helps to list the dependent variable last in the **graph matrix** variable list. That results in a neat row of dependent-versus-independent variable graphs across the bottom

Line Plots

Mechanically, line plots are scatterplots in which the points are connected by line segments. Like scatterplots, the various types of line plots belong to Stata's versatile **graph twoway** family. The scatterplot options that control axis labeling and markers work much the same with line plots, too. New options control the characteristics of the lines themselves.

Line plots tend to have different uses than scatterplots. For example, as time plots they depict changes in a variable over time. Dataset *cod.dta* contains time-series data reflecting the unhappy story of Newfoundland's Northern Cod fishery. This fishery, which had been among the world's richest, collapsed in 1992 primarily due to overfishing.

```
Contains data from C:\data\cod.dta
  obs:            38                       Newfoundland's Northern Cod
                                           fishery, 1960-1997
  vars:            5                       4 Jul 2005 15:02
  size:          684 (99.9% of memory free)
-------------------------------------------------------------------------
              storage  display   value
variable name  type    format    label    variable label
-------------------------------------------------------------------------
year           int     %8.0g              Year
cod            float   %8.0g              Total landings, 1000t
canada         int     %8.0g              Canadian landings, 1000t
TAC            int     %8.0g              Total Allowable Catch, 1000t
biomass        float   %9.0g              Estimated biomass, 1000t
-------------------------------------------------------------------------
Sorted by:   year
```

A simple time plot showing Canadian and total landings can be constructed by drawing line graphs of both variables against *year*. Figure 3.12 does this, showing the "killer spike" of international overfishing in the late 1960s, followed by a decade of Canadian fishing pressure in the 1980s, leading up to the 1992 collapse of the Northern Cod.

. **graph twoway line cod canada year**

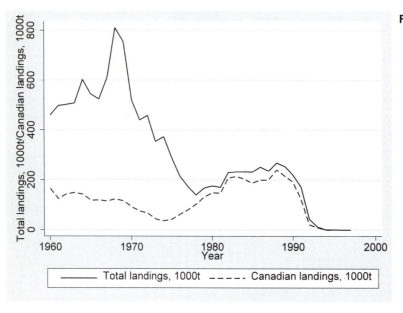

Figure 3.12

In Figure 3.12, Stata automatically chose a solid line for the first-named *y* variable, *cod*, and a dashed line for the second, *canada*. A legend at the bottom explains these meanings. We could improve this graph by rearranging the legend, and suppressing the redundant *y*-axis title, as illustrated in Figure 3.13.

```
. graph twoway line cod canada year, legend(label (1 "all nations")
       label(2 "Canada") position(2) ring(0) rows(2)) ytitle("")
```

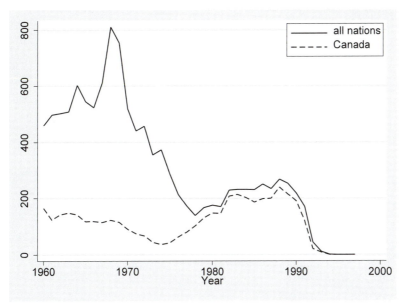

Figure 3.13

The **legend** option for Figure 3.13 breaks down as follows. Note that all of these suboptions occur *within the parentheses* following **legend** .

label(1 "all nations")	label first-named *y* variable "all nations"
label(2 "Canada")	label second-named *y* variable "Canada"
position(2)	place the legend at 2 o'clock position (upper right)
ring(0)	place the legend within the plot space
rows(2)	organize the legend to have two rows

By shortening the legend labels and placing them within the plot space, we leave more room to show the data and create a more attractive, readable figure. **legend** works similarly for other graph styles that have legends. Type **help legend_option** to see a list of the many suboptions available.

Figures 3.12 and 3.13 simply connect each data point with line segments. Several other connecting styles are possible, using the **connect** option. For example,

> **connect(stairstep)**

or equivalently,

> **connect(J)**

will cause points to be connected in stairstep (flat, then vertical) fashion. Figure 3.14 illustrates with a stairstep time plot of the government-set Total Allowable Catch (*TAC*) variable from *cod.dta*.

. `graph twoway line` *TAC year*, `connect(stairstep)`

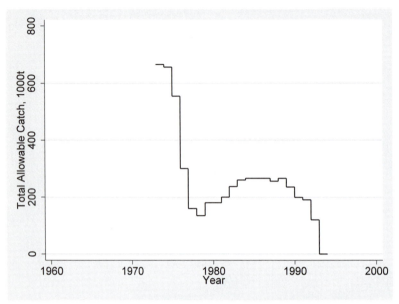

Figure 3.14

Other **connect** choices are listed below. The default, straight line segments, corresponds to **connect(direct)** or **connect(l)**. For more details, see **help connectstyle**.

connect( )	Abbreviation	Description
none	**i**	do not connect
direct	**l** (letter "el")	connect with straight lines
ascending	**L**	direct, but only if $x[i+1] > x[i]$
stairstep	**J**	flat, then vertical
stepstair		vertical, then flat

Figure 3.15 (on the following page) repeats this stairstep plot of *TAC*, but with some enhancements of axis labels and titles. The option **xtitle("")** requests no *x*-axis title (because "year" is obvious). We added tick marks at two-year intervals to the *x* axis, labeled the *y* axis at intervals of 100, and printed *y*-axis labels horizontally instead of vertically (the default).

```
. graph twoway line TAC year, connect(stairstep)  xtitle("")
     xtick(1960(2)2000)  ytitle("Thousands of tons")
     ylabel(0(100)800, angle(horizontal))  xtitle("")
     clpattern(dash)
```

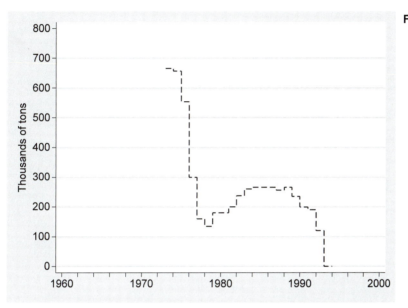

Figure 3.15

Instead of letting Stata determine the line patterns (solid, dashed, etc.) in Figure 3.15, we used the **clpattern(dash)** option to call for a dashed line. Possible line pattern choices are listed in the table below (also see **help linepatternstyle**).

clpattern( )	Description
solid	solid line
dash	dashed line
dot	dotted line
dash_dot	dash then dot
shortdash	short dash
shortdash_dot	short dash followed by dot
longdash	long dash
longdash_dot	long dash followed by dot
blank	invisible line
formula	for example, **clpattern(-.)** or **clpattern(—..)**

Before we move on to other examples and types, Figure 3.16 unites the three variables discussed in this section to create a single graphic showing the tragedy of the Northern Cod. Note how the **connect( )**, **clpattern( )**, and **legend( )** options work in this three-variable context.

```
. graph twoway line cod canada TAC year, connect(line line stairstep)
     clpattern(solid longdash dash) xtitle("") xtick(1960(2)2000)
     ytitle("Thousands of tons") ylabel(0(100)800, angle(horizontal))
     xtitle("") legend(label (1 "all nations") label(2 "Canada")
     label(3 "TAC") position(2) ring(0) rows(3))
```

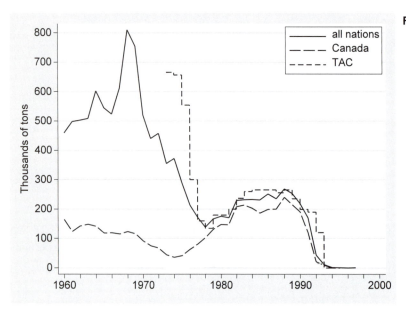

Figure 3.16

Connected-Line Plots

In the line plots of the previous section, data points are invisible and we see only the connecting lines. The **graph twoway connected** command creates connected-line plots in which the data points are marked by scatterplot symbols. The marker-symbol options described earlier for **graph twoway scatter**, and also the line-connecting options described for **graph twoway line**, both apply to **graph twoway connected** as well. Figure 3.17 shows a default example, a connected-line time plot of the cod biomass variable (*bio*) from *cod.dta*.

```
. graph twoway connected bio year
```

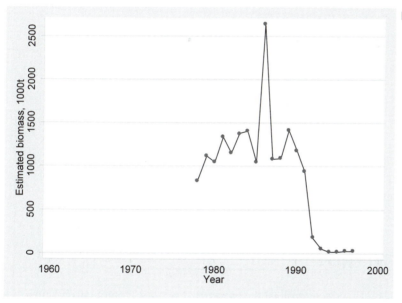

Figure 3.17

The dataset contains only biomass values for 1978 through 1997, resulting in much empty space in Figure 3.17. **if** qualifiers allow us to restrict the range of years. Figure 3.18, on the following page, does this. It also dresses up the image to show control of marker symbols, line patterns, axes, and legends. With cod landings and biomass both in the same image, we see that the biomass began its crash in the late 1980s, several years before a crisis was officially recognized.

```
. graph twoway connected bio cod year if year > 1977 & year < 1999,
    msymbol(T Oh) clpattern(dash solid) xlabel(1978(2)1996)
    xtick(1979(2)1997) ytitle("Thousands of tons") xtitle("")
    ylabel(0(500)2500, angle(horizontal))
    legend(label(1 "Estimated biomass") label(2 "Total landings")
    position(2) rows(2) ring(0))
```

Figure 3.18

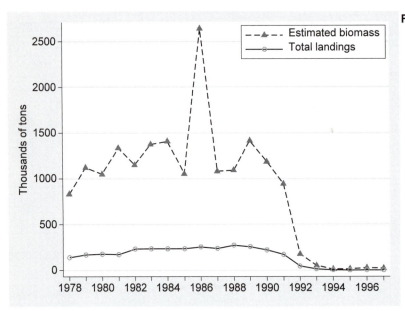

Other Twoway Plot Types

In addition to basic line plots and scatterplots, the **graph twoway** command encompasses a wide variety of other types. The following table lists the possibilities.

graph twoway	Description
scatter	scatterplot
line	line plot
connected	connected-line plot
scatteri	scatter with immediate arguments (data given *in* the command line)
area	line plot with shading
bar	twoway bar plot (different from **graph bar**)
spike	twoway spike plot
dropline	dropline plot (spikes dropped vertically or horizontally to given value)
dot	twoway dot plot (different from **graph dot**)
rarea	range plot, shading the area between high and low values

rbar	range plot with bars between high and low values
rspike	range plot with spikes between high and low values
rcap	range plot with capped spikes
rcapsym	range plot with spikes capped with symbols
rscatter	range plot with scatterplot marker symbols
rline	range plot with lines
rconnected	range plot with lines and markers
pcspike	paired-coordinate plot with spikes
pccapsym	paired-coordinate plot with spikes capped with symbols
pcarrow	paired-coordinate plot with arrows
pcbarrow	paired-coordinate plot with arrows having two heads
pcscatter	paired-coordinate plot with markers
pci	pcspike with immediate arguments
pcarrowi	pcarrow with immediate arguments
tsline	time-series plot
tsrline	time-series range plot
mband	straight line segments connect the (x, y) cross-medians within bands
mspline	cubic spline curve connects the (x, y) cross-medians within bands
lowess	LOWESS (locally weighted scatterplot smoothing) curve
lfit	linear regression line
qfit	quadratic regression curve
fpfit	fractional polynomial plot
lfitci	linear regression line with confidence band
qfitci	quadratic regression curve with confidence band
fpfitci	fractional polynomial plot with confidence band
function	line plot of function
histogram	histogram plot
kdensity	kernel density plot

The usual options to control line patterns, marker symbols, and so forth work where appropriate with all **twoway** commands. For more information about a particular command, type **help twoway_mband**, **help twoway_function**, etc. (using any of the names above). Note that **graph twoway bar** is a different command from **graph bar**. Similarly, **graph twoway dot** differs from **graph dot**. The **twoway** versions provide various methods for plotting a measurement *y* variable against a measurement *x* variable, analogous to a scatterplot or a line plot. The non-twoway versions, on the other hand, provide ways to plot summary statistics (such as means or medians) of one or more measurement *y* variables against categories of one or more *x* variables. The **twoway** versions thus are comparatively specialized, although (as with all **twoway** plots) they can be overlaid with other **twoway** plots for more complex graphical effects.

Many of these plot types are most useful in composite figures, constructed by overlaying two or more simple plots as described later in this chapter. Others produce nice stand-alone graphs. For example, Figure 3.19 shows an area plot of the Newfoundland cod landings.

```
. graph twoway area cod canada year, ytitle("")
```

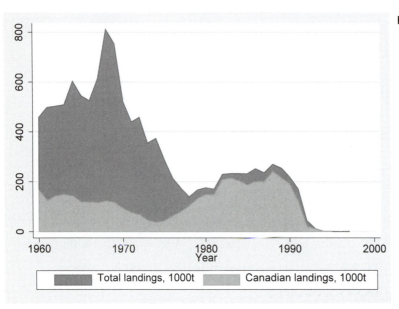

Figure 3.19

The shading in area graphs and other types with shaded regions can be controlled through the option **bcolor**. Type **help colorstyle** for a list of the available colors, which include gray scales. The darkest gray, gs0, is actually black. The lightest gray, gs16, is white. Other values are in between. For example, Figure 3.20 shows a light-gray version of this graph.

. **graph twoway area** *cod canada year*, **ytitle("") bcolor(gs12 gs14)**

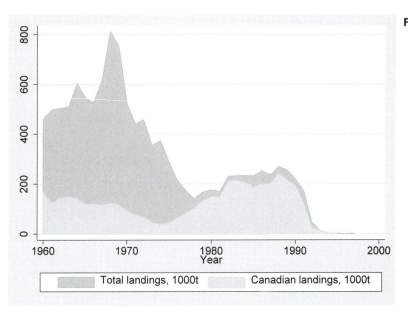

Figure 3.20

Unusually cold atmosphere/ocean conditions played a secondary role in Newfoundland's fisheries disaster, which involved not only the Northern Cod but also other species and populations. For example, key fish species in the neighboring Gulf of St. Lawrence declined during this period as well (Hamilton, Haedrich and Duncan 2003). Dataset *gulf.dta* describes environment and Northern Gulf cod catches (raw data from DFO 2003).

```
Contains data from C:\data\gulf.dta
  obs:             56                          Gulf of St. Lawrence
                                               environment and cod fishery
  vars:             7                          10 Jul 2005 11:51
  size:         1,344 (99.9% of memory free)
-------------------------------------------------------------------------------
              storage  display    value
variable name   type   format     label      variable label
-------------------------------------------------------------------------------
winter          int    %8.0g                  Winter
minarea         float  %9.0g                  Minimum ice area, 1000 km^2
maxarea         float  %9.0g                  Maximum ice area, 1000 km^2
mindays         byte   %8.0g                  Minimum ice days
maxdays         byte   %8.0g                  Maximum ice days
cil             float  %9.0g                  Cold Intermediate Layer
                                                temperature minimum, C
cod             float  %9.0g                  N. Gulf cod catch, 1000 tons
-------------------------------------------------------------------------------
Sorted by:  winter
```

The maximum annual ice cover averaged 173,017 km^2 during these years.

. **summarize** *maxarea*

```
    Variable |      Obs        Mean    Std. Dev.       Min        Max
-------------+--------------------------------------------------------
     maxarea |       38    173.0172    37.18623     47.8901   220.1905
```

Figure 3.21 uses this mean (173 thousand) as the base for a spike plot, in which spikes above and below the line show above and below-average ice cover, respectively. The **yline(173)** option draws a horizontal line at 173.

```
. graph twoway spike maxarea winter if winter > 1963, base(173)
     yline(173) ylabel(40(20)220, angle(horizontal))
     xlabel(1965(5)2000)
```

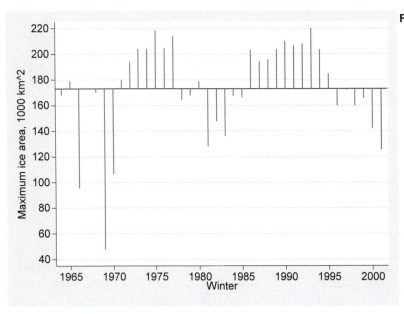

Figure 3.21

The **base()** format of Figure 3.21 emphasizes the succession of unusually harsh winters (above-average maximum ice cover) during the late 1980s and early 1990s, around the time of Newfoundland's fisheries crisis. We also see an earlier spell of mild winters in the early 1980s, and hints of a recent warming trend.

A different view of the same data, in Figure 3.22, employs lowess regression to smooth the time series. The bandwidth option, **bwidth(.4)**, specifies a curve based on smoothed data points that are calculated from weighted regressions within a moving band containing 40% of the sample. Lower bandwidths such as **bwidth(.2)**, or 20% of the data, would give us a more jagged, less smoothed curve that more closely resembles the raw data. Higher bandwidths such as **bwidth(.8)**, the default, will smooth more radically. Regardless of the bandwidth chosen, smoothed points towards either extreme of the *x* values must be calculated from increasingly narrow bands, and therefore will show less smoothing. Chapter 8 contains more about lowess smoothing.

```
. graph twoway lowess maxarea winter if winter > 1963, bwidth(.4)
     yline(173) ylabel(40(20)220, angle(horizontal))
     xlabel(1965(5)2000)
```

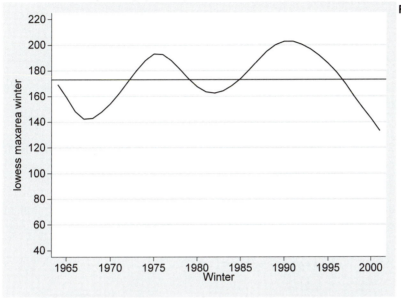

Figure 3.22

Range plots connect high and low *y* values at each level of *x*, using bars, spikes, or shaded areas. Daily stock market prices are often graphed in this way. Figure 3.23 shows a capped-spike range plot using the minimum and maximum ice cover variables from *gulf.dta*.

```
. graph twoway rcap minarea maxarea winter if winter > 1963,
     ylabel(0(20)220, angle(horizontal)) ytitle("Ice area, 1000 km^2")
     xlabel(1965(5)2000)
```

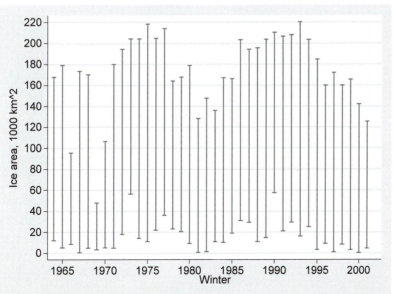

Figure 3.23

These examples by no means exhaust the possibilities for twoway graphs. Other applications appear throughout the book. Later in this chapter, we will see examples involving overlays of two or more twoway graphs, forming a single image.

Box Plots

Box plots convey information about center, spread, symmetry, and outliers at a glance. To obtain a single box plot, type a command of the form

```
. graph box y
```

If several different variables have roughly similar scales, we can visually compare their distributions through commands of the form

```
. graph box w x y z
```

One of the most common applications for box plots involves comparing the distribution of one variable over categories of a second. Figure 3.24 compares the distribution of *college* across states of four U.S. regions, from dataset *states.dta*.

```
. graph box college, over(region) yline(19.1)
```

Figure 3.24

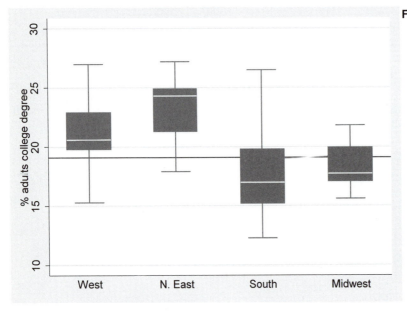

The median proportion of adults with college degrees tends to be highest in the Northeast, and lowest in the South. On the other hand, southern states are more variable. Regional medians (lines within boxes) in Figure 3.24 can be compared visually to the 50-state median indicated by the **yline(19.1)** option.. This median was obtained by typing

```
. summarize college if region < ., detail
```

Chapter 4 describes the **summarize, detail** command. The **if region < .** qualifier above restricted our analysis to observations that have nonmissing values of *region*; that is, to every place except Washington DC.

The box in a box plot extends from approximate first to third quartiles, a distance called the interquartile range (IQR). It therefore contains approximately the middle 50% of the data. Outliers, defined as observations more than 1.5IQR beyond the first or third quartile, are plotted individually in a box plot. No outliers appear among the four distributions in Figure 3.24. Stata's box plots define quartiles in the same manner as **summarize, detail**. This is not the same approximation used to calculate "fourths" for letter-value displays, **lv** (Chapter 4). See Frigge, Hoaglin, and Iglewicz (1989) and Hamilton (1992b) for more about quartile approximations and their role in identifying outliers.

Numerous options control the appearance, shading and details of boxes in a box plot; see **help graph_box** for a list. Figure 3.25 demonstrates some of these options, and also the horizontal arrangement of **graph hbox**, using per capita energy consumption from *states.dta*. The option **over(region, sort(1))** calls for boxes sorted in ascending order according to their medians on the first-named (and in this case, the only) *y* variable. **intensity(30)** controls the intensity of shading in the boxes, setting this somewhat lower (less dark) than the default seen in Figure 3.24. Counterintuitively, the vertical line marking the overall median (320) in Figure 3.25 requires a **yline** option, rather than **xline**.

. **graph hbox** *energy*, **over(***region***, sort(1)) yline(320) intensity(30)**

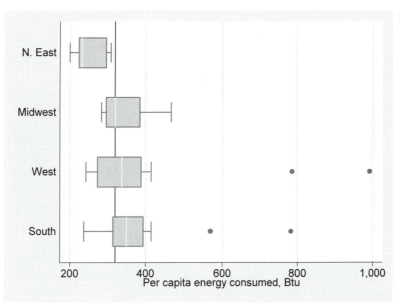

Figure 3.25

The energy box plots in Figure 3.25 make clear not only the differences among medians, but also the presence outliers — four very high-consumption states in the West and South. With a bit of further investigation, we find that these are oil-producing states: Wyoming, Alaska, Texas, and Louisiana. Box plots excel at drawing attention to outliers, which are easily overlooked (and often cause trouble) in other steps of statistical analysis.

Pie Charts

Pie charts are popular tools for "presentation graphics," although they have little value for analytical work. Stata's basic pie chart command has the form

```
. graph pie w x y z, pie
```

where the variables *w*, *x*, *y*, and *z* all measure quantities of something in similar units (for example, all are in dollars, hours, or people).

Dataset *AKethnic.dta*, on the ethnic composition of Alaska's population, provides an illustration. Alaska's indigenous Native population divides into three broad cultural/linguistic groups: Aleut, Indian (including Athabaska, Tlingit, and Haida), and Eskimo (Yupik and Inupiat). The variables *aleut*, *indian*, *eskimo*, and *nonnativ* are population counts for each group, taken from the 1990 U.S. Census. This dataset contains only three observations, representing three types or sizes of communities: cities of 10,000 people or more; towns of 1,000 to 10,000; and villages with fewer than 1,000 people.

```
Contains data from C:\data\AKethnic.dta
  obs:            3                          Alaska ethnicity 1990
  vars:           7                          4 Jul 2005 12:06
  size:          63 (99.9% of memory free)
-------------------------------------------------------------------------
              storage  display    value
variable name   type   format     label      variable label
-------------------------------------------------------------------------
comtype        byte    %8.0g      popcat     Community type (size)
pop            float   %9.0g                 Population
n              int     %8.0g                 number of communities
aleut          int     %8.0g                 Aleut
indian         int     %8.0g                 Indian
eskimo         int     %8.0g                 Eskimo
nonnativ       float   %9.0g                 Non-Native
-------------------------------------------------------------------------
Sorted by:
```

The majority of the state's population is non-Native, as clearly seen in a pie chart (Figure 3.26). The option **pie(3, explode)** causes the third-named variable, *eskimo*, to be "exploded" from the pie for emphasis. The fourth-named variable, *nonnativ*, is shaded a light gray color, **pie(4, color(gs13))**, for contrast with the smaller Native groups. (In this monochrome book, our examples use only gray-scale colors, but keep in mind that other possiblities such as **color(blue)** or **color(cranberry)** exist. Type **help colorstyle** for the list. **plabel(3 percent, gap(20))** causes a percentage label to be printed by the *eskimo* (variable 3) slice, with a gap of 20 relative radial units from the center. We see that about 8% of Alaska's population is Eskimo (Inupiat or Yupik). The **legend** option calls for a four-row box placed at the 11 o'clock position within the plot space.

```
. graph pie aleut indian eskimo nonnativ, pie(3, explode)
     pie(4, color(gs13)) plabel(3 percent, gap(20))
     legend(position(11) rows(4) ring(0))
```

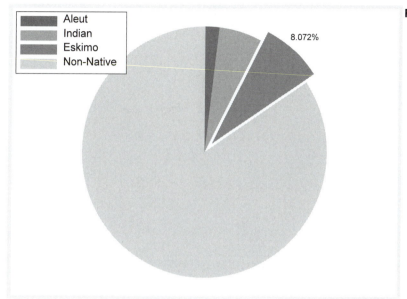

Figure 3.26

Non-Natives are the dominant group in Figure 3.26, but if we draw separate pies for each type of community by adding a **by(*comtype*)** option, new details emerge (Figure 3.27, next page). The option **angle0()** specifies the angle of the first slice of pie. Setting this first-slice angle at 0 (horizontal) orients the pies in Figure 3.27 in such a way that the labels are more readable. The figure shows that whereas Natives are only a small fraction of the population in Alaska cities, they constitute the majority among those living in villages. In particular, Eskimos make up a large fraction of villagers — 35% across all villages, and more than 90% in some. This gives Alaska villages a different character from Alaska cities.

```
. graph pie aleut indian eskimo nonnativ, pie(3, explode)
     pie(4, color(gs13)) plabel(3 percent, gap(8))
     legend(rows(1)) by(comtype) angle0(0)
```

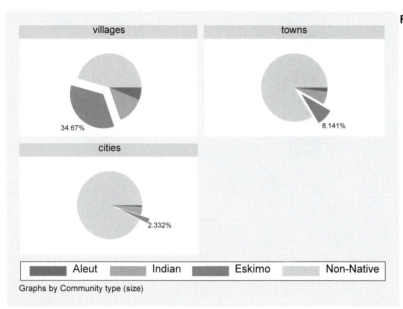

Figure 3.27

Bar Charts

Although they contain less information than box plots, bar charts provide simple and versatile displays for comparing sets of summary statistics such as means, medians, sums, or counts. To obtain vertical bars showing the mean of *y* across categories of *x*, for example, type

```
. graph bar (mean) y, over(x)
```

For horizontal bars showing the sum of *y* across categories of *x1* within categories of *x2*, type

```
. graph hbar (sum) y, over(x1) over(x2)
```

The bar chart could display any of the following statistics:

`mean`	Means (the default; used if the type of statistic is not specified)
`sd`	Standard deviations
`sum`	Sums
`rawsum`	Sums ignoring optionally specified weight
`count`	Numbers of nonmissing observations
`max`	Maximums
`min`	Minimums
`median`	Medians
`p1`	1st percentiles

| p2 | 2nd percentiles (and so forth to p99) |
| iqr | Interquartile ranges |

This list of available summary statistics is the same as that for the **collapse** command (see Chapter 2), and also for a number of other commands including **graph dot** (next section) and **table** (Chapter 4).

Dataset *statehealth.dta* contains further data on the U.S. states, combining socioeconomic measures from the 1990 Census with several health-risk indicators from the Centers for Disease Control (2003), averaged over 1994–98.

```
Contains data from C:\data\statehealth.dta
  obs:            51                    Health indicators 1994-98 (CDC)
  vars:           12                    9 Jul 2005 11:56
  size:        3,315 (99.9% of memory free)
-------------------------------------------------------------------------
              storage  display   value
variable name  type    format    label    variable label
-------------------------------------------------------------------------
state         str20   %20s                US State
region        byte    %9.0g     region    Geographical region
income        long    %10.0g              Median household income, 1990
income2       float   %11.0g    income2   Median income low or high
high          float   %9.0g               % adults HS diploma, 1990
college       float   %9.0g               % adults college degree, 1990
overweight    float   %9.0g               % overweight
inactive      float   %9.0g               % inactive in leisure time
smokeM        float   %9.0g               % male adults smoking
smokeF        float   %9.0g               % female adults smoking
smokeT        float   %9.0g               % adults smoking
motor         float   %9.0g               Age-adjusted motor-vehicle
                                             related deaths/100,000
-------------------------------------------------------------------------
Sorted by:  state
```

Figure 3.28 graphs the median percent of population inactive in leisure time (*inactive*) across four geographical regions (*region*). We see a pronounced regional difference: inactivity rates are highest in the South (36%), and lowest in the West (21%). Note that the vertical axis has automatically been labeled "p 50 of inactive," meaning the 50th percentile or median. The **blabel(bar)** option labels the bar heights (20.9, etc.). **bar(1, bcolor(gs(10))** specifies that bars for the first-named *y* variable (*inactive*; there is only one) should be filled with a medium-light gray color.

```
. graph bar (median) inactive, over(region) blabel(bar)
    bar(1, bcolor(gs10))
```

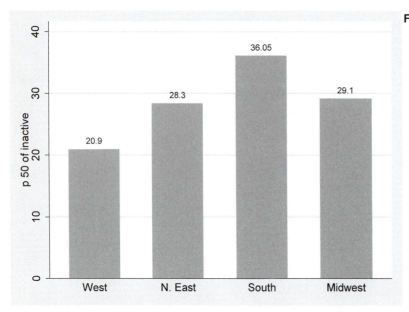

Figure 3.28

Figure 3.29 (following page) elaborates this idea by adding a second variable, *overweight*, and coloring its bars a darker gray. The bar labels are **size(medium)** in Figure 3.29, making them larger than the defaults, **size(small)**, used in Figure 3.28. Other possibilities for **size( )** suboptions include labels that are **tiny** , **medsmall** , **medlarge** , or **large**. See **help textsizestyle** for a complete list. Figure 3.29 shows that regional differences in the prevalence of overweight individuals are less pronounced than differences in inactivity, although both variables' medians are highest in the South and Midwest.

```
. graph bar (median) inactive overweight, over(region)
     blabel(bar, size(medium))
     bar(1, bcolor(gs10)) bar(2, bcolor(gs7))
```

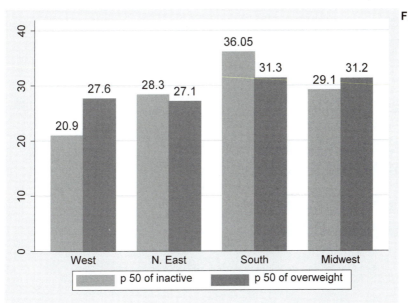

Figure 3.29

The risk indicators in *statehealth.dta* include motor-vehicle related fatalities per 100,000 population (*motor*). On the next page, Figure 3.30 breaks these down regionally, and then into subgroups of low- and high-income states (states having median household incomes below or above the national median), revealing a striking correlation with wealth. Within each region, the low-income states exhibit higher mean fatality rates. Across both income categories, fatality rates are higher in the South, and lower in the Northeast. The order of the two **over** options in the command controls their order in organizing the chart. For this example we chose a horizontal bar chart or **hbar**. In such horizontal charts, **ytitle**, **yline**, etc. refer to the horizontal axis. **yline(17.2)** marks the overall mean.

```
. graph hbar (mean) motor, over(income2) over(region) yline(17.2)
     ytitle("Mean motor-vehicle related fatalities/100,000")
```

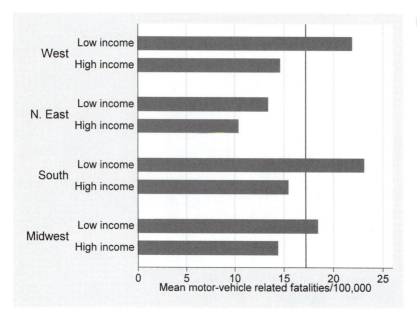

Figure 3.30

Bars also can be stacked, as shown in Figure 3.31. This plot, based on the Alaska ethnicity data (*AKethnic.dta*), employs all the defaults to display ethnic composition by type of community (village, town, or city).

```
. graph bar (sum) nonnativ aleut indian eskimo, over(comtyp) stack
```

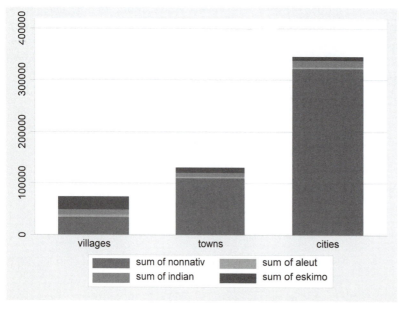

Figure 3.31

Figure 3.32 redraws this plot with better legend and axis labels. The **over** option now includes suboptions that relabel the community types so the horizontal axis is more informative. The **legend** option specifies four rows in the same vertical order as the bars themselves, and placed in the 11 o'clock position inside the plot space. It also improves legend labels. **ytitle**, **ylabel**, and **ytitle** options format the vertical axis.

```
. graph bar (sum) nonnativ aleut indian eskimo,
      over(comtyp, relabel(1 "Villages <1,000" 2 "Towns 1,000-10,000"
        3 "Cities >10,000"))
      legend(rows(4) order(4 3 2 1) position(11) ring(0)
        label(1  "Non-native") label(2 "Aleut")
        label(3 "Indian") label(4 "Eskimo"))
      stack ytitle(Population)
      ylabel(0(100000)300000) ytick(50000(100000)350000)
```

Figure 3.32

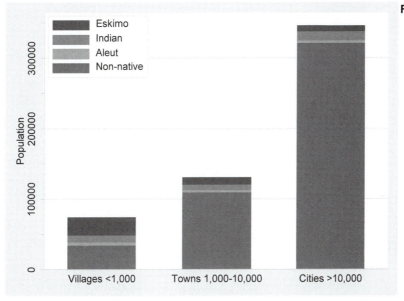

Figure 3.32 plots the same variables as the pie chart in Figure 3.27, but displays them quite differently. Whereas the pie charts show relative sizes (percentages) of ethnic groups within each community type, this bar chart shows their absolute sizes. Consequently, Figure 3.32 tells us something that Figure 3.27 could not: the majority of Alaska's Eskimo (Yupik and Inupiat) population lives in villages.

Dot Plots

Dot plots serve much the same purpose as bar charts: visually comparing statistical summaries of one or more measurement variables. The organization and Stata options for the two types of plot are broadly similar, including the choices of statistical summaries. To see a dot plot comparing the medians of variables *x*, *y*, *z*, and *w*, type

```
. graph dot (median) x y z w
```

For a dot plot comparing the mean of *y* across categories of *x*, type

```
. graph dot (mean) y, over(x)
```

Figure 3.33 shows a dot plot of male and female smoking rates by region, from *statehealth.dta*. The **over** option includes a suboption, **sort(smokeM)**, which calls for the regions to be sorted in order of their mean values of *smokeM* — that is, from lowest to highest smoking rates. We also specify a solid triangle as the marker symbol for *smokeM*, and hollow circle for *smokeF*.

```
. graph dot (mean) smokeM smokeF, over(region, sort(smokeM))
        marker(1, msymbol(T)) marker(2, msymbol(Oh))
```

Figure 3.33

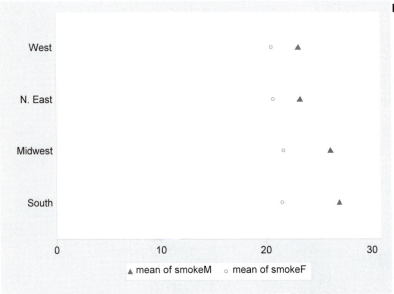

Although Figure 3.33 displays only eight means, it does so in a way that facilitates several comparisons. We see that smoking rates are generally higher for males; that among both sexes they are higher in the South and Midwest; and that regional variations are substantially greater for the male smoking rates. Bar charts could convey the same information, but one advantage of dot plots is their compactness. Dot plots (particularly when rows are sorted by the statistic of interest, as in Figure 3.33) remain easily readable even with a dozen or more rows.

Symmetry and Quantile Plots

Box plots, bar charts, and dot plots summarize measurement variable distributions, hiding individual data points to clarify overall patterns. Symmetry and quantile plots, on the other hand, include points for every observation in a distribution. They are harder to read than summary graphs, but convey more detailed information.

A histogram of per-capita energy consumption in the 50 U.S. states (from *states.dta*) appears in Figure 3.34. The distribution includes a handful of very high-consumption states, which happen to be oil producers. A superimposed normal (Gaussian) curve indicates that *energy* has a lighter-than-normal left tail, and a heavier-than-normal right tail — the definition of positive skew.

```
. histogram energy, start(100) width(100) xlabel(0(100)1000)
        frequency norm
```

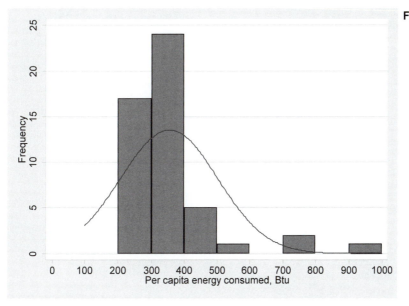

Figure 3.34

Figure 3.35 depicts this distribution as a symmetry plot. It plots the distance of the *i*th observation above the median (vertical) against the distance of the *i*th observation below the median. All points would lie on the diagonal line if this distribution were symmetrical. Instead, we see that distances above the median grow steadily larger than corresponding distances below the median, a symptom of positive skew. Unlike Figure 3.34, Figure 3.35 also reveals that the energy-consumption distribution is approximately symmetrical near its center.

```
. symplot energy
```

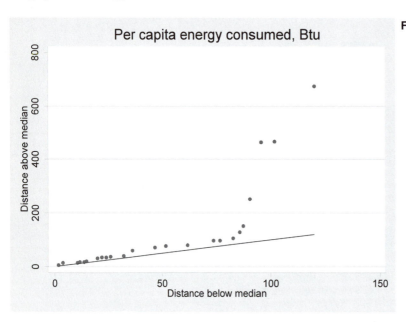

Figure 3.35

Quantiles are values below which a certain fraction of the data lie. For example, a .3 quantile is that value higher than 30% of the data. If we sort *n* observations in ascending order, the *i*th value forms the $(i - .5)/n$ quantile. The following commands would calculate quantiles of variable *energy*:

```
. drop if energy >= .
```

```
. sort energy
```

```
. generate quant = (_n - .5)/_N
```

As mentioned in Chapter 2, _n and _N are Stata system variables, always unobtrusively present when there are data in memory. _n represents the current observation number, and _N the total number of observations.

Quantile plots automatically calculate what fraction of the observations lie below each data value, and display the results graphically as in Figure 3.36. Quantile plots provide a graphic reference for someone who does not have the original data at hand. From well-labeled quantile plots, we can estimate order statistics such as median (.5 quantile) or quartiles (.25 and .75 quantiles). The IQR equals the rise between .25 and .75 quantiles. We could also read a quantile plot to estimate the fraction of observations falling below a given value.

. **quantile** *energy*

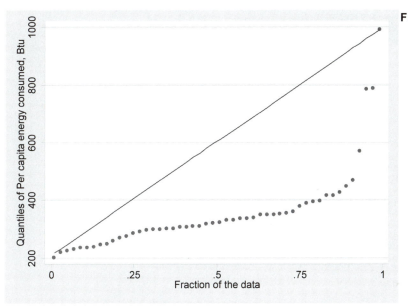

Figure 3.36

Quantile–normal plots, also called normal probability plots, compare quantiles of a variable's distribution with quantiles of a theoretical normal distribution having the same mean and standard deviation. They allow visual inspection for departures from normality in every part of a distribution, which can help guide decisions regarding normality assumptions and efforts to find a normalizing transformation. Figure 3.37, a quantile–normal plot of *energy*, confirms the severe positive skew that we had already observed. The **grid** option calls for a set of lines marking the .05, .10, .25 (first quartile), .50 (median), .75 (third quartile), .90, and .95 quantiles of both distributions. The .05, .50, and .95 quantile values are printed along the top and right-hand axes.

```
. qnorm energy, grid
```

Figure 3.37

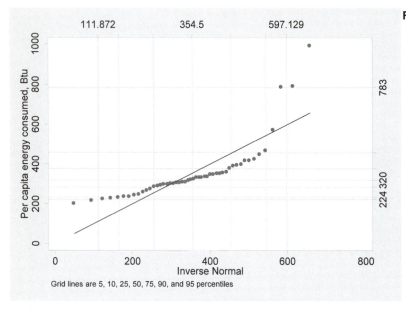

Quantile–quantile plots resemble quantile–normal plots, but they compare quantiles (ordered data points) of two empirical distributions instead of comparing one empirical distribution with a theoretical normal distribution. On the following page, Figure 3.38 shows a quantile–quantile plot of the mean math SAT score versus the mean verbal SAT score in 50 states and the District of Columbia. If the two distributions were identical, we would see points along the diagonal line. Instead, data points form a straight line roughly parallel to the diagonal, indicating that the two variables have different means but similar shapes and standard deviations.

. qqplot *msat vsat*

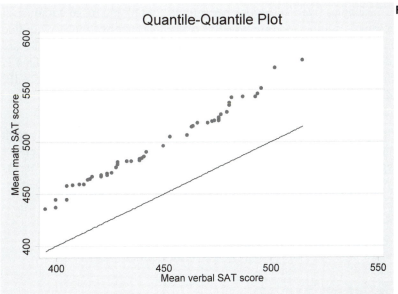

Figure 3.38

Regression with Graphics (Hamilton 1992a) includes an introduction to reading quantile-based plots. Chambers et al. (1983) provide more details. Related Stata commands include **pnorm** (standard normal probability plot), **pchi** (chi-squared probability plot), and **qchi** (quantile–chi-squared plot).

Quality Control Graphs

Quality control charts help to monitor output from a repetitive process such as industrial production. Stata offers four basic types: c chart, p chart, R chart, and $\bar{x}$ chart. A fifth type, called Shewhart after the inventor of these methods, consists of vertically-aligned $\bar{x}$ and R charts. Iman (1994) provides a brief introduction to R and $\bar{x}$ charts, including the tables used in calculating their control limits. The *Base Reference Manual* gives the command details and formulas used by Stata. Basic outlines of these commands are as follows:

. cchart *defects unit*
Constructs a c chart with the number of nonconformities or defects (*defects*) graphed against the unit number (*unit*). Upper and lower control limits, based on the assumption that number of nonconformities per unit follows a Poisson distribution, appear as horizontal lines in the chart. Observations with values outside these limits are said to be "out of control."

. pchart *rejects unit ssize*
Constructs a p chart with the proportion of items rejected (*rejects / ssize*) graphed against the unit number (*unit*). Upper and lower control limit lines derive from a normal approximation, taking sample size (*ssize*) into account. If *ssize* varies across units, the control limits will vary too, unless we add the option **stabilize** .

. `rchart x1 x2 x3 x4 x5, connect(1)`

Constructs an R (range) chart using the replicated measurements in variables *x1* through *x5* — that is, in this example, five replications per sample. Graph the range within each sample against the sample number, and (optionally) connect successive ranges with line segments. Horizontal lines indicate the mean range and control limits. Control limits are estimated from the sample size if the process standard deviation is unknown. When σ is known we can include this information in the command. For example, assuming σ = 10,

. `rchart x1 x2 x3 x4 x5, connect(1) std(10)`

. `xchart x1 x2 x3 x4 x5, connect(1)`

Constructs an $\bar{x}$ (mean) chart using the replicated measurements in variables *x1* through *x5*. Graphs the mean within each sample against the sample number and connect successive means with line segments. The mean range is estimated from the mean of sample means and control limits from sample size, unless we override these defaults. For example, if we know that the process actually has μ = 50 and σ = 10,

. `xchart x1 x2 x3 x4 x5, connect(1) mean(50) std(10)`

Alternatively, we could specify particular upper and lower control limits:

. `xchart x1 x2 x3 x4 x5, connect(1) mean(50) lower(40)`
    `upper(60)`

. `shewhart x1 x2 x3 x4 x5, mean(50) std(10)`

In one figure, vertically aligns an $\bar{x}$ chart with an R chart.

To illustrate a p chart, we turn to the quality inspection data in *quality1.dta*.

```
Contains data from C:\data\quality1.dta
  obs:             16                          Quality control example 1
  vars:             3                          4 Jul 2005 12:07
  size:           112  (99.9% of memory free)
-------------------------------------------------------------------------------
                storage   display    value
variable name    type     format     label      variable label
-------------------------------------------------------------------------------
day              byte     %9.0g                  Day sampled
ssize            byte     %9.0g                  Number of units sampled
rejects          byte     %9.0g                  Number of units rejected
-------------------------------------------------------------------------------
Sorted by:
```

. `list in 1/5`

```
     +----------------------+
     | day   ssize   rejects |
     |----------------------|
  1. |  58      53       10 |
  2. |   7      53       12 |
  3. |  26      52       12 |
  4. |  21      52       10 |
  5. |   6      51       10 |
     +----------------------+
```

Note that sample size varies from unit to unit, and that the units (days) are not in order. **pchart** handles these complications automatically, creating the graph with changing control limits seen in Figure 3.39. (For constant control limits despite changing sample sizes, add the **stabilize** option.)

. pchart *rejects day ssize*

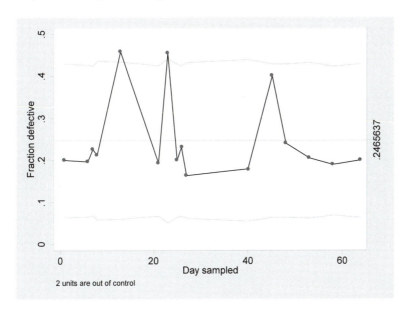

Figure 3.39

Dataset *quality2.dta*, borrowed from Iman (1994:662), serves to illustrate **rchart** and **xchart** . Variables *x1* through *x4* represent repeated measurements from an industrial production process; 25 units with four replications each form the dataset.

```
Contains data from C:\data\quality2.dta
  obs:            25                          Quality control (Iman 1994:662)
  vars:            4                          4 Jul 2005 12:07
  size:          500 (99.9% of memory free)
-------------------------------------------------------------------------------
              storage  display     value
variable name   type   format      label      variable label
-------------------------------------------------------------------------------
x1             float   %9.0g
x2             float   %9.0g
x3             float   %9.0g
x4             float   %9.0g
-------------------------------------------------------------------------------
Sorted by:
```

. list in 1/5

```
     +-----------------------+
     |  x1     x2     x3    x4 |
     |-----------------------|
  1. | 4.6      2      4   3.6 |
  2. | 6.7    3.8    5.1   4.7 |
  3. | 4.6    4.3    4.5   3.9 |
  4. | 4.9      6    4.8   5.7 |
  5. | 7.6    6.9    2.5   4.7 |
     +-----------------------+
```

Figure 3.40, an R chart, graphs variation in the process range over the 25 units. **rchart** informs us that one unit's range is "out of control."

```
. rchart x1 x2 x3 x4, connect(l)
```

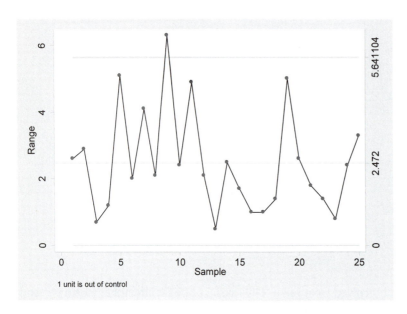

Figure 3.40

Figure 3.41, an $\bar{x}$ chart, shows variation in the process mean. None of these 25 means falls outside the control limits.

```
. xchart x1 x2 x3 x4, connect(l)
```

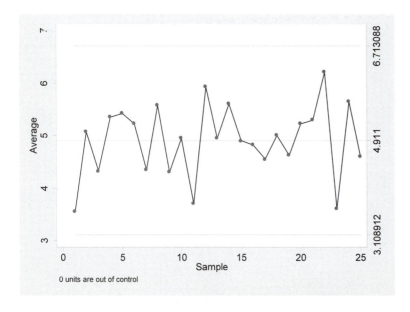

Figure 3.41

Adding Text to Graphs

Titles, captions, and notes can be added to make graphs more self-explanatory. The default versions of titles and subtitles appear above the plot space; notes (which might document the data source, for instance) and captions appear below. These defaults can be overridden, of course. Type **help title_options** for more information about placement of titles, or **help textbox_options** for details concerning their content. Figure 3.42 demonstrates the default versions of these four options in a scatterplot of the prevalence of smoking and college graduates among U.S. states, using *statehealth.dta*. Figure 3.42 also includes titles for both the left and right *y* axes, **yaxis(1 2)**, and top and bottom *x* axes, **xaxis(1 2)**. Subsequent **ytitle** and **xtitle** options refer to the second axes specifically, by including the **axis(2)** suboption. *y* axis 2 is not necessarily on the right, and *x* axis 2 is not necessarily on the left, as we will see later; but these are their default positions.

```
. graph twoway scatter smokeT college, yaxis(1 2) xaxis(1 2)
    title("This is the TITLE") subtitle("This is the SUBTITLE")
    caption("This is the CAPTION") note("This is the NOTE")
    ytitle("Percent adults smoking")
    ytitle("This is Y AXIS 2", axis(2))
    xtitle("Percent adults with Bachelor's degrees or higher")
    xtitle("This is X AXIS 2", axis(2))
```

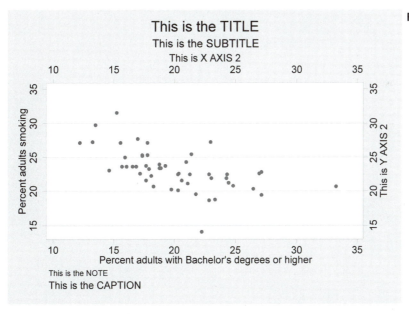

Figure 3.42

Titles add text boxes outside of the plot space. We can also add text boxes at specified coordinates within the plot space. Several outliers stand out in this scatterplot. Upon investigation, they turn out to be Washington DC (highest *college* value, at far right), Utah (lowest *smokeT* value, at bottom center), and Nevada (highest *smokeT* value, at upper left). Text boxes provide a way for us to identify these observations within our graph, as demonstrated in Figure 3.43. The option **text(15.5 22.5 "Utah")** places the word "Utah" at position *y* = 15.5, *x* = 22.5 in the scatterplot, directly above Utah's data point.

Similarly, we place the word "Nevada" at $y = 33.5$, $x = 15$, and draw a box (with small margins; see **help marginstyle**) around that state's name. Three lines of left-justified text are placed next to Washington DC (each line specified in its own set of quotation marks). Any text box or title can have multiple lines in this fashion; we specify each line individually in its own set of quotations, then specify justification or other suboptions. The "Nevada" box uses a default shaded background, whereas for the "Washington DC" box we chose a white background color (see **help textbox_options** and **help colorstyle**).

```
. graph twoway scatter smokeT college, yaxis(1 2) xaxis(1 2)
      title("This is the TITLE") subtitle("This is the SUBTITLE")
      caption("This is the CAPTION") note("This is the NOTE")
      ytitle("Percent adults smoking")
      ytitle("This is Y AXIS 2", axis(2))
      xtitle("Percent adults with Bachelor's degrees or higher")
      xtitle("This is X AXIS 2", axis(2))
      text(15.5 22.5 "Utah")
      text(33.5 15 "Nevada", box margin(small))
      text(23.5 32 "Washington DC" "is not actually" "a state",
         box justification(left) margin(small) bfcolor(white))
```

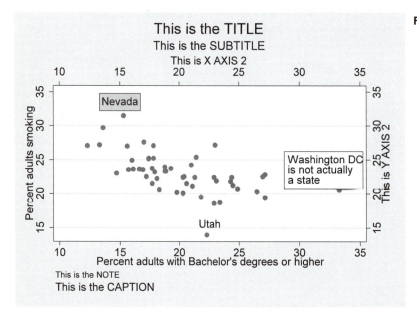

Figure 3.43

Overlaying Multiple Twoway Plots

Two or more plots from the versatile **graph twoway** family can be overlaid, one atop the other, to form a single unified image. Figure 1.1 in Chapter 1 gave a simple example. The **twoway** family includes several model-based types such as **lfit** (linear regression line), **qfit** (quadratic regression curve), and so forth. By themselves, such plots provide minimal information. For example, Figure 3.44 depicts the linear regression line, with 95% confidence bands for the conditional mean, from the regression of *smokeT* on *college* (*states.dta*).

```
. graph twoway lfitci smokeT college
```

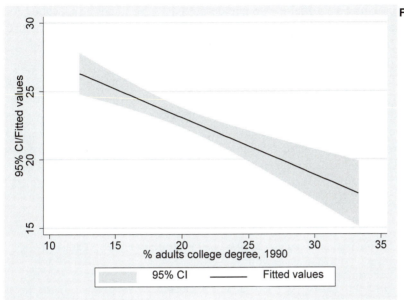

Figure 3.44

A more informative graph results when we overlay a scatterplot on top of the regression line plot, as seen in Figure 3.45. To do this, we essentially give two distinct graphing commands, separated by " || ".

```
. graph twoway lfitci smokeT college
     ||   scatter smokeT college
```

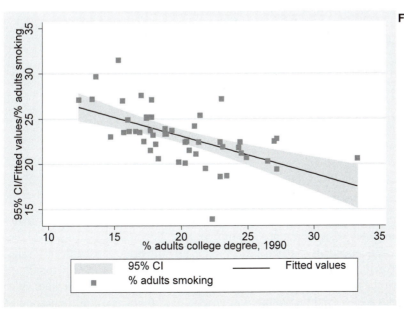

Figure 3.45

The second plot (scatterplot) overprints the first in Figure 3.45. This order has consequences for the default line style (solid, dashed, etc.) and also for the marker symbols (squares, circles, etc.) used by each sub-plot. More importantly, it superimposes the scatterplot points on the confidence bands so the points remain visible. Try reversing the order of the two plots in the command, to see how this works.

Figure 3.46 takes this idea a step further, improving the image through axis labeling and legend options. Because these options apply to the graph as a whole, not just to one of the subplots, the options are placed after a second **||** separator, followed by a comma. Most of these options resemble those used in previous examples. The **order(2 1)** option here does something new: it omits one of the three legend items, so that only two of them (2, the regression line, followed by 1, the confidence interval) appear in the figure. Compare this legend with Figure 3.45 to see the difference. Although we list only two legend items in Figure 3.46, it is still necessary to specify a **rows(3)** legend format as if all three were retained.

```
. graph twoway lfitci smokeT college
     || scatter smokeT college
     || , xlabel(12(2)34) ylabel(14(2)32, angle(horizontal))
     xtitle("Percent adults with Bachelor's degrees or higher")
     ytitle("Percent adults smoking")
     note("Data from CDC and US Census")
     legend(order(2 1) label(1 "95% c.i.") label(2 "regression line")
         rows(3) position(1) ring(0))
```

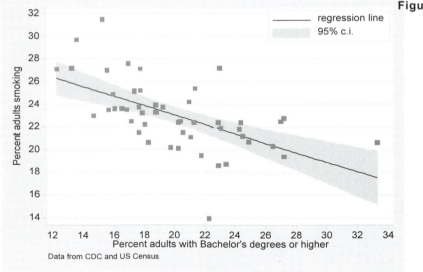

Figure 3.46

The two separate plots (**lfitci** and **scatter**) overlaid in Figure 3.46 share the same *y* and *x* scales, so a single set of axes applies to both. When the variables of interest have different scales, we need independently scaled axes. Figure 3.47 illustrates this with an overlay of two line plots based on the Gulf of St. Lawrence environmental data in *gulf.dta*. This figure combines time series of the minimum mean temperature of the Gulf's cold intermediate layer waters (*cil*), in degrees Celsius, and maximum winter ice cover (*maxarea*), in thousands of square kilometers. The *cil* plot makes use of **yaxis(1)**, which by default is on the left. The *maxarea* plot makes use of **yaxis(2)**, which by default is on the right. The various **ylabel** , **ytitle** , **yline** , and **yscale** options each include an **axis(1)** or **axis(2)** suboption, declaring which *y* axis they refer to. Extra spaces inside the quotation marks for **ytitle** provided a quick way to place the words of these titles where we want them, near the numerical labels. (For a different approach, see Figure 3.48.) The text box containing "Northern Gulf fisheries decline and collapse" is drawn with medium-wide margins around the text; see **help marginstyle** for other choices. **yscale(range( ))** options give both *y* axes a range wider than their data, with specific values chosen after experimenting to find the best vertical separation between the two series.

```
. graph twoway line cil winter,  yaxis(1)  yscale(range(-1,3) axis(1))
       ytitle("Degrees C                              ", axis(1))
       yline(0)  ylabel(-1(.5)1.5, axis(1) angle(horizontal) nogrid)
       text(1 1992 "Northern Gulf" "fisheries decline" "and collapse"
         , box margin(medium))
       ||  line maxarea winter,
       yaxis(2) ylabel(50(50)200, axis(2) angle(horizontal))
       yscale(range(-100,221) axis(2))
       ytitle("                              1000s of km^2", axis(2))
       yline(173.6, axis(2) lpattern(dot))
       ||  if winter > 1949,
       xtitle("")  xlabel(1950(10)2000) xtick(1950(2)2002)
       legend(position(11) ring(0) rows(2) order(2 1)
          label(1 "Max ice area") label(2 "Min CIL temp"))
       note("Source:  Hamilton, Haedrich and Duncan (2003); data from
          DFO (2003)")
```

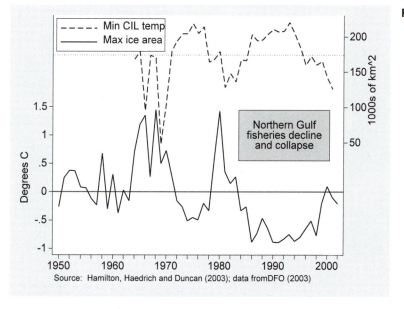

Figure 3.47

The text box on the right in Figure 3.47 marks the late-1980s and early-1990s period when key fisheries including the Northern Gulf cod declined or collapsed. As the graph shows, the fisheries declines coincided with the most sustained cold and ice conditions on record.

To place cod catches in the same graph with temperature and ice, we need three independent vertical scales. Figure 3.48 involves three overlaid plots, with all *y* axes at left (default). The basic forms of the three component plots are as follows:

connected *maxarea winter*

A connected-line plot of *maxarea* vs. *winter*, using *y* axis 3 (which will be leftmost in our final graph). The *y* axis scale ranges from –300 to +220, with no grid of horizontal lines. Its title is "Ice area, 1000 km^2." This title is placed in the "northwest" position, **placement(nw)**.

line *cil winter*

A line plot of *cil* vs. *winter*, using *y* axis 2. *y* scale ranges from –4 to +3, with default labels.

connected *cod winter*

A connected-line plot of *cod* vs. *winter*, using *y* axis 1. The title placement is "southwest,"

placement(sw).

Bringing these three component plots together, the full command for Figure 3.48 appears on the next page. *y* ranges for each of the overlaid plots were chosen by experimenting to find the "right" amount of vertical separation among the three series. Options applied to the whole graph restrict the analysis to years since 1959, specify legend and *x* axis labeling, and request vertical grid lines.

```
. graph twoway connected maxarea winter, yaxis(3)
    yscale(range(-300,220) axis(3)) ylabel(50(50)200, nogrid axis(3))
    ytitle("Ice area, 1000 km^2", axis(3) placement(nw))
    clpattern(dash)

    || line cil winter, yaxis(2) yscale(range(-4,3) axis(2))
    ylabel(, nogrid axis(2))
    ytitle("CIL temperature, degrees C", axis(2)) clpattern(solid)

    || connected cod winter, yaxis(1) yscale(range(0,200) axis(1))
    ylabel(, nogrid axis(1))
    ytitle("Cod catch, 1000 tons", axis(1) placement(sw))

    || if winter > 1959,
    legend(ring(0) position(7) label(1 "Max ice area")
        label(2 "Min CIL temp") label(3 "Cod catch") rows(3))

    xtitle("") xlabel(1960(5)2000, grid)
```

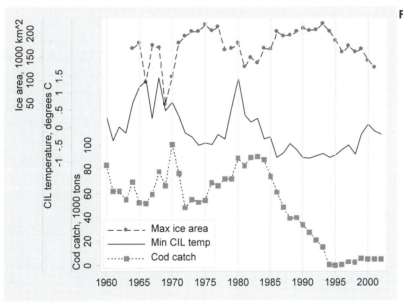

Figure 3.48

Graphing with Do-Files

Complicated graphics like Figure 3.48 require **graph** commands that are many physical lines long (although Stata views the whole command as one logical line). Do-files, introduced in Chapter 2, help in writing such multi-line commands. They also make it easy to save the command for future re-use, in case we later want to modify the graph or draw it again.

The following commands, typed into Stata's Do-file Editor and saved with the file name *fig03_48.do*, become a new do-file for drawing Figure 3.48. Typing

```
. do fig03_48
```

then causes the do-file to execute, redrawing the graph and saving it in two formats.

```
#delimit ;
use c:\data\gulf.dta, clear ;
graph twoway connected maxarea winter, yaxis(3)
   yscale(range(-300,220) axis(3)) ylabel(50(50)200, nogrid axis(3))
   ytitle("Ice area, 1000 km^2", axis(3) placement(nw))
   clpattern(dash)
   || line cil winter, yaxis(2) yscale(range(-4,3) axis(2))
   ylabel(, nogrid axis(2))
   ytitle("CIL temperature, degrees C", axis(2)) clpattern(solid)
   || connected cod winter, yaxis(1) yscale(range(0,200) axis(1))
   ylabel(, nogrid axis(1))
   ytitle("Cod catch, 1000 tons", axis(1) placement(sw))
   || if winter > 1959,
   legend(ring(0) position(7) label(1 "Max ice area")
     label(2 "Min CIL temp") label(3 "Cod catch") rows(3))
   xtitle("") xlabel(1960(5)2000, grid)
   saving(c:\data\fig03_48.gph, replace) ;
graph export c:\data\fig03_48.eps, replace ;
#delimit cr
```

The first line of this do-file sets the semicolon (;) as end-of-line delimiter. Thereafter, Stata does not consider a line finished until it encounters a semicolon. The second line simply retrieves the dataset (*gulf.dta*) needed to draw Figure 3.48; note the semicolon that finishes this line. The long `graph twoway` command occupies the next 15 lines on this page, but Stata treats this all as one logical line that ends with the semicolon after the `saving()` option. This option saves the graph in Stata's .gph format.

Next, the `graph export` command creates a second version of the same graph in Encapsulated Postscript format, as indicated by the .eps suffix in the filename *fig04_48.eps*. (Type **help graph_export** to learn more about this command, which is particularly useful for writing programs or do-files that will create graphs repeatedly.)

The do-file's final `#delimit cr` command re-sets a carriage return as the end-of-line delimiter, going back to Stata's usual mode. Although it is not visible on paper, the line `#delimit cr` must itself end with a carriage return (hit the Enter key), creating one last blank line at the end of the do-file.

Retrieving and Combining Graphs

Any graph saved in Stata's "live" .gph format can subsequently be retrieved into memory by the **graph use** command. For example, we could retrieve Figure 3.48 by typing

```
. graph use fig03_48
```

Once the graph is in memory, it is displayed onscreen and can be printed or saved again with a different name or format. From a graph saved earlier in .gph format, we could subsequently save versions in other formats such as Postscript (.ps), Portable Network Graphics (.png), or Enhanced Windows metafile (.emf). We also could change the color scheme, either through menus or directly in the **graph use** command. *fig03_48.gph* was saved in the s2 monochrome scheme, but we could see how it looks in the s1 color scheme by typing

```
. graph use fig03_47, scheme(s1color)
```

Graphs saved on disk can also be combined by the **graph combine** command. This provides a way to bring multiple plots into the same image. For illustration, we return to the Gulf of St. Lawrence data shown earlier in Figure 3.48. The following commands draw three simple time plots (not shown), saving them with the names *fig03_49a.gph*, *fig03_49b.gph*, and *fig03_49c.gph*. The **margin(medium)** suboptions specify the margin width for title boxes within each plot.

```
. graph twoway line maxarea winter if winter > 1964, xtitle("")
    xlabel(1965(5)2000, grid) ylabel(50(50)200, nogrid)
    title("Maximum winter ice area", position(4) ring(0) box
      margin(medium))
    ytitle("1000 km^2") saving(fig03_49a)
. graph twoway line cil winter if winter > 1964, xtitle("")
    xlabel(1965(5)2000, grid) ylabel(-1(.5)1.5, nogrid)
    title("Minimum CIL temperature", position(1) ring(0) box
      margin(medium))
    ytitle("Degrees C") saving(fig03_49b)
. graph twoway line cod winter if winter > 1964,  xtitle("")
    xlabel(1965(5)2000, grid) ylabel(0(20)100, nogrid)
    title("Northern Gulf cod catch", position(1) ring(0) box
      margin(medium))
    ytitle("1000 tons") saving(fig03_49c)
```

To combine these plots, we type the following command. Because the three plots have identical *x* scales, it makes sense to align the graphs vertically, in three rows. The **imargin** option specifies "very small" margins around the individual plots of Figure 3.49.

```
. graph combine fig03_49a.gph fig03_49b.gph fig03_49c.gph,
    imargin(vsmall) rows(3)
```

Figure 3.49

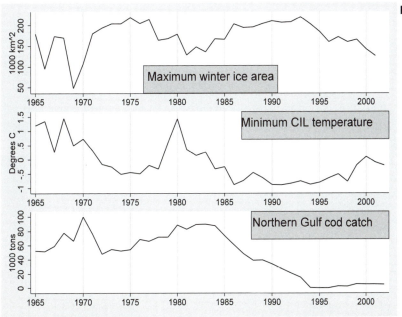

Type **help graph_combine** for more information on this command. Options control details including the number of rows or columns, the size of text and markers (which otherwise become smaller as the number of plots increases), and the margins between individual plots. They can also specify whether *x* or *y* axes of twoway plots have common scales, or assign all components a common color scheme. Titles can be added to the combined graph, which can be printed, saved, retrieved, or for that matter combined again in the usual ways.

Our final example illustrates several of these **graph combine** options, and a way to build graphs with unequal-sized components. Suppose we want a scatterplot similar to the smoking vs. college grads plot seen earlier in Figure 3.42, but with box plots of the *y* and *x* variables drawn beside their respective axes. Using *statehealth.dta*, we might first try to do this by drawing a vertical box plot of *smokeT*, a scatterplot of *smokeT* vs. *college*, and a horizontal box plot of *college*, and then combining the three plots into one image (not shown) with the following commands.

```
. graph box smokeT, saving(wrong1)

. graph twoway scatter smokeT college, saving(wrong2)

. graph hbox college, saving(wrong2)

. graph combine wrong1.gph wrong2.gph wrong3.gph
```

The combined graph produced by the commands above would look wrong, however. We would end up with two fat box plots, each the size of the whole scatterplot, and none of the axes aligned. For a more satisfactory version, we need to start by creating a thin vertical box plot of *smokeT*. The **fxsize(20)** option in the following command fixes the plot's *x* (horizontal) size at 20% of normal, resulting in a normal height but only 20% width plot. Two empty caption lines are included for spacing reasons that will be apparent in the final graph.

```
. graph box smokeT, fxsize(20) caption("" "")
     ytitle("") ylabel(none) ytick(15(5)35, grid) saving(fig03_50a)
```

For the second component, we create a straightforward scatterplot of *smokeT* vs. *college*.

```
. graph twoway scatter smokeT college,
     ytitle("Percent adults smoking")
     xtitle("Percent adults with Bachelor's degrees or higher")
     ylabel(, grid) xlabel(, grid) saving(fig03_50b)
```

The third component is a thin horizontal box plot of *college*. This plot should have normal width, but a *y* (vertical) size fixed at 20% of normal. For spacing reasons, two empty left titles are included.

```
. graph hbox college, fysize(20) l1title("") l2title("")
     ylabel(none) ytick(10(5)35, grid) ytitle("") saving(fig03_50c)
```

These three components come together in Figure 3.50. The **graph combine** command's **cols(2)** option arranges the plots in two columns, like a 2-by-2 table with one empty cell. The **holes(3)** option specifies that the empty cell should be the third one, so our three component graphs fill positions 1, 2, and 4. **iscale(1.05)** enlarges marker symbols and text by about 5%, for readability. The empty captions or titles we built into the original box plots compensate for the two lines of text (title and label) on each axis of the scatterplot, so the box plots align (although not quite perfectly) with the scatterplot axes.

```
.  graph combine fig03_50a.gph fig03_50b.gph fig03_50c.gph,
       cols(2) holes(3) iscale(1.05)
```

Figure 3.50

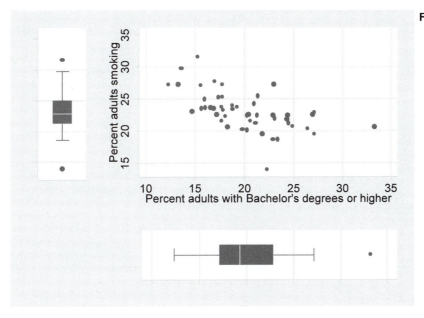

4

Summary Statistics and Tables

The **summarize** command finds simple descriptive statistics such as medians, means, and standard deviations of measurement variables. More flexible arrangements of summary statistics are available through the command **tabstat**. For categorical or ordinal variables, **tabulate** obtains frequency distribution tables, cross-tabulations, assorted tests, and measures of association. **tabulate** can also construct one- or two-way tables of means and standard deviations across categories of other variables. A general table-making command, **table**, produces as many as seven-way tables in which the cells contain statistics such as frequencies, sums, means, or medians. Finally, we review further one-variable procedures including normality tests, transformations, and displays for exploratory data analysis (EDA). Most of the analyses covered in this chapter can be accomplished either through the commands shown or through menu selections under Statistics – Summaries, tables & tests.

In addition to such general-purpose analyses, Stata provides many tables of particular interest to epidemiologistsl. These are not described in this chapter, but can be viewed by typing **help epitab**. Selvin (1996) introduces the topic.

Example Commands

. **summarize y1 y2 y3**
   Calculates simple summary statistics (means, standard deviations, minimum and maximum values, and numbers of observations) for the variables listed.

. **summarize y1 y2 y3, detail**
   Obtains detailed summary statistics including percentiles, median, mean, standard deviation, variance, skewness, and kurtosis.

. **summarize y1 if x1 > 3 & x2 < .**
   Finds summary statistics for $y1$ using only those observations for which variable $x1$ is greater than 3, and $x2$ is not missing.

. **summarize y1 [fweight = w], detail**
   Calculates detailed summary statistics for $y1$ using the frequency weights in variable w.

. **tabstat y1, stats(mean sd skewness kurtosis n)**
   Calculates only the specified summary statistics for variable $y1$.

. **tabstat y1, stats(min p5 p25 p50 p75 p95 max) by(x1)**
   Calculates the specified summary statistics (minimum, 5th percentile, 25th percentile, etc.) for measurement variable $y1$, within categories of $x1$.

. `tabulate x1`

Displays a frequency distribution table for all nonmissing values of variable *x1*.

. `tabulate x1, sort miss`

Displays a frequency distribution of *x1*, including the missing values. Rows (values) are sorted from most to least frequent.

. `tab1 x1 x2 x3 x4`

Displays a series of frequency distribution tables, one for each of the variables listed.

. `tabulate x1 x2`

Displays a two-variable cross-tabulation with *x1* as the row variable, and *x2* as the columns.

. `tabulate x1 x2, chi2 nof column`

Produces a cross-tabulation and Pearson χ^2 test of independence. Does not show cell frequencies, but instead gives the column percentages in each cell.

. `tabulate x1 x2, missing row all`

Produces a cross-tabulation that includes missing values in the table and in the calculation of percentages. Calculates "all" available statistics (Pearson and likelihood χ^2, Cramer's *V*, Goodman and Kruskal's gamma, and Kendall's τ_b).

. `tab2 x1 x2 x3 x4`

Performs all possible two-way cross-tabulations of the listed variables.

. `tabulate x1, summ(y)`

Produces a one-way table showing the mean, standard deviation, and frequency of *y* values within each category of *x1*.

. `tabulate x1 x2, summ(y) means`

Produces a two-way table showing the mean of *y* at each combination of *x1* and *x2* values.

. `by x3, sort:   tabulate x1 x2, exact`

Creates a three-way cross-tabulation, with subtables for *x1* (row) by *x2* (column) at each value of *x3*. Calculates Fisher's exact test for each subtable. **by varname, sort:** works as a prefix for almost any Stata command where it makes sense. The **sort** option is unnecessary if the data already are sorted on *varname*.

. `table y x2 x3, by(x4 x5) contents(freq)`

Creates a five-way cross-tabulation, of *y* (row) by *x2* (column) by *x3* (supercolumn), by *x4* (superrow 1) by *x5* (superrow 2). Cells contain frequencies.

. `table x1 x2, contents(mean y1 median y2)`

Creates a two-way table of *x1* (row) by *x2* (column). Cells contain the mean of *y1* and the median of *y2*.

Summary Statistics for Measurement Variables

Dataset *VTtown.dta* contains information from residents of a town in Vermont. A survey was conducted soon after routine state testing had detected trace amounts of toxic chemicals in the town's water supply. Higher concentrations were found in several private wells and near the public schools. Worried citizens held meetings to discuss possible solutions to this problem.

```
Contains data from C:\data\VTtown.dta
  obs:           153                          VT town survey (Hamilton 1985)
  vars:            7                          11 Jul 2005 18:05
  size:        1,683 (99.9% of memory free)
-------------------------------------------------------------------------------
              storage   display    value
variable name   type    format     label       variable label
-------------------------------------------------------------------------------
gender         byte     %8.0g      sexlbl      Respondent's gender
lived          byte     %8.0g                  Years lived in town
kids           byte     %8.0g      kidlbl      Have children <19 in town?
educ           byte     %8.0g                  Highest year school completed
meetings       byte     %8.0g      kidlbl      Attended meetings on pollution
contam         byte     %8.0g      contamlb    Believe own property/water
                                                 contaminated
school         byte     %8.0g      close       School closing opinion
-------------------------------------------------------------------------------
Sorted by:
```

To find the mean and standard deviation of the variable *lived* (years the respondent had lived in town), type

. **summarize** *lived*

```
    Variable |     Obs        Mean    Std. Dev.       Min         Max
-------------+---------------------------------------------------------
       lived |     153    19.26797    16.95466          1          81
```

This table also gives the number of nonmissing observations and the variable's minimum and maximum values. If we had simply typed **summarize** with no variable list, we would obtain means and standard deviations for every numerical variable in the dataset.

To see more detailed summary statistics, type

. **summarize** *lived*, **detail**

```
                        Years lived in town
-------------------------------------------------------------
          Percentiles     Smallest
 1%            1               1
 5%            2               1
10%            3               1         Obs               153
25%            5               1         Sum of Wgt.       153

50%           15                         Mean          19.26797
                           Largest       Std. Dev.     16.95466
75%           29              65
90%           42              65         Variance      287.4606
95%           55              68         Skewness      1.208804
99%           68              81         Kurtosis      4.025642
```

This **summarize, detail** output includes basic statistics plus the following:

Percentiles: Notably the first quartile (25th percentile), median (50th percentile), and third quartile (75th percentile). Because many samples do not divide evenly into quarters or other standard fractions, these percentiles are approximations.

Four smallest and four largest values, where outliers might show up.

Sum of weights: Stata understands four types of weights: analytical weights (**aweight**), frequency weights (**fweight**), importance weights (**iweight**), and sampling weights (**pweight**). Different procedures allow, and make sense with, different kinds of weights. **summarize, detail** , for example, permits **aweight** or **fweight**. For explanations see **help weights**.

Variance: Standard deviation squared (more properly, standard deviation equals the square root of variance).

Skewness: The direction and degree of asymmetry. A perfectly symmetrical distribution has skewness = 0. Positive skew (heavier right tail) results in skewness > 0; negative skew (heavier left tail) results in skewness < 0.

Kurtosis: Tail weight. A normal (Gaussian) distribution is symmetrical and has kurtosis = 3. If a symmetrical distribution has heavier-than-normal tails (that is, is sharply peaked), it will have kurtosis > 3. Kurtosis < 3 indicates lighter-than-normal tails.

The **tabstat** command provides a more flexible alternative to **summarize**. We can specify just which summary statistics we want to see. For example,

```
. tabstat lived, stats(mean range skewness)

    variable |      mean      range   skewness
-------------+------------------------------
       lived |  19.26797         80   1.208804
---------------------------------------------
```

With a **by(varname)** option, **tabstat** constructs a table containing summary statistics for each value of *varname*. The following example contains means, standard deviations, medians, interquartile ranges, and number of nonmissing observations of *lived*, for each category of *gender*. The means and medians both indicate that, on average, the women in this sample had lived in town for fewer years than the men. Note that the median column is labeled "p50", meaning 50th percentile.

```
. tabstat lived, stats(mean sd median iqr n) by(gender)

Summary for variables: lived
     by categories of: gender (Respondent's gender)

gender |      mean         sd        p50        iqr          N
-------+------------------------------------------------------
  male |  23.48333   19.69125       19.5         28         60
female |  16.54839   14.39468         13         19         93
-------+------------------------------------------------------
 Total |  19.26797   16.95466         15         24        153
--------------------------------------------------------------
```

Statistics available for the **stats()** option of **tabstat** include:

mean	Mean
count	Count of nonmissing observations
n	Same as **count**
sum	Sum
max	Maximum

min	Minimum
range	Range = max – min
sd	Standard deviation
var	Variance
cv	Coefficient of variation = sd / mean
semean	Standard error of mean = sd / sqrt(n)
skewness	Skewness
kurtosis	Kurtosis
median	Median (same as **p50**)
p1	1st percentile (similarly, **p5** , **p10** , **p25** , **p50** , **p75** , **p95** , or **p99**)
iqr	Interquartile range = p75 – p25
q	Quartiles; equivalent to specifying **p25 p50 p75**

Further **tabstat** options give control over the table layout and labeling. Type **help tabstat** to see a complete list.

The statistics produced by **summarize** or **tabstat** describe the sample at hand. We might also want to draw inferences about the population, for example, by constructing a 99% confidence interval for the mean of *lived*:

```
. ci lived, level(99)

    Variable |     Obs        Mean    Std. Err.      [99% Conf. Interval]
-------------+-------------------------------------------------------------
       lived |     153    19.26797    1.370703       15.69241    22.84354
```

Based on this sample, we could be 99% confident that the population mean lies somewhere in the interval from 15.69 to 22.84 years. Here we used a **level( )** option to specify a 99% confidence interval. If we omit this option, **ci** defaults to a 95% confidence interval.

Other options allow **ci** to calculate exact confidence intervals for variables that follow binomial or Poisson distributions. A related command, **cii** , calculates normal, binomial, or Poisson confidence intervals directly from summary statistics, such as we might encounter in a published article. It does not require the raw data. Type **help ci** for details about both commands.

Exploratory Data Analysis

Statistician John Tukey invented a toolkit of methods for exploratory data analysis (EDA), which involves analyzing data in an exploratory and skeptical way without making unneeded assumptions (see Tukey 1977; also Hoaglin, Mosteller, and Tukey 1983, 1985). Box plots, introduced in Chapter 3, are one of Tukey's best-known innovations. Another is the stem-and-leaf display, a graphical arrangement of ordered data values in which initial digits form the "stems" and following digits for each observation make up the "leaves."

```
. stem lived

Stem-and-leaf plot for lived (Years lived in town)

  0* | 1111111222223333333344444444
  0. | 5555555555556666666667777889999
  1* | 0000001122223333334
  1. | 55555567788899
  2* | 0000000111112224444
  2. | 56778899
  3* | 00000124
  3. | 5555666789
  4* | 0012
  4. | 59
  5* | 00134
  5. | 556
  6* |
  6. | 5558
  7* |
  7. |
  8* | 1
```

stem automatically chose a double-stem version here, in which `1*` denotes first digits of 1 and second digits of 0–4 (that is, respondents who had lived in town 10–14 years). `1.` denotes first digits of 1 and second digits of 5 to 9 (15–19 years). We can control the number of lines per initial digit with the **lines( )** option. For example, a five-stem version in which the `1*` stem hold leaves of 0–1, `1t` leaves of 2–3, `1f` leaves of 4–5, `1s` leaves of 6–7, and `1.` leaves of 8–9 could be obtained by typing

```
. stem lived, lines(5)
```

Type **help stem** for information about other options.

Letter-value displays (**lv**) use order statistics to dissect a distribution.

```
. lv lived
```

#	153			Years lived in town			spread	pseudosigma
M	77	|		15		|		
F	39	|	5	17	29	|	24	17.9731
E	20	|	3	21	39	|	36	15.86391
D	10.5	|	2	27	52	|	50	16.62351
C	5.5	|	1	30.75	60.5	|	59.5	16.26523
B	3	|	1	33	65	|	64	15.15955
A	2	|	1	34.5	68	|	67	14.59762
Z	1.5	|	1	37.75	74.5	|	73.5	15.14113
	1	|	1	41	81	|	80	15.32737
		|				|		
		|				|	# below	# above
inner fence	|		-31		65	|	0	5
outer fence	|		-67		101	|	0	0

`M` denotes the median, and `F` the "fourths" (quartiles, using a different approximation than the quartile approximation used by **summarize, detail** and **tabsum**). `E`, `D`, `C`, ... denote cutoff points such that roughly 1/8, 1/16, 1/32, ... of the distribution remains outside in the tails. The second column of numbers gives the "depth," or distance from nearest extreme, for each letter value. Within the center box, the middle column gives "midsummaries," which are averages of the two letter values. If midsummaries drift away from the median, as they do for *lived*, this tells us that the distribution becomes progressively more skewed as we move farther out into the tails. The "spreads" are differences between pairs of letter values. For instance, the spread between F 's equals the approximate interquartile range. Finally, "pseudosigmas" in the right-hand column estimate what the standard deviation should be if these letter values described a Gaussian population. The F pseudosigma, sometimes called a "pseudo standard deviation" (*PSD*), provides a simple and outlier-resistant check for approximate normality in symmetrical distributions:

1. Comparing mean with median diagnoses overall skew:
    mean > median   positive skew
    mean = median   symmetry
    mean < median   negative skew

2. If the mean and median are similar, indicating symmetry, then a comparison between standard deviation and *PSD* helps to evaluate tail normality:
    standard deviation > *PSD*   heavier-than-normal tails
    standard deviation = *PSD*   normal tails
    standard deviation < *PSD*   lighter-than-normal tails
    Let F_1 and F_3 denote 1st and 3rd fourths (approximate 25th and 75th percentiles). Then the interquartile range, *IQR*, equals $F_3 - F_1$, and $PSD = IQR / 1.349$.

lv also identifies mild and severe outliers. We call an *x* value a "mild outlier" when it lies outside the inner fence, but not outside the outer fence:

$$F_1 - 3IQR \le x < F_1 - 1.5IQR \quad \text{or} \quad F_3 + 1.5IQR < x \le F_3 + 3IQR$$

The value of *x* is a "severe outlier" if it lies outside the outer fence:

$$x < F_1 - 3IQR \quad \text{or} \quad x > F_3 + 3IQR$$

lv gives these cutoffs and the number of outliers of each type. Severe outliers, values beyond the outer fences, occur sparsely (about two per million) in normal populations. Monte Carlo simulations suggest that the presence of any severe outliers in samples of $n = 15$ to about 20,000 should be sufficient evidence to reject a normality hypothesis at $\alpha = .05$ (Hamilton 1992b). Severe outliers create problems for many statistical techniques.

summarize, **stem**, and **lv** all confirm that *lived* has a positively skewed sample distribution, not at all resembling a theoretical normal curve. The next section introduces more formal normality tests, and transformations that can reduce a variable's skew.

Normality Tests and Transformations

Many statistical procedures work best when applied to variables that follow normal distributions. The preceding section described exploratory methods to check for approximate normality, extending the graphical tools (histograms, box plots, symmetry plots, and quantile–normal plots) presented in Chapter 3. A skewness–kurtosis test, making use of the skewness and kurtosis statistics shown by **summarize, detail**, can more formally evaluate the null hypothesis that the sample at hand came from a normally-distributed population.

```
. sktest lived
```

```
                  Skewness/Kurtosis tests for Normality
                                                ------- joint ------
    Variable |  Pr(Skewness)   Pr(Kurtosis)  adj chi2(2)    Prob>chi2
-------------+---------------------------------------------------------
       lived |     0.000          0.028         24.79        0.0000
```

sktest here rejects normality: *lived* appears significantly nonnormal in skewness ($P =$ .000), kurtosis ($P = .028$), and in both statistics considered jointly ($P = .0000$). Stata rounds off displayed probabilities to three or four decimals; "0.0000" really means $P < .00005$.

Other normality or log-normality tests include Shapiro–Wilk W (**swilk**) and Shapiro–Francia W' (**sfrancia**) methods. Type **help sktest** to see the options.

Nonlinear transformations such as square roots and logarithms are often employed to change distributions' shapes, with the aim of making skewed distributions more symmetrical and perhaps more nearly normal. Transformations might also help linearize relationships between variables (Chapter 8). Table 4.1 shows a progression called the "ladder of powers" (Tukey 1977) that provides guidance for choosing transformations to change distributional shape. The variable *lived* exhibits mild positive skew, so its square root might be more symmetrical. We could create a new variable equal to the square root of *lived* by typing

```
. generate srlived = lived ^.5
```

Instead of **lived ^.5**, we could equally well have written **sqrt(lived)** .

Logarithms are another transformation that can reduce positive skew. To generate a new variable equal to the natural (base *e*) logarithm of *lived*, type

```
. generate loglived = ln(lived)
```

In the ladder of powers and related transformation schemes such as Box–Cox, logarithms take the place of a "0" power. Their effect on distribution shape is intermediate between .5 (square root) and –.5 (reciprocal root) transformations.

Table 4.1: Ladder of Powers

Transformation	Formula	Effect
cube	**new = old ^3**	reduce severe negative skew
square	**new = old ^2**	reduce mild negative skew
raw	**old**	no change (raw data)
square root	**new = old ^.5**	reduce mild positive skew
$\log_e$	**new = ln(old)**	reduce positive skew
(or $\log_{10}$)	**new = log10(old)**	
negative reciprocal root	**new = -(old ^-.5)**	reduce severe positive skew
negative reciprocal	**new = -(old ^-1)**	reduce very severe positive skew
negative reciprocal square	**new = -(old ^-2)**	"
negative reciprocal cube	**new = -(old ^-3)**	"

When raising to a power less than zero, we take negatives of the result to preserve the original order — the highest value of *old* becomes transformed into the highest value of *new*, and so forth. When *old* itself contains negative or zero values, it is necessary to add a constant before transformation. For example, if *arrests* measures the number of times a person has been arrested (0 for many people), then a suitable log transformation could be

```
. generate larrests = ln(arrests + 1)
```

The **ladder** command combines the ladder of powers with **sktest** tests for normality. It tries each power on the ladder, and reports whether the result is significantly nonnormal. This can be illustrated using the severely skewed variable *energy*, per capita energy consumption, from *states.dta*.

```
. ladder energy
```

Transformation	formula	chi2(2)	P(chi2)
cube	energy^3	53.74	0.000
square	energy^2	45.53	0.000
raw	energy	33.25	0.000
square-root	sqrt(energy)	25.03	0.000
log	log(energy)	15.88	0.000
reciprocal root	1/sqrt(energy)	7.36	0.025
reciprocal	1/energy	1.32	0.517
reciprocal square	1/(energy^2)	4.13	0.127
reciprocal cube	1/(energy^3)	11.56	0.003

It appears that the reciprocal transformation, $1/energy$ (or $energy^{-1}$), most closely resembles a normal distribution. Most of the other transformations (including the raw data) are significantly nonnormal. Figure 4.1 (produced by the **gladder** command) visually supports this conclusion by comparing histograms of each transformation to normal curves.

```
. gladder energy
```

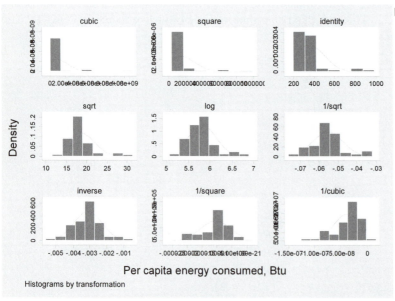

Figure 4.1

Histograms by transformation

Figure 4.2 shows a corresponding set of quantile–normal plots for these ladder of powers transformations, obtained by the "quantile ladder" command **qladder**. To make the tiny plots more readable in this example we scale the labels and marker symbols up by 25% with the **scale(1.25)** option. The axis labels (which would be unreadable and crowded) are suppressed by the options **ylabel(none) xlabel(none)**.

. **qladder** *energy*, **scale(1.25) ylabel(none) xlabel(none)**

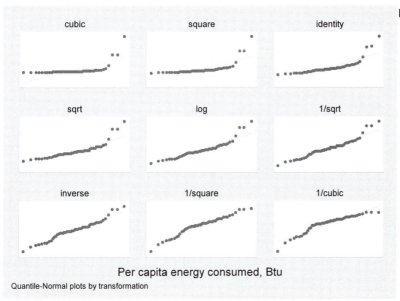

Figure 4.2

Per capita energy consumed, Btu

Quantile-Normal plots by transformation

An alternative technique called Box–Cox transformation offers finer gradations between transformations and automates the choice among them (easier for the analyst, but not always a good thing). The command **bcskew0** finds a value of λ (lambda) for the Box–Cox transformations

$$y^{(\lambda)} = \{y^\lambda - 1\} / \lambda \qquad \lambda > 0 \text{ or } \lambda < 0$$

or

$$y^{(\lambda)} = \ln(y) \qquad \lambda = 0$$

such that $y^{(\lambda)}$ has approximately 0 skewness. Applying this to *energy*, we obtain the transformed variable *benergy*:

. **bcskew0** *benergy* = *energy*, **level(95)**

```
      Transform |         L     [95% Conf. Interval]      Skewness
----------------+-------------------------------------------------
 (energy^L-1)/L |  -1.246052    -2.052503  -.6163383       .000281
(1 missing value generated)
```

That is, *benergy* = (*energy*$^{-1.246}$ – 1)/(–1.246) is the transformation that comes closest to symmetry (as defined by the skewness statistic). The Box–Cox parameter $\lambda = -1.246$ is not far from our ladder-of-powers choice, the –1 power. The confidence interval for λ,

$$-2.0525 < \lambda < -.6163$$

allows us to reject some other possibilities including logarithms ($\lambda = 0$) or square roots ($\lambda = .5$). Chapter 8 describes a Box–Cox approach to regression modeling.

Frequency Tables and Two-Way Cross-Tabulations

The methods described above apply to measurement variables. Categorical variables require other approaches, such as tabulation. Returning to the survey data in *VTtown.dta*, we could find the percentage of respondents who attended meetings concerning the pollution problem by tabulating the categorical variable *meetings*:

```
. tabulate meetings

   Attended |
meetings on |
  pollution |      Freq.     Percent        Cum.
------------+-----------------------------------
         no |        106       69.28       69.28
        yes |         47       30.72      100.00
------------+-----------------------------------
      Total |        153      100.00
```

tabulate can produce frequency distributions for variables that have thousands of values. To construct a manageable frequency distribution table for a variable with many values, however, you might first want to group those values by applying **generate** with its **recode** or **autocode** options (see Chapter 2 or **help generate**).

   tabulate followed by two variable names creates a two-way cross-tabulation. For example, here is a cross-tabulation of *meetings* by *kids* (whether respondent has children under 19 living in town):

```
. tabulate meetings kids

   Attended |
   meetings | Have children <19 in
         on |         town?
  pollution |         no        yes |      Total
------------+----------------------+----------
         no |         52         54 |        106
        yes |         11         36 |         47
------------+----------------------+----------
      Total |         63         90 |        153
```

The first-named variable forms the rows, and the second forms columns in the resulting table. We see that only 11 of these 153 people were non-parents who attended the meetings.

   tabulate has a number of options that are useful with frequency tables:

all   Equivalent to the options **chi2 lrchi2 gamma taub V**. Not all of these options will be equally appropriate for a given table. **gamma** and **taub** assume that both variables have ordered categories, whereas **chi2** , **lrchi2** , and **V** do not.

cchi2   Displays the contribution to Pearson χ^2 (chi-squared) in each cell of a two-way table.

cell   Shows total percentages for each cell.

chi2   Pearson χ^2 test of hypothesis that row and column variables are independent.

clrchi2   Displays the contribution to likelihood-ratio χ^2 in each cell of a two-way table.

`column`	Shows column percentages for each cell.
`exact`	Fisher's exact test of the independence hypothesis. Superior to **chi2** if the table contains thin cells with low expected frequencies. Often too slow to be practical in large tables, however.
`expected`	Displays the expected frequency under the assumption of independence in each cell of a two-way table.
`gamma`	Goodman and Kruskal's γ (gamma), with its asymptotic standard error (ASE). Measures association between ordinal variables, based on the number of concordant and discordant pairs (ignoring ties). $-1 \leq \gamma \leq 1$.
`generate(new)`	Creates a set of dummy variables named *new1*, *new2*, and so on to represent the values of the tabulated variable.
`lrchi2`	Likelihood-ratio χ^2 test of independence hypothesis. Not obtainable if the table contains any empty cells.
`matcell(matname)`	Saves the reported frequencies in *matname*.
`matcol(matname)`	Saves the numeric values of the $1 \times c$ column stub in *matname*.
`matrow(matname)`	Saves the numeric values of the $r \times 1$ row stub in *matname*.
`missing`	Includes "missing" as one row and/or column of the table.
`nofreq`	Does not show cell frequencies.
`nokey`	Suppresses the display of a key above two-way tables. The default is to display the key if more than one cell statistic is requested and otherwise to omit it. Specifying **key** forces the display of the key.
`nolabel`	Shows numerical values rather than value labels of labeled numeric variables.
`plot`	Produces a simple bar chart of the relative frequencies in a one-way table.
`replace`	Indicates that the immediate data specified as arguments to the **tabi** command are to be left as the current data in memory, replacing whatever data were there.
`row`	Shows row percentages for each cell.
`sort`	Displays the rows in descending order of frequency (and ascending order of the variable within equal values of frequency).
`subpop(varname)`	Excludes observations for which *varname* = 0 in tabulating frequencies. The identities of the rows and columns will be determined from all the data, including the *varname* = 0 group, and so there may be entries in the table with frequency 0.
`taub`	Kendall's τ_b (tau-b), with its asymptotic standard error (ASE). Measures association between ordinal variables. **taub** is similar to **gamma**, but uses a correction for ties. $-1 \leq \tau_b \leq 1$.
`v`	Cramer's V (note capitalization), a measure of association for nominal variables. In 2×2 tables, $-1 \leq V \leq 1$. In larger tables, $0 \leq V \leq 1$.

wrap   Requests that Stata take no action on wide, two-way tables to make them readable. Unless **wrap** is specified, wide tables are broken into pieces for readability.

To get the column percentages (because the column variable, *kids*, is the independent variable) and a χ^2 test for the cross-tabulation of *meetings* by *kids*, type

```
. tabulate meetings kids, column chi2
```

```
+-------------------+
| Key               |
|-------------------|
|     frequency     |
| column percentage |
+-------------------+

Attended |
meetings | Have children <19 in
      on |        town?
pollution |       no       yes |     Total
-----------+----------------------+----------
       no |       52        54 |       106
          |    82.54     60.00 |     69.28
-----------+----------------------+----------
      yes |       11        36 |        47
          |    17.46     40.00 |     30.72
-----------+----------------------+----------
    Total |       63        90 |       153
          |   100.00    100.00 |    100.00

      Pearson chi2(1) =   8.8464   Pr = 0.003
```

Forty percent of the respondents with children attended meetings, compared with about 17% of the respondents without children. This association is statistically significant ($P = .003$).

Occasionally we might need to re-analyze a published table, without access to the original raw data. A special command, **tabi** ("immediate" tabulation), accomplishes this. Type the cell frequencies on the command line, with table rows separated by " \ ". For illustration, here is how **tabi** could reproduce the previous χ^2 analysis, given only the four cell frequencies:

```
. tabi 52 54 \ 11 36, column chi2
```

```
+-------------------+
| Key               |
|-------------------|
|     frequency     |
| column percentage |
+-------------------+

          |         col
      row |        1         2 |     Total
-----------+----------------------+----------
        1 |       52        54 |       106
          |    82.54     60.00 |     69.28
-----------+----------------------+----------
        2 |       11        36 |        47
          |    17.46     40.00 |     30.72
-----------+----------------------+----------
    Total |       63        90 |       153
          |   100.00    100.00 |    100.00

      Pearson chi2(1) =   8.8464   Pr = 0.003
```

Unlike **tabulate**, **tabi** does not require or refer to any data in memory. By adding the **replace** option, however, we can ask **tabi** to replace whatever data are in memory with the new cross-tabulation. Statistical options (**chi2**, **exact**, **nofreq**, and so forth) work the same for **tabi** as they do with **tabulate**.

Multiple Tables and Multi-Way Cross-Tabulations

With surveys and other large datasets, we sometimes need frequency distributions of many different variables. Instead of asking for each table separately, for example by typing **tabulate** *meetings*, then **tabulate** *gender*, and finally **tabulate** *kids*, we could simply use another specialized command, **tab1**:

```
. tab1 meetings gender kids
```

Or, to produce one-way frequency tables for each variable from *gender* through *school* in this dataset (the maximum is 30 variables at one time), type

```
. tab1 gender-school
```

Similarly, **tab2** creates multiple two-way tables. For example, the following command cross-tabulates every two-way combination of the listed variables:

```
. tab2 meetings gender kids
```

tab1 and **tab2** offer the same options as **tabulate**.

To form multi-way contingency tables, one approach uses the ordinary **tabulate** command with a **by** prefix. Here is a three-way cross-tabulation of *meetings* by *kids* by *contam* (respondent believes his or her own property or water contaminated), with χ^2 tests for the independence of *meetings* and *kids* within each level of *contam*:

```
. by contam, sort: tabulate meetings kids, nofreq col chi2
-----------------------------------------------------------------------
-> contam = no

  Attended |
  meetings | Have children <19 in
        on |        town?
 pollution |        no        yes |      Total
-----------+----------------------+----------
        no |     91.30      68.75 |      78.18
       yes |      8.70      31.25 |      21.82
-----------+----------------------+----------
     Total |    100.00     100.00 |     100.00

           Pearson chi2(1) =    7.9814    Pr = 0.005
```

```
-> contam = yes

   Attended |
   meetings | Have children <19 in
         on |         town?
  pollution |      no        yes |      Total
-----------+----------------------+----------
        no |     58.82      38.46 |      46.51
       yes |     41.18      61.54 |      53.49
-----------+----------------------+----------
     Total |    100.00     100.00 |     100.00

        Pearson chi2(1) =    1.7131    Pr = 0.191
```

Parents were more likely to attend meetings, among both the contaminated and uncontaminated groups. Only among the larger uncontaminated group is this "parenthood effect" statistically significant, however. As multi-way tables separate the data into smaller subsamples, the size of these subsamples has noticeable effects on significance-test outcomes.

This approach can be extended to tabulations of greater complexity. For example, to get a four-way cross-tabulation of *gender* by *contam* by *meetings* by *kids,* with χ^2 tests for each *meetings* by *kids* subtable (results not shown), type the command

. **by *gender contam*, sort: tabulate *meetings kids*, column chi2**

A better way to produce multi-way tables, if we do not need percentages or statistical tests, is through Stata's general table-making command, **table**. This versatile command has many options, only a few of which are illustrated here. To construct a simple frequency table of *meetings*, type

. **table *meetings*, contents(freq)**

```
----------------------
Attended   |
meetings   |
on         |
pollution  |       Freq.
-----------+-----------
       no  |        106
      yes  |         47
----------------------
```

For a two-way frequency table or cross-tabulation, type

. **table *meetings kids*, contents(freq)**

```
----------------------
           |    Have
Attended   |  children
meetings   |   <19 in
on         |    town?
pollution  |   no    yes
-----------+-----------
       no  |   52     54
      yes  |   11     36
----------------------
```

If we specify a third categorical variable, it forms the "supercolumns" of a three-way table:

```
. table meetings kids contam, contents(freq)

-----------------------------------
          |       Believe own
          |      property/water
Attended  |   contaminated and Have
meetings  |   children <19 in town?
on        | --- no ---     --- yes --
pollution |  no    yes      no    yes
----------+------------------------
      no  |  42     44      10     10
     yes  |   4     20       7     16
-----------------------------------
```

More complicated tables require the **by( )** option, which allows up to four "supperrow" variables. **table** thus can produce up to seven-way tables: one row, one column, one supercolumn, and up to four superrows. Here is a four-way example:

```
. table meetings kids contam, contents(freq) by(gender)

-----------------------------------
Responden |
t's       |
gender    |
and       |       Believe own
Attended  |      property/water
meetings  |   contaminated and Have
on        |   children <19 in town?
pollution | --- no ---     --- yes --
          |  no    yes      no    yes
----------+------------------------
male      |
      no  |  18     18       3      3
     yes  |   2      7       3      6
----------+------------------------
female    |
      no  |  24     26       7      7
     yes  |   2     13       4     10
-----------------------------------
```

The **contents( )** option of **table** specifies what statistics the table's cells contain:

contents(freq)	Frequency
contents(mean *varname*)	Mean of *varname*
contents(sd *varname*)	Standard deviation of *varname*
contents(sum *varname*)	Sum of *varname*
contents(rawsum *varname*)	Sums ignoring optionally specified weight
contents(count *varname*)	Count of nonmissing observations of *varname*
contents(n *varname*)	Same as **count**
contents(max *varname*)	Maximum of *varname*
contents(min *varname*)	Minimum of *varname*
contents(median *varname*)	Median of *varname*
contents(iqr *varname*)	Interquartile range (IQR) of *varname*

    `contents(p1 varname)`   1st percentile of *varname*

    `contents(p2 varname)`   2nd percentile of *varname* (so forth to `p99`)

The next section illustrates several more of these options.

Tables of Means, Medians, and Other Summary Statistics

`tabulate` readily produces tables of means and standard deviations within categories of the tabulated variable. For example, to form a one-way table with means of *lived* within each category of *meetings*, type

. `tabulate meetings, summ(lived)`

```
   Attended |
meetings on |    Summary of Years lived in town
   pollution |       Mean   Std. Dev.        Freq.
------------+-----------------------------------
        no |   21.509434   17.743809          106
       yes |   14.212766   13.911109           47
------------+-----------------------------------
     Total |   19.267974   16.954663          153
```

Meetings attenders appear to be relative newcomers, averaging 14.2 years in town, compared with 21.5 years for those who did not attend.

    We can also use **tabulate** to form a two-way table of means by typing

. `tabulate meetings kids, sum(lived) means`

```
                       Means of Years lived in town

   Attended |
   meetings |    Have children <19
        on |        in town?
   pollution |       no        yes |    Total
-----------+-------------------------+----------
        no |  28.307692  14.962963 |  21.509434
       yes |  23.363636  11.416667 |  14.212766
-----------+-------------------------+----------
     Total |  27.444444  13.544444 |  19.267974
```

Both parents and nonparents among the meeting attenders tend to have lived fewer years in town, so the newcomer/oldtimer division noticed in the previous table is not a spurious reflection of the fact that parents with young children were more likely to attend.

    The **means** option used above called for a table containing only means. Otherwise we get a bulkier table with means, standard deviations, and frequencies in each cell. Chapter 5 describes statistical tests for hypotheses about subgroup means.

    Although it performs no tests, **table** nicely builds up to seven-way tables containing means, standard deviations, sums, medians, or other statistics (see the option list in previous section). Here is a one-way table showing means of *lived* within categories of *meetings*:

```
. table meetings, contents(mean lived)

----------------------
Attended   |
meetings   |
on         |
pollution  | mean(lived)
-----------+-----------
      no   |    21.5094
     yes   |    14.2128
----------------------
```

A two-way table of means is a straightforward extension:

```
. table meetings kids, contents(mean lived)

--------------------------
Attended   |
meetings   |Have children <19
on         |    in town?
pollution  |    no      yes
-----------+--------------
      no   | 28.3077   14.963
     yes   | 23.3636   11.4167
--------------------------
```

Table cells can contain more than one statistic. Suppose we want a two-way table with both means and medians of the variable *lived*:

```
. table meetings kids, contents(mean lived median lived)

--------------------------
Attended   |
meetings   |Have children <19
on         |    in town?
pollution  |    no      yes
-----------+--------------
      no   | 28.3077   14.963
           |    27.5     12.5
           |
     yes   | 23.3636   11.4167
           |      21        6
--------------------------
```

The medians in the table above confirm our earlier conclusion based on means: the meeting attenders, both parents and nonparents, tended to have lived fewer years in town than their non-attending counterparts. Medians within each cell are less than the means, reflecting the positive skew (means pulled up by a few long-time residents) of the variable *lived*.

The cell contents shown by **table** could be means, medians, sums, or other summary statistics for two or more different variables.

Using Frequency Weights

`summarize`, `tabulate`, `table`, and related commands can be used with frequency weights that indicate the number of replicated observations. For example, file *sextab2.dta* contains results from a British survey of sexual behavior (Johnson et al. 1992). It apparently has 48 observations:

```
Contains data from C:\data\sextab2.dta
  obs:            48                          British sex survey (Johnson 92)
  vars:            4                          11 Jul 2005 18:05
  size:          432 (99.9% of memory free)
-------------------------------------------------------------------------------
              storage  display    value
variable name  type    format     label        variable label
-------------------------------------------------------------------------------
age            byte    %8.0g      age          Age
gender         byte    %8.0g      gender       Gender
lifepart       byte    %8.0g      partners     # heterosex partners lifetime
count          int     %8.0g                   Number of individuals
-------------------------------------------------------------------------------
Sorted by:  age  lifepart  gender
```

One variable, *count*, indicates the number of individuals with each combination of characteristics, so this small dataset actually contains information from over 18,000 respondents. For example, 405 respondents were male, ages 16 to 24, and reported having no heterosexual partners so far in their lives.

```
. list in 1/5

     +------------------------------------+
     |  age    gender   lifepart   count  |
     |------------------------------------|
  1. | 16-24     male      none     405   |
  2. | 16-24   female      none     465   |
  3. | 16-24     male       one     323   |
  4. | 16-24   female       one     606   |
  5. | 16-24     male       two     194   |
     +------------------------------------+
```

We use *count* as a frequency weight to create a cross-tabulation of *lifepart* by *gender*:

```
. tabulate lifepart gender [fw = count]
```

# heterosex partners lifetime	Gender male	female	Total
none	544	586	1130
one	1734	4146	5880
two	887	1777	2664
3-4	1542	1908	3450
5-9	1630	1364	2994
10+	2048	708	2756
Total	8385	10489	18874

The usual `tabulate` options work as expected with frequency weights. Here is the same table showing column percentages instead of frequencies:

```
. tabulate lifepart gender [fweight = count], column nof
```

```
        # |
heterosex |
 partners |        Gender
 lifetime |    male    female |     Total
----------+----------------------+----------
     none |    6.49      5.59 |      5.99
      one |   20.68     39.53 |     31.15
      two |   10.58     16.94 |     14.11
      3-4 |   18.39     18.19 |     18.28
      5-9 |   19.44     13.00 |     15.86
      10+ |   24.42      6.75 |     14.60
----------+----------------------+----------
    Total |  100.00    100.00 |    100.00
```

Other types of weights such as probability or analytical weights do not work as well with **tabulate** because their meanings are unclear regarding the command's principal options.

A different application of frequency weights can be demonstrated with **summarize**. File *college1.dta* contains information on a random sample consisting of 11 U.S. colleges, drawn from *Barron's Compact Guide to Colleges* (1992).

```
Contains data from C:\data\college1.dta
  obs:           11                          Colleges sample 1 (Barron's 92)
  vars:           5                          11 Jul 2005 18:05
  size:         429 (99.9% of memory free)
-------------------------------------------------------------------------------
              storage   display    value
variable name   type    format     label      variable label
-------------------------------------------------------------------------------
school          str28   %28s                   College or university
enroll          int     %8.0g                  Full-time students 1991
pctmale         byte    %8.0g                  Percent male 1991
msat            int     %8.0g                  Average math SAT
vsat            int     %8.0g                  Average verbal SAT
-------------------------------------------------------------------------------
Sorted by:
```

The variables include *msat*, the mean math Scholastic Aptitude Test score at each of the 11 schools.

```
. list school enroll msat
```

```
     +------------------------------------------------+
     |                       school   enroll   msat |
     |------------------------------------------------|
  1. |            Brown University     5550    680 |
  2. |                 U. Scranton     3821    554 |
  3. | U. North Carolina/Asheville     2035    540 |
  4. |            Claremont College     849    660 |
  5. |            DePaul University    6197    547 |
     |------------------------------------------------|
  6. |       Thomas Aquinas College     201    570 |
  7. |            Davidson College     1543    640 |
  8. |         U. Michigan/Dearborn    3541    485 |
  9. |        Mass. College of Art      961    482 |
 10. |             Oberlin College     2765    640 |
     |------------------------------------------------|
 11. |          American University    5228    587 |
     +------------------------------------------------+
```

We can easily find the mean *msat* value among these 11 schools by typing

```
. summarize msat
```

```
    Variable |       Obs        Mean    Std. Dev.       Min        Max
-------------+--------------------------------------------------------
        msat |        11    580.4545    67.63189        482        680
```

This summary table gives each school's mean math SAT score the same weight. DePaul University, however, has 30 times as many students as Thomas Aquinas College. To take the different enrollments into account we could weight by *enroll*,

```
. summarize msat [fweight = enroll]
```

```
    Variable |       Obs        Mean    Std. Dev.       Min        Max
-------------+--------------------------------------------------------
        msat |     32691     583.064    63.10665        482        680
```

Typing

```
. summarize msat [freq = enroll]
```

would accomplish the same thing.

The enrollment-weighted mean, unlike the unweighted mean, is equivalent to the mean for the 32,691 students at these colleges (assuming they all took the SAT). Note, however, that we could not say the same thing about the standard deviation, minimum, or maximum. Apart from the mean, most individual-level statistics cannot be calculated simply by weighting data that already are aggregated. Thus, we need to use weights with caution. They might make sense in the context of one particular analysis, but seldom do for the dataset as a whole, when many different kinds of analyses are needed.

ANOVA and Other Comparison Methods

Analysis of variance (ANOVA) encompasses a set of methods for testing hypotheses about differences between means. Its applications range from simple analyses where we compare the means of y across categories of x, to more complicated situations with multiple categorical and measurement x variables. t tests for hypotheses regarding a single mean (one-sample) or a pair of means (two-sample) correspond to elementary forms of ANOVA.

Rank-based "nonparametric" tests, including sign, Mann–Whitney, and Kruskal–Wallis, take a different approach to comparing distributions. These tests make weaker assumptions about measurement, distribution shape, and spread. Consequently, they remain valid under a wider range of conditions than ANOVA and its "parametric" relatives. Careful analysts sometimes use parametric and nonparametric tests together, checking to see whether both point toward similar conclusions. Further troubleshooting is called for when parametric and nonparametric results disagree.

anova is the first of Stata's model-fitting commands to be introduced in this book. Like the others, it has considerable flexibility encompassing a wide variety of models. **anova** can fit one-way and N-way ANOVA or analysis of covariance (ANCOVA) for balanced and unbalanced designs, including designs with missing cells. It can also fit factorial, nested, mixed, or repeated-measures designs. One follow-up command, **predict**, calculates predicted values, several types of residuals, and assorted standard errors and diagnostic statistics after **anova**. Another followup command, **test**, obtains tests of user-specified null hypotheses. Both **predict** and **test** work similarly with other Stata model-fitting commands, such as **regress** (Chapter 6).

The following menu choices give access to most operations described in this chapter:

Statistics – Summaries, tables, & tests – Classical tests of hypotheses

Statistics – Summaries, tables, & tests – Nonparametric tests of hypotheses

Statistics – ANOVA/MANOVA

Statistics – General post-estimation – Obtain predictions, residuals, etc., after estimation

Graphics – Overlaid twoway graphs

Example Commands

. `anova y x1 x2`

Performs two-way ANOVA, testing for differences among the means of *y* across categories of *x1* and *x2*.

. `anova y x1 x2 x1*x2`

Performs a two-way factorial ANOVA, including both the main and interaction (*x1*x2*) effects of categorical variables *x1* and *x2*.

. `anova y x1 x2 x3 x1*x2 x1*x3 x2*x3 x1*x2*x3`

Performs a three-way factorial ANOVA, including the three-way interaction *x1*x2*x3*, as well as all two-way interactions and main effects.

. `anova reading curriculum / teacher|curriculum`

Fits a nested model to test the effects of three types of curriculum on students' reading ability (*reading*). *teacher* is nested within *curriculum* (`teacher|curriculum`) because several different teachers were assigned to each curriculum. The *Base Reference Manual* provides other nested ANOVA examples, including a split-plot design.

. `anova headache subject medication, repeated(medication)`

Fits a repeated-measures ANOVA model to test the effects of three types of headache medication (*medication*) on the severity of subjects' headaches (*headache*). The sample consists of 20 subjects who report suffering from frequent headaches. Each subject tried each of the three medications at separate times during the study.

. `anova y x1 x2 x3 x4 x2*x3, continuous(x3 x4) regress`

Performs analysis of covariance (ANCOVA) with four independent variables, two of them (*x1* and *x2*) categorical and two of them (*x3* and *x4*) measurements. Includes the *x2*x3* interaction, and shows results in the form of a regression table instead of the default ANOVA table.

. `kwallis y, by(x)`

Performs a Kruskal–Wallis test of the null hypothesis that *y* has identical rank distributions across the *k* categories of *x* (*k* > 2).

. `oneway y x`

Performs a one-way analysis of variance (ANOVA), testing for differences among the means of *y* across categories of *x*. The same analysis, with a different output table, is produced by `anova y x`.

. `oneway y x, tabulate scheffe`

Performs one-way ANOVA, including a table of sample means and Scheffé multiple-comparison tests in the output.

. `ranksum y, by(x)`

Performs a Wilcoxon rank-sum test (also known as a Mann–Whitney *U* test) of the null hypothesis that *y* has identical rank distributions for both categories of dichotomous variable *x*. If we assume that both rank distributions possess the same shape, this amounts to a test for whether the two medians of *y* are equal.

. `serrbar ymean se x, scale(2)`

Constructs a standard-error-bar plot from a dataset of means. Variable *ymean* holds the group means of *y*; *se* the standard errors; and *x* the values of categorical variable *x*. `scale(2)` asks for bars extending to ±2 standard errors around each mean (default is ±1 standard error).

. `signrank y1 = y2`

Performs a Wilcoxon matched-pairs signed-rank test for the equality of the rank distributions of *y1* and *y2*. We could test whether the median of *y1* differs from a constant such as 23.4 by typing the command `signrank y1 = 23.4`.

. `signtest y1 = y2`

Tests the equality of the medians of *y1* and *y2* (assuming matched data; that is, both variables measured on the same sample of observations). Typing `signtest y1 = 5` would perform a sign test of the null hypothesis that the median of *y1* equals 5.

. `ttest y = 5`

Performs a one-sample *t* test of the null hypothesis that the population mean of *y* equals 5.

. `ttest y1 = y2`

Performs a one-sample (paired difference) *t* test of the null hypothesis that the population mean of *y1* equals that of *y2*. The default form of this command assumes that the data are paired. With unpaired data (*y1* and *y2* are measured from two independent samples), add the option `unpaired`.

. `ttest y, by(x) unequal`

Performs a two-sample *t* test of the null hypothesis that the population mean of *y* is the same for both categories of variable *x*. Does not assume that the populations have equal variances. (Without the `unequal` option, `ttest` does assume equal variances.)

One-Sample Tests

One-sample *t* tests have two seemingly different applications:

1. Testing whether a sample mean $\bar{y}$ differs significantly from an hypothesized value μ_0.
2. Testing whether the means of y_1 and y_2, two variables measured over the same set of observations, differ significantly from each other. This is equivalent to testing whether the mean of a "difference score" variable created by subtracting y_1 from y_2 equals zero.

We use essentially the same formulas for either application, although the second starts with information on two variables instead of one.

The data in *writing.dta* were collected to evaluate a college writing course based on word processing (Nash and Schwartz 1987). Measures such as the number of sentences completed in timed writing were collected both before and after students took the course. The researchers wanted to know whether the post-course measures showed improvement.

```
.  describe
```

```
Contains data from C:\data\writing.dta
   obs:            24                          Nash and Schwartz (1987)
  vars:             9                          12 Jul 2005 10:16
  size:           312 (99.9% of memory free)
-------------------------------------------------------------------------------
              storage  display    value
variable name   type   format     label      variable label
-------------------------------------------------------------------------------
id              byte   %8.0g      slbl       Student ID
preS            byte   %8.0g                 # of sentences (pre-test)
preP            byte   %8.0g                 # of paragraphs (pre-test)
preC            byte   %8.0g                 Coherence scale 0-2 (pre-test)
preE            byte   %8.0g                 Evidence scale 0-6 (pre-test)
postS           byte   %8.0g                 # of sentences (post-test)
postP           byte   %8.0g                 # of paragraphs (post-test)
postC           byte   %8.0g                 Coherence scale 0-2 (post-test)
postE           byte   %8.0g                 Evidence scale 0-6 (post-test)
-------------------------------------------------------------------------------
Sorted by:
```

Suppose that we knew that students in previous years were able to complete an average of 10 sentences. Before examining whether the students in *writing.dta* improved during the course, we might want to learn whether at the start of the course they were essentially like earlier students — in other words, whether their pre-test (*preS*) mean differs significantly from the mean of previous students (10). To see a one-sample t test of $H_0 : \mu = 10$, type

```
.  ttest preS = 10
```

```
One-sample t test
```

```
-------------------------------------------------------------------------------
Variable |     Obs        Mean    Std. Err.    Std. Dev.   [95% Conf. Interval]
---------+---------------------------------------------------------------------
    preS |      24    10.79167    .9402034     4.606037     8.846708    12.73663
-------------------------------------------------------------------------------
Degrees of freedom: 23
```

```
                           Ho: mean(preS) = 10

    Ha: mean < 10               Ha: mean != 10                 Ha: mean > 10
      t =   0.8420                t =   0.8420                    t =   0.8420
  P < t =   0.7958            P > |t| =   0.4084              P > t =   0.2042
```

The notation P > t means "the probability of a greater value of t"— that is, the one-tail test probability. The two-tail probability of a greater absolute t appears as P > |t| = .4084 . Because this probability is high, we have no reason to reject $H_0 : \mu = 10$. Note that **ttest** automatically provides a 95% confidence interval for the mean. We could get a different confidence interval, such as 90%, by adding a **level(90)** option to this command.

A nonparametric counterpart, the sign test, employs the binomial distribution to test hypotheses about single medians. For example, we could test whether the median of *preS* equals 10. **signtest** gives us no reason to reject that null hypothesis either.

```
. signtest preS = 10
```

Sign test

```
        sign |      observed      expected
-------------+------------------------------
    positive |            12            11
    negative |            10            11
        zero |             2             2
-------------+------------------------------
         all |            24            24
```

One-sided tests:
  Ho: median of preS - 10 = 0 vs.
  Ha: median of preS - 10 > 0
    Pr(#positive >= 12) =
      Binomial(n = 22, x >= 12, p = 0.5) =   0.4159

  Ho: median of preS - 10 = 0 vs.
  Ha: median of preS - 10 < 0
    Pr(#negative >= 10) =
      Binomial(n = 22, x >= 10, p = 0.5) =   0.7383

Two-sided test:
  Ho: median of preS - 10 = 0 vs.
  Ha: median of preS - 10 != 0
    Pr(#positive >= 12 or #negative >= 12) =
      min(1, 2*Binomial(n = 22, x >= 12, p = 0.5)) =   0.8318

Like **ttest**, **signtest** includes right-tail, left-tail, and two-tail probabilities. Unlike the symmetrical t distributions used by **ttest**, however, the binomial distributions used by **signtest** have different left- and right-tail probabilities. In this example, only the two-tail probability matters because we were testing whether the *writing.dta* students "differ" from their predecessors.

Next, we can test for improvement during the course by testing the null hypothesis that the mean number of sentences completed before and after the course (that is, the means of *preS* and *postS*) are equal. The **ttest** command accomplishes this as well, finding a significant improvement.

```
. ttest postS = preS
```

Paired t test

```
------------------------------------------------------------------------------
Variable |     Obs         Mean    Std. Err.    Std. Dev.   [95% Conf. Interval]
---------+--------------------------------------------------------------------
   postS |      24       26.375    1.693779     8.297787    22.87115    29.87885
    preS |      24     10.79167    .9402034     4.606037    8.846708    12.73663
---------+--------------------------------------------------------------------
    diff |      24     15.58333    1.383019     6.775382    12.72234    18.44433
------------------------------------------------------------------------------
```

Ho: mean(postS - preS) = mean(diff) = 0

Ha: mean(diff) < 0	Ha: mean(diff) != 0	Ha: mean(diff) > 0
t = 11.2676	t = 11.2676	t = 11.2676
P < t = 1.0000	P > |t| = 0.0000	P > t = 0.0000

Because we expect "improvement," not just "difference" between the *preS* and *postS* means, a one-tail test is appropriate. The displayed one-tail probability rounds off four decimal places to zero ("0.0000" really means $P < .00005$). Students' mean sentence completion does significantly improve. Based on this sample, we are 95% confident that it improves by between 12.7 and 18.4 sentences.

　　t tests assume that variables follow a normal distribution. This assumption usually is not critical because the tests are moderately robust. When nonnormality involves severe outliers, however, or occurs in small samples, we might be safer turning to medians instead of means and employing a nonparametric test that does not assume normality. The Wilcoxon signed-rank test, for example, assumes only that the distributions are symmetrical and continuous. Applying a signed-rank test to these data yields essentially the same conclusion as **ttest** , that students' sentence completion significantly improved. Because both tests agree on this conclusion, we can assert it with more assurance.

```
. signrank postS = preS

Wilcoxon signed-rank test

        sign |      obs    sum ranks     expected
-------------+-------------------------------------
    positive |       24          300          150
    negative |        0            0          150
        zero |        0            0            0
-------------+-------------------------------------
         all |       24          300          300

unadjusted variance      1225.00
adjustment for ties        -1.63
adjustment for zeros        0.00
                        ----------
adjusted variance        1223.38

Ho: postS = preS
             z =      4.289
    Prob > |z| =     0.0000
```

Two-Sample Tests

The remainder of this chapter draws examples from a survey of college undergraduates by Ward and Ault (1990) (*student2.dta*).

```
. describe

Contains data from C:\data\student2.dta
  obs:           243                          Student survey (Ward & Ault 1990)
  vars:           19                          12 Jul 2005 10:16
  size:         6,561 (99.9% of memory free)
-------------------------------------------------------------------------------
              storage   display    value
variable name   type    format     label     variable label
-------------------------------------------------------------------------------
id              int     %8.0g                 Student ID
year            byte    %8.0g      year       Year in college
age             byte    %8.0g                 Age at last birthday
gender          byte    %9.0g      s          Gender (male)
major           byte    %8.0g                 Student major
relig           byte    %8.0g      v4         Religious preference
drink           byte    %9.0g                 33-point drinking scale
gpa             float   %9.0g                 Grade Point Average
grades          byte    %8.0g      grades     Guessed grades this semester
```

```
belong          byte    %8.0g    belong      Belong to fraternity/sorority
live            byte    %8.0g    v10         Where do you live?
miles           byte    %8.0g                How many miles from campus?
study           byte    %8.0g                Avg. hours/week studying
athlete         byte    %8.0g    yes         Are you a varsity athlete?
employed        byte    %8.0g    yes         Are you employed?
allnight        byte    %8.0g    allnight    How often study all night?
ditch           byte    %8.0g    times       How many class/month ditched?
hsdrink         byte    %9.0g                High school drinking scale
aggress         byte    %9.0g                Aggressive behavior scale
--------------------------------------------------------------------------
Sorted by:  id
```

About 19% of these students belong to a fraternity or sorority:

```
. tabulate belong

  Belong to |
 fraternity/ |
   sorority |      Freq.     Percent        Cum.
------------+-----------------------------------
     member |         47       19.34       19.34
  nonmember |        196       80.66      100.00
------------+-----------------------------------
      Total |        243      100.00
```

Another variable, *drink*, measures how often and heavily a student drinks alcohol, on a 33-point scale. Campus rumors might lead one to suspect that fraternity/sorority members tend to differ from other students in their drinking behavior. Box plots comparing the median *drink* values of members and nonmembers, and a bar chart comparing their means, both appear consistent with these rumors. Figure 5.1 combines these two separate plot types in one image.

```
. graph box drink, over(belong) ylabel(0(5)35) saving(fig05_01a)

. graph bar (mean) drink, over(belong) ylabel(0(5)35) saving(fig05_01b)

. graph combine fig05_01a.gph fig05_01b.gph, col(2) iscale(1.05)
```

Figure 5.1

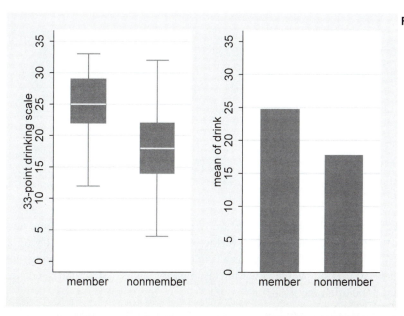

The **ttest** command, used earlier for one-sample and paired-difference tests, can perform two-sample tests as well. In this application its general syntax is **ttest** *measurement*, **by(***categorical***)**. For example,

`. ttest drink, by(belong)`

```
Two-sample t test with equal variances

------------------------------------------------------------------------------
   Group |     Obs        Mean     Std. Err.    Std. Dev.   [95% Conf. Interval]
---------+--------------------------------------------------------------------
  member |      47     24.7234     .7124518     4.884323    23.28931    26.1575
nonmembe |     196     17.7602     .4575013     6.405018    16.85792   18.66249
---------+--------------------------------------------------------------------
combined |     243      19.107     .431224      6.722117    18.25756   19.95643
---------+--------------------------------------------------------------------
    diff |               6.9632    .9978608                  4.997558   8.928842
------------------------------------------------------------------------------
Degrees of freedom: 241

          Ho: mean(member) - mean(nonmembe) = diff = 0

   Ha: diff < 0                Ha: diff != 0                Ha: diff > 0
      t =   6.9781                t =   6.9781                t =   6.9781
  P < t =   1.0000            P > |t| =   0.0000            P > t =   0.0000
```

As the output notes, this *t* test rests on an equal-variances assumption. But the fraternity and sorority members' sample standard deviation appears somewhat lower — they are more alike than nonmembers in their reported drinking behavior. To perform a similar test without assuming equal variances, add the option **unequal** :

`. ttest drink, by(belong) unequal`

```
Two-sample t test with unequal variances

------------------------------------------------------------------------------
   Group |     Obs        Mean     Std. Err.    Std. Dev.   [95% Conf. Interval]
---------+--------------------------------------------------------------------
  member |      47     24.7234     .7124518     4.884323    23.28931    26.1575
nonmembe |     196     17.7602     .4575013     6.405018    16.85792   18.66249
---------+--------------------------------------------------------------------
combined |     243      19.107     .431224      6.722117    18.25756   19.95643
---------+--------------------------------------------------------------------
    diff |               6.9632    .8466965                  5.280627   8.645773
------------------------------------------------------------------------------
Satterthwaite's degrees of freedom:     88.22

          Ho: mean(member) - mean(nonmembe) = diff = 0

   Ha: diff < 0                Ha: diff != 0                Ha: diff > 0
      t =   8.2240                t =   8.2240                t =   8.2240
  P < t =   1.0000            P > |t| =   0.0000            P > t =   0.0000
```

Adjusting for unequal variances does not alter our basic conclusion that members and nonmembers are significantly different. We can further check this conclusion by trying a nonparametric Mann–Whitney *U* test, also known as a Wilcoxon rank-sum test. Assuming that the rank distributions have similar shape, the rank-sum test here indicates that we can reject the null hypothesis of equal population medians.

```
. ranksum drink, by(belong)

Two-sample Wilcoxon rank-sum (Mann-Whitney) test

      belong |      obs    rank sum     expected
-------------+---------------------------------
      member |       47        8535         5734
   nonmember |      196       21111        23912
-------------+---------------------------------
    combined |      243       29646        29646

unadjusted variance     187310.67
adjustment for ties       -472.30
                        ----------
adjusted variance       186838.36

Ho: drink(belong==member) = drink(belong==nonmember)
           z =    6.480
    Prob > |z| =    0.0000
```

One-Way Analysis of Variance (ANOVA)

Analysis of variance (ANOVA) provides another way, more general than t tests, to test for differences among means. The simplest case, one-way ANOVA, tests whether the means of y differ across categories of x. One-way ANOVA can be performed by a **oneway** command with the general form **oneway** *measurement categorical*. For example,

```
. oneway drink belong, tabulate

  Belong to |
 fraternity/ | Summary of 33-point drinking scale
   sorority |      Mean    Std. Dev.        Freq.
------------+-----------------------------------
     member |  24.723404   4.8843233           47
  nonmember |  17.760204   6.4050179          196
------------+-----------------------------------
      Total |  19.106996   6.7221166          243

                    Analysis of Variance
    Source          SS        df      MS             F     Prob > F
------------------------------------------------------------------------
Between groups   1838.08426      1   1838.08426     48.69    0.0000
Within groups    9097.13385    241   37.7474433
------------------------------------------------------------------------
    Total       10935.2181     242   45.1868517

Bartlett's test for equal variances:  chi2(1) =    4.8378  Prob>chi2 = 0.028
```

The **tabulate** option produces a table of means and standard deviations in addition to the analysis of variance table itself. One-way ANOVA with a dichotomous x variable is equivalent to a two-sample t test, and its F statistic equals the corresponding t statistic squared. **oneway** offers more options and processes faster, but it lacks **ttest**'s **unequal** option for abandoning the equal-variances assumption.

oneway formally tests the equal-variances assumption, using Bartlett's χ^2. A low Bartlett's probability implies that ANOVA's equal-variance assumption is implausible, in which case we should not trust the ANOVA *F* test results. In the **oneway** *drink belong* example above, Bartlett's *P* = .028 casts doubt on the ANOVA's validity.

ANOVA's real value lies not in two-sample comparisons, but in more complicated comparisons of three or more means. For example, we could test whether mean drinking behavior varies by year in college:

```
. oneway drink year, tabulate scheffe
```

```
    Year in | Summary of 33-point drinking scale
    college |        Mean    Std. Dev.        Freq.
------------+-----------------------------------------
   Freshman |      18.975    6.9226033           40
  Sophomore |   21.169231    6.5444853           65
     Junior |   19.453333    6.2866081           75
     Senior |   16.650794    6.6409257           63
------------+-----------------------------------------
      Total |   19.106996    6.7221166          243
```

```
                       Analysis of Variance
      Source              SS          df        MS              F     Prob > F
---------------------------------------------------------------------------------
Between groups       666.200518        3    222.066839         5.17     0.0018
Within groups        10269.0176      239    42.9666008
---------------------------------------------------------------------------------
     Total           10935.2181      242    45.1868517
```

```
Bartlett's test for equal variances:   chi2(3) =    0.5103   Prob>chi2 = 0.917
```

```
              Comparison of 33-point drinking scale by Year in college
                                  (Scheffe)
Row Mean-|
Col Mean |   Freshman    Sophomor      Junior
---------+----------------------------------------
Sophomor |    2.19423
         |      0.429
         |
  Junior |    .478333     -1.7159
         |      0.987       0.498
         |
  Senior |   -2.32421     -4.51844    -2.80254
         |      0.382       0.002       0.103
```

We can reject the hypothesis of equal means (*P* = .0018), but not the hypothesis of equal variances (*P* = .917). The latter is "good news" regarding the ANOVA's validity.

The box plots in Figure 5.2 (next page) support this conclusion, showing similar variation within each category. This figure, which combines separate box plots and dot plots, shows that differences among medians and among means follow similar patterns.

```
. graph hbox drink, over(year) ylabel(0(5)35) saving(fig05_02a)

. graph dot (mean) drink, over(year) ylabel(0(5)35, grid)
      marker(1, msymbol(S)) saving(fig05_02b)

. graph combine fig05_02a.gph fig05_02b.gph, row(2) iscale(1.05)
```

Figure 5.2

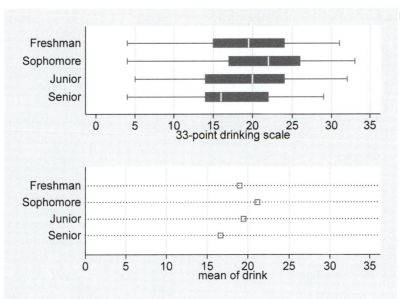

The **scheffe** option (Scheffé multiple-comparison test) produces a table showing the differences between each pair of means. The freshman mean equals 18.975 and the sophomore mean equals 21.16923, so the sophomore–freshman difference is $21.16923 - 18.975 = 2.19423$, not statistically distinguishable from zero ($P = .429$). Of the six contrasts in this table, only the senior–sophomore difference, $16.6508 - 21.1692 = -4.5184$, is significant ($P = .002$). Thus, our overall conclusion that these four groups' means are not the same arises mainly from the contrast between seniors (the lightest drinkers) and sophomores (the heaviest).

oneway offers three multiple-comparison options: **scheffe**, **bonferroni**, and **sidak** (see *Base Reference Manual* for definitions). The Scheffé test remains valid under a wider variety of conditions, although it is sometimes less sensitive.

The Kruskal–Wallis test (**kwallis**), a K-sample generalization of the two-sample rank-sum test, provides a nonparametric alternative to one-way ANOVA. It tests the null hypothesis of equal population medians.

```
. kwallis drink, by(year)

Test: Equality of populations (Kruskal-Wallis test)

    +----------------------------+
    |      year | Obs | Rank Sum |
    |-----------+-----+----------|
    |  Freshman |  40 |  4914.00 |
    | Sophomore |  65 |  9341.50 |
    |    Junior |  75 |  9300.50 |
    |    Senior |  63 |  6090.00 |
    +----------------------------+

chi-squared =     14.453 with 3 d.f.
probability =      0.0023

chi-squared with ties =     14.490 with 3 d.f.
probability =      0.0023
```

Here, the **kwallis** results ($P = .0023$) agree with our **oneway** findings of significant differences in *drink* by year in college. Kruskal–Wallis is generally safer than ANOVA if we have reason to doubt ANOVA's equal-variances or normality assumptions, or if we suspect problems caused by outliers. **kwallis**, like **ranksum**, makes the weaker assumption of similar-shaped distributions within each group. In principle, **ranksum** and **kwallis** should produce similar results when applied to two-sample comparisons, but in practice this is true only if the data contain no ties. **ranksum** incorporates an exact method for dealing with ties, which makes it preferable for two-sample problems.

Two- and *N*-Way Analysis of Variance

One-way ANOVA examines how the means of measurement variable *y* vary across categories of one other variable *x*. *N*-way ANOVA generalizes this approach to deal with two or more categorical *x* variables. For example, we might consider how drinking behavior varies not only by fraternity or sorority membership, but also by gender. We start by examining a two-way table of means:

```
. table belong gender, contents(mean drink) row col

----------------------------------------
Belong to |
fraternit |
y/sororit |        Gender (male)
y         |   Female     Male      Total
----------+-----------------------------
   member | 22.44444  26.13793   24.7234
nonmember | 16.51724   19.5625   17.7602
          |
    Total | 17.31343  21.31193    19.107
----------------------------------------
```

It appears that in this sample, males drink more than females and members drink more than nonmembers. The member–nonmember difference appears similar among males and females. Stata's *N*-way ANOVA command, **anova**, can test for significant differences among these means attributable to belonging to a fraternity or sorority, gender, or the interaction of belonging and gender (written *belong*gender*).

```
. anova drink belong gender belong*gender

                        Number of obs =       243     R-squared     =  0.2221
                        Root MSE      = 5.96592     Adj R-squared =  0.2123

          Source |  Partial SS    df       MS              F     Prob > F
    -------------+----------------------------------------------------
           Model |  2428.67237     3   809.557456        22.75    0.0000
                 |
          belong |   1406.2366     1    1406.2366         39.51    0.0000
          gender |  408.520097     1   408.520097         11.48    0.0008
    belong*gender |  3.78016612     1   3.78016612         0.11    0.7448
                 |
        Residual |  8506.54574   239   35.5922416
    -------------+----------------------------------------------------
           Total |  10935.2181   242   45.1868517
```

In this example of "two-way factorial ANOVA," the output shows significant main effects for *belong* (*P* = .0000) and *gender* (*P* = .0008), but their interaction contributes little to the model (*P* = .7448). This interaction cannot be distinguished from zero, so we might prefer to fit a simpler model without the interaction term (results not shown):

`. anova drink belong gender`

To include any interaction term with **anova**, specify the variable names joined by *. Unless the number of observations with each combination of *x* values is the same (a condition called "balanced data"), it can be hard to interpret the main effects in a model that also includes interactions. This does not mean that the main effects in such models are unimportant, however. Regression analysis might help to make sense of complicated ANOVA results, as illustrated in the following section.

Analysis of Covariance (ANCOVA)

Analysis of Covariance (ANCOVA) extends *N*-way ANOVA to encompass a mix of categorical and continuous *x* variables. This is accomplished through the **anova** command if we specify which variables are continuous. For example, when we include *gpa* (college grade point average) among the independent variables, we find that it, too, is related to drinking behavior.

`. anova drink belong gender gpa, continuous(gpa)`

```
                 Number of obs =      218     R-squared     =  0.2970
                 Root MSE      = 5.68939     Adj R-squared =  0.2872

     Source |   Partial SS     df       MS              F     Prob > F
------------+----------------------------------------------------------
      Model |   2927.03087      3   975.676958        30.14     0.0000
            |
     belong |   1489.31999      1   1489.31999        46.01     0.0000
     gender |   405.137843      1   405.137843        12.52     0.0005
        gpa |     407.0089      1     407.0089        12.57     0.0005
            |
   Residual |   6926.99206    214   32.3691218
------------+----------------------------------------------------------
      Total |   9854.02294    217   45.4102439
```

From this analysis we know that a significant relationship exists between *drink* and *gpa* when we control for *belong* and *gender*. Beyond their *F* tests for statistical significance, however, ANOVA or ANCOVA ordinarily do not provide much descriptive information about how variables are related. Regression, with its explicit model and parameter estimates, does a better descriptive job. Because ANOVA and ANCOVA amount to special cases of regression, we could restate these analyses in regression form. Stata does so automatically if we add the **regress** option to **anova**. For instance, we might want to see regression output in order to understand results from the following ANCOVA.

```
.  anova drink belong gender belong*gender gpa,  continuous(gpa)
      regress

      Source |       SS        df       MS              Number of obs =      218
-------------+------------------------------           F(  4,    213) =    22.57
       Model |  2933.45823      4   733.364558          Prob > F      =   0.0000
    Residual |   6920.5647    213   32.4909141          R-squared     =   0.2977
-------------+------------------------------           Adj R-squared =   0.2845
       Total |  9854.02294    217   45.4102439          Root MSE      =   5.7001


------------------------------------------------------------------------------
       drink        Coef.   Std. Err.       t    P>|t|     [95% Conf. Interval]
------------------------------------------------------------------------------
       _cons     27.47676    2.439962    11.26   0.000      22.6672    32.28633
belong
          1      6.925384    1.286774     5.38   0.000     4.388942    9.461826
          2     (dropped)
gender
          1     -2.629057     .8917152   -2.95   0.004    -4.386774    -.8713407
          2     (dropped)
gpa           -3.054633     .8593498    -3.55   0.000    -4.748552   -1.360713
belong*gender
       1  1     -.8656158    1.946211    -0.44   0.657    -4.701916    2.970685
       1  2     (dropped)
       2  1     (dropped)
       2  2     (dropped)
------------------------------------------------------------------------------
```

With the **regress** option, we get the **anova** output formatted as a regression table. The top part gives the same overall F test and R^2 as a standard ANOVA table. The bottom part describes the following regression:

> We construct a separate dummy variable {0,1} representing each category of each x variable, except for the highest categories, which are dropped. Interaction terms (if specified in the variable list) are constructed from the products of every possible combination of these dummy variables. Regress y on all these dummy variables and interactions, and also on any continuous variables specified in the command line.

The previous example therefore corresponds to a regression of *drink* on four x variables:

1. a dummy coded 1 = fraternity/sorority member, 0 otherwise (highest category of *belong*, nonmember, gets dropped);

2. a dummy coded 1 = female, 0 otherwise (highest category of *gender*, male, gets dropped);

3. the continuous variable *gpa*;

4. an interaction term coded 1 = sorority female, 0 otherwise.

Interpret the individual dummy variables' regression coefficients as effects on predicted or mean y. For example, the coefficient on the first category of *gender* (female) equals -2.629057. This informs us that the mean drinking scale levels for females are about 2.63 points lower than those of males with the same grade point average and membership status. And we know that among students of the same gender and membership status, mean drinking scale values decline by 3.054633 with each one-point increase in grades. Note also that we have confidence intervals and individual t tests for each coefficient; there is much more information in the **anova, regress** output than in the ANOVA table alone.

Predicted Values and Error-Bar Charts

After **anova** , the followup command **predict** calculates predicted values, residuals, or standard errors and diagnostic statistics. One application for such statistics is in drawing graphical representations of the model's predictions, in the form of error-bar charts. For a simple illustration, we return to the one-way ANOVA of *drink* by *year*:

```
. anova drink year
```

```
                          Number of obs =      243     R-squared     =  0.0609
                          Root MSE      = 6.55489     Adj R-squared =  0.0491

       Source |   Partial SS     df        MS               F     Prob > F
    ----------+----------------------------------------------------------
        Model |   666.200518      3   222.066839            5.17     0.0018
              |
         year |   666.200518      3   222.066839            5.17     0.0018
              |
     Residual |   10269.0176    239   42.9666008
    ----------+----------------------------------------------------------
        Total |   10935.2181    242   45.1868517
```

To calculate predicted means from the recent **anova**, type **predict** followed by a new variable name:

```
. predict drinkmean
(option xb assumed; fitted values)
```

```
. label variable drinkmean "Mean drinking scale"
```

With the **stdp** option, **predict** calculates standard errors of the predicted means:

```
. predict SEdrink, stdp
```

Using these new variables, we apply the **serrbar** command to create an error-bar chart. The **scale(2)** option tells **serrbar** to draw error bars of plus and minus two standard errors, from

$$drinkmean - 2 \times SEdrink$$

to

$$drinkmean + 2 \times SEdrink.$$

In a **serrbar** command, the first-listed variable should be the means or *y* variable; the second-listed, the standard error or standard deviation (depending on which you want to show); and the third-listed variable defines the *x* axis. The **plot( )** option for **serrbar** can specify a second plot to overlay on the standard-error bars. In Figure 5.3, we overlay a line plot that connects the *drinkmean* values with solid line segments.

```
. serrbar drinkmean SEdrink year, scale(2)
     plot(line drinkmean year, clpattern(solid)) legend(off)
```

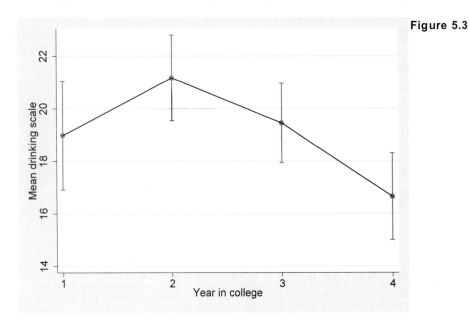

Figure 5.3

For a two-way factorial ANOVA, error-bar charts help us to visualize main and interaction effects. Although the usual error-bar command **serrbar** can, with effort, be adapted for this purpose, an alternative approach using the more flexible **graph twoway** family will be illustrated below. First, we perform ANOVA, obtain group means (predicted values) and their standard errors, then generate new variables equal to the group means plus or minus two standard errors. The example examines the relationship between students' aggressive behavior (*aggress*), gender, and year in college. Both the main effects of *gender* and *year*, and their interaction, are statistically significant.

```
. anova aggress gender year gender*year
```

	Number of obs =	243	R-squared	=	0.2503
	Root MSE =	1.45652	Adj R-squared	=	0.2280

Source	Partial SS	df	MS	F	Prob > F
Model	166.482503	7	23.7832147	11.21	0.0000
gender	94.3505972	1	94.3505972	44.47	0.0000
year	19.0404045	3	6.34680149	2.99	0.0317
gender*year	24.1029759	3	8.03432529	3.79	0.0111
Residual	498.538073	235	2.12143861		
Total	665.020576	242	2.74801891		

```
. predict aggmean
(option xb assumed; fitted values)

. label variable aggmean "Mean aggressive behavior scale"

. predict SEagg, stdp

. gen agghigh = aggmean + 2 * SEagg

. gen agglow = aggmean - 2 * SEagg

. graph twoway connected aggmean year
       || rcap agghigh agglow year
       || , by(gender, legend(off) note(""))
       ytitle("Mean aggressive behavior scale")
```

Figure 5.4

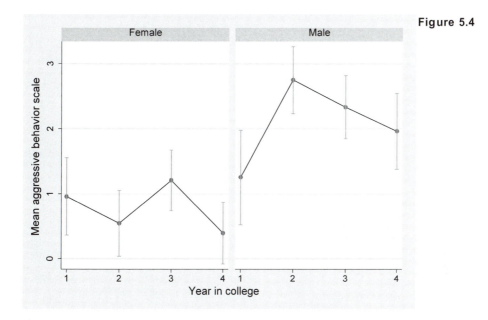

Figure 5.4 built error-bar charts by overlaying two pairs of plots. The first pair are female and male connected-line plots, connecting the group means of *aggress* (which we calculated using **predict**, and saved as the variable *aggmean*). The second pair are female and male capped-spike range plots (**twoway rcap**) in which the vertical spikes connecting variables *agghigh* (group means of *aggress* plus two standard errors) and *agglow* (group means of *aggress* minus two standard errors). The **by(*gender*)** option produced sub-plots for females and males. Notice that to suppress legends and notes in a graph that uses a **by( )** option, **legend(off)** and **note("")** must appear as suboptions within **by( )**.

The resulting error-bar chart (Figure 5.4) shows female means on the aggressive-behavior scale fluctuating at comparatively low levels during the four years of college. Male means are higher throughout, with a sophomore-year peak that resembles the pattern seen earlier for drinking (Figures 5.2 and 5.3). Thus, the relationship between *aggress* and *year* is different for males and females. This graph helps us to understand and explain the significant interaction effect.

predict works the same way with regression analysis (**regress**) as it does with **anova** because the two share a common mathematical framework. A list of some other

predict options appears in Chapter 6, and further examples using these options are given in Chapter 7. The options include residuals that can be used to check assumptions regarding error distributions, and also a suite of diagnostic statistics (such as leverage, Cook's *D*, and *DFBETA*) that measure the influence of individual observations on model results. The Durbin–Watson test (**dwstat**), described in Chapter 13, can also be used after **anova** to test for first-order autocorrelation. Conditional effect plotting (Chapter 7) provides a graphical approach that can aid interpretation of more complicated regression, ANOVA, or ANCOVA models.

Linear Regression Analysis

Stata offers an exceptionally broad range of regression procedures. A partial list of the possibilities can be seen by typing **help regress**. This chapter introduces **regress** and related commands that perform simple and multiple ordinary least squares (OLS) regression. One followup command, **predict**, calculates predicted values, residuals, and diagnostic statistics such as leverage or Cook's *D*. Another followup command, **test**, performs tests of user-specified hypotheses. **regress** can accomplish other analyses including weighted least squares and two-stage least squares. Regression with dummy variables, interaction effects, polynomial terms, and stepwise variable selection are covered briefly in this chapter, along with a first look at residual analysis.

The following menus access most of the operations discussed:

Statistics – Linear regression and related – Linear regression

Statistics – Linear regression and related – Regression diagnostics

Statistics – General post-estimation – Obtain predictions, residuals, etc., after estimation

Graphics – Overlaid twoway graphs

Statistics – Cross-sectional time series

Example Commands

. `regress y x`

Performs ordinary least squares (OLS) regression of variable *y* on one predictor, *x*.

. `regress y x if ethnic == 3 & income > 50`

Regresses *y* on *x* using only that subset of the data for which variable *ethnic* equals 3 and *income* is greater than 50.

. `predict yhat`

Generates a new variable (here arbitrarily named *yhat*) equal to the predicted values from the most recent regression.

. `predict e, resid`

Generates a new variable (here arbitrarily named *e*) equal to the residuals from the most recent regression.

. `graph twoway lfit y x  ||  scatter y x`

Draws the simple regression line (**lfit** or linear fit) with a scatterplot of *y* vs. *x*.

. `graph twoway mspline` *yhat* `x  ||  scatter` *y* `x`

Draws a simple regression line with a scatterplot of *y* vs. *x* by connecting (with a smooth cubic spline curve) the regression's predicted values (in this example named *yhat*).

Note: There are many alternative ways to draw regression lines or curves in Stata. These alternatives include the **twoway** graph types **mspline** (illustrated above), **mband**, **line**, **lfit**, **lfitci**, **qfit**, and **qfitci**, each of which has its own advantages and options. Usually we combine (overlay) the regression line or curve with a scatterplot. If the scatterplot comes second in our **graph twoway** command, as in the example above, then scatterplot points will print on top of the regression line. Placing the scatterplot first in the command causes the line to print on top of the scatter. Examples throughout this and the following chapters illustrate some of these different possibilities.

. `rvfplot`

Draws a residual versus fitted (predicted values) plot, automatically based on the most recent regression.

. `graph twoway scatter` *e yhat*`, yline(0)`

Draws a residual versus predicted values plot using the variables *e* and *yhat*.

. `regress` *y x1 x2 x3*

Performs multiple regression of *y* on three predictor variables, *x1*, *x2*, and *x3*.

. `regress` *y x1 x2 x3*`, robust`

Calculates robust (Huber/White) estimates of standard errors. See the *User's Guide* for details. The **robust** option works with many other model fitting commands as well.

. `regress` *y x1 x2 x3*`, beta`

Performs multiple regression and includes standardized regression coefficients ("beta weights") in the output table.

. `correlate` *x1 x2 x3 y*

Displays a matrix of Pearson correlations, using only observations with no missing values on all of the variables specified. Adding the option **covariance** produces a variance–covariance matrix instead of correlations.

. `pwcorr` *x1 x2 x3 y*`, sig`

Displays a matrix of Pearson correlations, using pairwise deletion of missing values and showing probabilities from *t* tests of H_0:$\rho = 0$ on each correlation.

. `graph matrix` *x1 x2 x3 y*`, half`

Draws a scatterplot matrix. Because their variable lists are the same, this example yields a scatterplot matrix having the same organization as the correlation matrix produced by the preceding **pwcorr** command. Listing the dependent (*y*) variable last creates a matrix in which the bottom row forms a series of *y*-versus-*x* plots.

. `test` *x1 x2*

Performs an *F* test of the null hypothesis that coefficients on *x1* and *x2* both equal zero in the most recent regression model.

.`xi:   regress` *y x1 x2* `i.`*catvar*`*`*x2*

Performs "expanded interaction" regression of *y* on predictors *x1*, *x2*, a set of dummy variables created automatically to represent categories of *catvar*, and a set of interaction terms equal to those dummy variables times measurement variable *x2*. **help xi** gives more details.

. `sw regress y x1 x2 x3, pr(.05)`

Performs stepwise regression using backward elimination until all remaining predictors are significant at the .05 level. All listed predictors are entered on the first iteration. Thereafter, each iteration drops one predictor with the highest *P* value, until all predictors remaining have probabilities below the "probability to retain," `pr(.05)`. Options permit forward or hierarchical selection. Stepwise variants exist for many other model-fitting commands as well; type `help sw` for a list.

. `regress y x1 x2 x3 [aweight = w]`

Performs weighted least squares (WLS) regression of *y* on *x1*, *x2*, and *x3*. Variable *w* holds the analytical weights, which work as if we had multiplied each variable and the constant by the square root of *w*, and then performed an ordinary regression. Analytical weights are often employed to correct for heteroskedasticity when the *y* and *x* variables are means, rates, or proportions, and *w* is the number of individuals making up each aggregate observation (e.g., city or school) in the data. If the *y* and *x* variables are individual-level, and the weights indicate numbers of replicated observations, then use frequency weights `[fweight = w]` instead. See `help svy` if the weights reflect design factors such as disproportionate sampling.

. `regress y1 y2 x (x z)`
. `regress y2 y1 z (x z)`

Estimates the reciprocal effects of *y1* and *y2*, using instrumental variables *x* and *z*. The first parts of these commands specify the structural equations:

$$y1 = \alpha_0 + \alpha_1 y2 + \alpha_2 x + \epsilon_1$$
$$y2 = \beta_0 + \beta_1 y2 + \beta_2 w + \epsilon_2$$

The parentheses in the commands enclose variables that are exogenous to all of the structural equations. `regress` accomplishes two-stage least squares (2SLS) in this example.

. `svy: regress y x1 x2 x3`

Regresses *y* on predictors *x1*, *x2*, and *x3*, with appropriate adjustments for a complex survey sampling design. We assume that a `svyset` command has previously been used to set up the data, by specifying the strata, clusters, and sampling probabilities. `help svy` lists the many procedures available for working with complex survey data. `help regress` outlines the syntax of this particular command; follow references to the *User's Guide* and the *Survey Data Reference Manual* for details.

. `xtreg y x1 x2 x3 x4, re`

Fits a panel (cross-sectional time series) model with random effects by generalized least squares (GLS). An observation in panel data consists of information about unit *i* at time *t*, and there are multiple observations (times) for each unit. Before using `xtreg`, the variable identifying the units was specified by an `iis` ("*i* is") command, and the variable identifying time by `tis` ("*t* is"). Once the data have been saved, these definitions are retained for future analysis by `xtreg` and other `xt` procedures. `help xt` lists available panel estimation procedures. `help xtreg` gives the syntax of this command and references to the printed documentation. If your data include many observations for each unit, a time-series approach could be more appropriate. Stata's time series procedures (introduced in Chapter 13) provide further tools for analyzing panel data. Consult the *Longitudinal/Panel Data Reference Manual* for a full description.

```
. xtmixed population year  ||  city: year
```

Assume that we have yearly data on population, for a number of different cities. The **xtmixed population year** part specifies a "fixed-effect" model, similar to ordinary regression, which describes the average trend in population. The **|| city: year** part specifies a "random-effects" model, allowing unique intercepts and slopes (different starting points and growth rates) for each city.

```
. xtmixed SAT grades prepcourse  ||  district:  pctcollege  ||  region:
```

Fits a hierarchical (nested or multi-level) linear model predicting students's SAT scores as a function of the individual students' grades and whether they took a preparation course; the percent college graduates among their school district's adults; and region of the country (*region* affecting *y*-intercept only). See the *Longitudinal/Panel Data Reference Manual* for much more about the **xtmixed** command, which is new with Stata 9.

The Regression Table

File *states.dta* contains educational data on the U.S. states and District of Columbia:

```
. describe state csat expense percent income high college region
```

variable name	storage type	display format	value label	variable label
state	str20	%20s		State
csat	int	%9.0g		Mean composite SAT score
expense	int	%9.0g		Per pupil expenditures prim&sec
percent	byte	%9.0g		% HS graduates taking SAT
income	long	%10.0g		Median household income
high	float	%9.0g		% adults HS diploma
college	float	%9.0g		% adults college degree
region	byte	%9.0g	region	Geographic region

Political leaders occasionally use mean Scholastic Aptitude Test (SAT) scores to make pointed comparisons between the educational systems of different U.S. states. For example, some have raised the question of whether SAT scores are higher in states that spend more money on education. We might try to address this question by regressing mean composite SAT scores (*csat*) on per-pupil expenditures (*expense*). The appropriate Stata command has the form **regress y x**, where *y* is the predicted or dependent variable, and *x* the predictor or independent variable.

```
. regress csat expense
```

Source	SS	df	MS
Model	48708.3001	1	48708.3001
Residual	175306.21	49	3577.67775
Total	224014.51	50	4480.2902

Number of obs = 51
F(1, 49) = 13.61
Prob > F = 0.0006
R-squared = 0.2174
Adj R-squared = 0.2015
Root MSE = 59.814

csat	Coef.	Std. Err.	t	P>|t|	[95% Conf. Interval]
expense	-.0222756	.0060371	-3.69	0.001	-.0344077 -.0101436
_cons	1060.732	32.7009	32.44	0.000	995.0175 1126.447

This regression tells an unexpected story: the more money a state spends on education, the lower its students' mean SAT scores. Any causal interpretation is premature at this point, but the regression table does convey information about the linear statistical relationship between *csat* and *expense*. At upper right it gives an overall *F* test, based on the sums of squares at the upper left. This *F* test evaluates the null hypothesis that coefficients on all *x* variables in the model (here there is only one *x* variable, *expense*) equal zero. The *F* statistic, 13.61 with 1 and 49 degrees of freedom, leads easily to rejection of this null hypothesis ($P = .0006$). Prob > F means "the probability of a greater *F*" statistic if we drew samples randomly from a population in which the null hypothesis is true.

At upper right, we also see the coefficient of determination, $R^2 = .2174$. Per-pupil expenditures explain about 22% of the variance in states' mean composite SAT scores. Adjusted R^2, $R^2_a = .2015$, takes into account the complexity of the model relative to the complexity of the data. This adjusted statistic is often more informative for research.

The lower half of the regression table gives the fitted model itself. We find coefficients (slope and *y*-intercept) in the first column, here yielding the prediction equation

$$\text{predicted } csat = 1060.732 - .0222756 expense$$

The second column lists estimated standard errors of the coefficients. These are used to calculate *t* tests (columns 3–4) and confidence intervals (columns 5–6) for each regression coefficient. The *t* statistics (coefficients divided by their standard errors) test null hypotheses that the corresponding population coefficients equal zero. At the $\alpha = .05$ significance level, we could reject this null hypothesis regarding both the coefficient on *expense* ($P = .001$) and the *y*-intercept (".000", really meaning $P < .0005$). Stata's modeling commands print 95% confidence intervals routinely, but we can request other levels by specifying the **level( )** option, as shown in the following:

```
. regress csat expense, level(99)
```

Because these data do not represent a random sample from some larger population of U.S. states, hypothesis tests and confidence intervals lack their usual meanings. They are discussed in this chapter anyway for purposes of illustration.

The term _cons stands for the regression constant, usually set at one. Stata automatically includes a constant unless we tell it not to. The **nocons** option causes Stata to suppress the constant, performing regression through the origin. For example,

```
. regress y x,   nocons
```

or

```
. regress y x1 x2 x3, nocons
```

In certain advanced applications, you might need to specify your own constant. If the "independent variables" include a user-supplied constant (named *c*, for example), employ the **hascons** option instead of **nocons** :

```
. regress y c x, hascons
```

Using **nocons** in this situation would result in a misleading *F* test and R^2. Consult the *Base Reference Manual* or **help regress** for more about **hascons** .

Multiple Regression

Multiple regression allows us to estimate how *expense* predicts *csat*, while adjusting for a number of other possible predictor variables. We can incorporate other predictors of *csat* simply by listing these variables in the command

```
. regress csat expense percent income high college
```

```
      Source |       SS       df       MS              Number of obs =      51
-------------+------------------------------           F(  5,    45) =   42.23
       Model | 184663.309      5   36932.6617           Prob > F      =  0.0000
    Residual | 39351.2012     45   874.471137           R-squared     =  0.8243
-------------+------------------------------           Adj R-squared =  0.8048
       Total | 224014.51      50   4480.2902            Root MSE      =  29.571


        csat |     Coef.   Std. Err.       t    P>|t|     [95% Conf. Interval]
-------------+----------------------------------------------------------------
     expense |   .0033528   .0044709     0.75   0.457    -.005652    .0123576
     percent |  -2.618177   .2538491   -10.31   0.000    -3.129455   -2.106898
      income |   .0001056   .0011661     0.09   0.928    -.002243    .0024542
        high |   1.630841    .992247     1.64   0.107    -.367647    3.629329
     college |   2.030894   1.660118     1.22   0.228    -1.312756    5.374544
       _cons |   851.5649   59.29228    14.36   0.000     732.1441    970.9857
```

This yields the multiple regression equation

$$\text{predicted } csat = 851.56 + .00335expense - 2.618percent + .0001income + 1.63high + 2.03college$$

Controlling for four other variables weakens the coefficient on *expense* from –.0223 to .00335, which is no longer statistically distinguishable from zero. The unexpected negative relationship between *expense* and *csat* found in our earlier simple regression evidently can be explained by other predictors.

Only the coefficient on *percent* (percentage of high school graduates taking the SAT) attains significance at the .05 level. We could interpret this "fourth-order partial regression coefficient" (so called because its calculation adjusts for four other predictors) as follows.

$b_2 = -2.618$: Predicted mean SAT scores decline by 2.618 points, with each one-point increase in the percentage of high school graduates taking the SAT — if *expense*, *income*, *high*, and *college* do not change.

Taken together, the five *x* variables in this model explain about 80% of the variance in states' mean composite SAT scores ($R^2_a = .8048$). In contrast, our earlier simple regression with *expense* as the only predictor explained only 20% of the variance in *csat*.

To obtain standardized regression coefficients ("beta weights") with any regression, add the **beta** option. Standardized coefficients are what we would see in a regression where all the variables had been transformed into standard scores (means 0, standard deviations 1).

```
. regress csat expense percent income high college, beta

      Source |       SS       df       MS              Number of obs =      51
-------------+------------------------------           F( 5,    45) =    42.23
       Model | 184663.309      5  36932.6617           Prob > F      =   0.0000
    Residual | 39351.2012     45  874.471137           R-squared     =   0.8243
-------------+------------------------------           Adj R-squared =   0.8048
       Total |  224014.51     50  4480.2902            Root MSE      =   29.571


-------------------------------------------------------------------------------
        csat |     Coef.   Std. Err.       t    P>|t|                      Beta
-------------+-----------------------------------------------------------------
     expense |   .0033528   .0044709     0.75   0.457                   .070185
     percent |  -2.618177   .2538491   -10.31   0.000                 -1.024538
      income |   .0001056   .0011661     0.09   0.928                  .0101321
        high |   1.630841    .992247     1.64   0.107                  .1361672
     college |   2.030894   1.660118     1.22   0.228                  .1263952
       _cons |   851.5649   59.29228    14.36   0.000                         .
-------------------------------------------------------------------------------
```

The standardized regression equation is

$$\text{predicted } csat^* = .07 expense^* - 1.0245 percent^* + .01 income^* + .136 high^* + .126 college^*$$

where *csat**, *expense**, etc. denote these variables in standard-score form. We might interpret the standardized coefficient on *percent*, for example, as follows:

> $b_2^* = -1.0245$: Predicted mean SAT scores decline by 1.0245 standard deviations, with each one-standard-deviation increase in the percentage of high school graduates taking the SAT — if *expense*, *income*, *high*, and *college* do not change.

The *F* and *t* tests, R^2, and other aspects of the regression remain the same.

Predicted Values and Residuals

After any regression, the **predict** command can obtain predicted values, residuals, and other case statistics. Suppose we have just done a regression of composite SAT scores on their strongest single predictor:

```
. regress csat percent
```

Now, to create a new variable called *yhat* containing predicted *y* values from this regression, type

```
. predict yhat
. label variable yhat "Predicted mean SAT score"
```

Through the **resid** option, we can also create another new variable containing the residuals, here named *e*:

```
. predict e, resid
. label variable e "Residual"
```

We might instead have obtained the same predicted *y* and residuals through two **generate** commands:

```
. generate yhat0 = _b[_cons] + _b[percent]*percent
```

```
. generate e0 = csat - yhat0
```

Stata temporarily remembers coefficients and other details from the recent regression. Thus _b[*varname*] refers to the coefficient on independent variable *varname*. _b[_cons] refers to the coefficient on _cons (usually, the *y*-intercept). These stored values are useful in programming and some advanced applications, but for most purposes, **predict** saves us the trouble of generating *yhat0* and *e0* "by hand" in this fashion.

Residuals contain information about where the model fits poorly, and so are important for diagnostic or troubleshooting analysis. Such analysis might begin just by sorting and examining the residuals. Negative residuals occur when our model overpredicts the observed values. That is, in these states the mean SAT scores are lower than we would expect, based on what percentage of students took the test. To list the states with the five lowest residuals, type

```
. sort e

. list state percent csat yhat e in 1/5
```

```
     +----------------------------------------------------------+
     |          state   percent   csat       yhat          e |
     |----------------------------------------------------------|
  1. | South Carolina        58    832   894.3333   -62.3333 |
  2. |   West Virginia        17    926   986.0953  -60.09526 |
  3. | North Carolina        57    844   896.5714   -52.5714 |
  4. |          Texas        44    874   925.6666  -51.66666 |
  5. |         Nevada        25    919   968.1905  -49.19049 |
     +----------------------------------------------------------+
```

The four lowest residuals belong to southern states, suggesting that we might be able to improve our model, or better understand variation in mean SAT scores, by somehow taking region into account.

Positive residuals occur when actual *y* values are higher than predicted. Because the data already have been sorted by *e*, to list the five highest residuals we add the qualifier

```
     in -5/1
```

"−5" in this qualifier means the 5th-from-last observation, and the letter "el" (note that this is not the number "1") stands for the last observations. The qualifiers **in 47/1** or **in 47/51** could accomplish the same thing.

```
. list state percent csat yhat e in -5/1
```

```
     +----------------------------------------------------------+
     |          state   percent   csat       yhat          e |
     |----------------------------------------------------------|
 47. | Massachusetts        79    896   847.3333  48.66673 |
 48. |   Connecticut        81    897   842.8571  54.14292 |
 49. |  North Dakota         6   1073   1010.714  62.28567 |
 50. | New Hampshire        75    921   856.2856  64.71434 |
 51. |          Iowa         5   1093   1012.952  80.04758 |
     +----------------------------------------------------------+
```

predict also derives other statistics from the most recently-fitted model. Below are some **predict** options that can be used after **anova** or **regress** .

. predict *new*	Predicted values of *y*. predict *new*, **xb** means the same thing (referring to **Xb**, the vector of predicted *y* values).
. predict *new*, cooksd	Cook's *D* influence measures.
. predict new, covratio	*COVRATIO* influence measures; effect of each observation on the variance–covariance matrix of estimates.
. predict *DFx1*, dfbeta(*x1*)	*DFBETA*s measuring each observation's influence on the coefficient of predictor *x1*.
. predict new, dfits	*DFITS* influence measures.
. predict *new*, hat	Diagonal elements of hat matrix (leverage).
. predict *new*, resid	Residuals.
. predict *new*, rstandard	Standardized residuals.
. predict *new*, rstudent	Studentized (jackknifed) residuals.
. predict *new*, stdf	Standard errors 'of predicted individual *y*, sometimes called the standard errors of forecast or the standard errors of prediction.
. predict *new*, stdp	Standard errors of predicted mean *y*.
. predict *new*, stdr	Standard errors of residuals.
. predict *new*, welsch	Welsch's distance influence measures.

Further options obtain predicted probabilities and expected values; type **help regress** for a list. All **predict** options create case statistics, which are new variables (like predicted values and residuals) that have a value for each observation in the sample.

When using **predict** , substitute a new variable name of your choosing for *new* in the commands shown above. For example, to obtain Cook's *D* influence measures, type

. predict *D*, cooksd

Or you can find hat matrix diagonals by typing

. predict *h*, hat

The names of variables created by **predict** (such as *yhat*, *e*, *D*, *h*) are arbitrary and are invented by the user. As with other elements of Stata commands, we could abbreviate the options to the minimum number of letters it takes to identify them uniquely. For example,

. predict *e*, resid

could be shortened to

. pre *e*, re

Basic Graphs for Regression

This section introduces some elementary graphs you can use to represent a regression model or examine its fit. Chapter 7 describes more specialized graphs that aid post-regression diagnostic work.

In simple regression, predicted values lie on the line defined by the regression equation. By plotting and connecting predicted values, we can make that line visible. The **lfit** (linear fit) command automatically draws a simple regression line.

```
. graph twoway lfit csat percent
```

Ordinarily, it is more interesting to overlay a scatterplot on the regression line, as done in Figure 6.1.

```
. graph twoway lfit csat percent
       || scatter csat percent
       || , ytitle("Mean composite SAT score") legend(off)
```

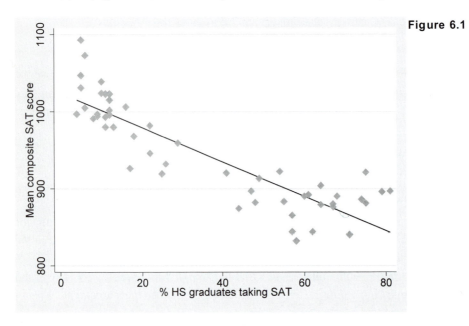

Figure 6.1

We could draw the same Figure 6.1 graph "by hand" using the predicted values (*yhat*) generated after the regression, and a command of the form

```
graph twoway mspline yhat percent, bands(50)
       || scatter csat percent
       || , legend(off) ytitle("Mean composite SAT score")
```

The second approach is more work, but offers greater flexibility for advanced applications such as conditional effect plots or nonlinear regression. Working directly with the predicted values also keeps the analyst closer to the data, and to what a regression model is doing. **graph twoway mspline** (cubic spline curve fit to 50 cross-medians) simply draws a straight line when applied to linear predicted values, but will equally well draw a smooth curve in the case of nonlinear predicted values.

Residual-versus-predicted-values plots provide useful diagnostic tools (Figure 6.2). After any regression analysis (also after some other models, such as ANOVA) we can automatically draw a residual-versus-fitted (predicted values) plot just by typing

```
. rvfplot, yline(0)
```

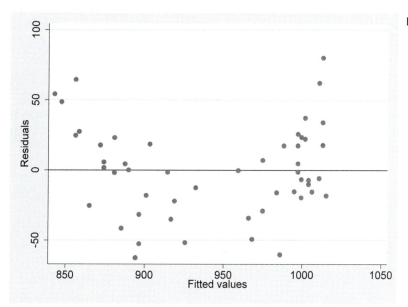

Figure 6.2

The "by-hand" alternative for drawing Figure 6.2 would be

```
. graph twoway scatter e yhat, yline(0)
```

Figure 6.2 reveals that our present model overlooks an obvious pattern in the data. The residuals or prediction errors appear to be mostly positive at first (due to too-high predictions), then mostly negative, followed by mostly positive residuals again. Later sections will seek a model that better fits these data.

predict can generate two kinds of standard errors for the predicted y values, which have two different applications. These applications are sometimes distinguished by the names "confidence intervals" and "prediction intervals": A "confidence interval" in this context expresses our uncertainty in estimating the conditional mean of y at a given x value (or a given combination of x values, in multiple regression). Standard errors for this purpose are obtained through

```
. predict SE, stdp
```

Select an appropriate t value. With 49 degrees of freedom, for 95% confidence we should use $t = 2.01$, found by looking up the t distribution or simply by asking Stata:

```
. display invttail(49,.05/2)
  2.0095752
```

Then the lower confidence limit is approximately

```
. generate low1 = yhat - 2.01*SE
```

and the upper confidence limit is

```
. generate high1 = yhat + 2.01*SE
```

Confidence bands in simple regression have an hourglass shape, narrowest at the mean of *x*. We could graph these using an overlaid **twoway** command such as the following.

```
. graph twoway mspline low1 percent, clpattern(dash) bands(50)
       ||   mspline high1 percent, clpattern(dash) bands(50)
       ||   mspline yhat percent, clpattern(solid) bands(50)
       ||   scatter csat percent
       ||   , legend(off) ytitle("Mean composite SAT score")
```

Shaded-area range plots (see **help twoway_rarea**) offer a different way to draw such graphs, shading the range between *low1* and *high1*. Alternatively, **lfitci** can do this automatically, and take care of the confidence-band calculations, as illustrated in Figure 6.3. Note the **stdp** option, calling for a conditional-mean confidence band (actually, the default).

```
. graph twoway lfitci csat percent, stdp
       ||   scatter csat percent, msymbol(O)
       ||   , ytitle("Mean composite SAT score")  legend(off)
       title("Confidence bands for conditional means (stdp)")
```

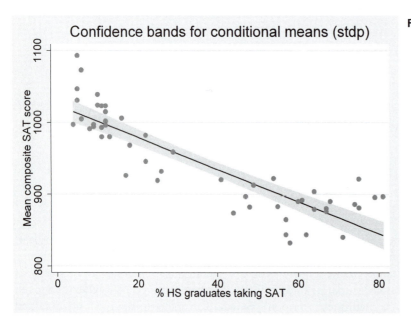

Figure 6.3

The second type of confidence interval for regression predictions is sometimes called a "prediction interval." This expresses our uncertainty in estimating the unknown value of *y* for an individual observation with known *x* value(s). Standard errors for this purpose are obtained by typing

```
. predict SEyhat, stdf
```

Figure 6.4 (next page) graphs this prediction band using **lfitci** with the **stdf** option. Predicting the *y* values of individual observations as done in Figure 6.4 inherently involves greater uncertainty, and hence wider bands, than does predicting the conditional mean of *y* (Figure 6.3). In both instances, the bands are narrowest at the mean of *x*.

```
. graph twoway lfitci csat percent, stdf
      || scatter csat percent, msymbol(O)
      || , ytitle("Mean composite SAT score")  legend(off)
      title("Confidence bands for individual-case predictions (stdf)")
```

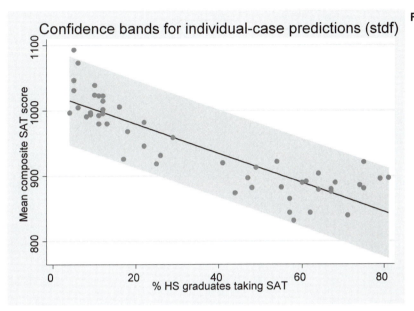

Figure 6.4

As with other confidence intervals and hypothesis tests in OLS regression, the standard errors and bands just described depend on the assumption of independent and identically distributed errors. Figure 6.2 has cast doubt on this assumption, so the results in Figures 6.3 and 6.4 could be misleading.

Correlations

correlate obtains Pearson product-moment correlations between variables.

```
. correlate csat expense percent income high college
```

(obs=51)

	csat	expense	percent	income	high	college
csat	1.0000					
expense	-0.4663	1.0000				
percent	-0.8758	0.6509	1.0000			
income	-0.4713	0.6784	0.6733	1.0000		
high	0.0858	0.3133	0.1413	0.5099	1.0000	
college	-0.3729	0.6400	0.6091	0.7234	0.5319	1.0000

correlate uses only a subset of the data that has no missing values on any of the variables listed (with these particular variables, that does not matter because no observations have missing values). In this respect, the **correlate** command resembles **regress**, and given the same variable list, they will use the same subset of the data. Analysts not employing regression or other multi-variable techniques, however, might prefer to find correlations based upon all of the observations available for each variable pair. The command **pwcorr** (pairwise correlation) accomplishes this, and can also furnish *t*-test probabilities for the null hypotheses that each individual correlation equals zero.

```
. pwcorr csat expense percent income high college, sig

             |     csat   expense   percent    income      high   college
-------------+------------------------------------------------------------
        csat |   1.0000
             |
             |
     expense |  -0.4663    1.0000
             |   0.0006
             |
     percent |  -0.8758    0.6509    1.0000
             |   0.0000    0.0000
             |
      income |  -0.4713    0.6784    0.6733    1.0000
             |   0.0005    0.0000    0.0000
             |
        high |   0.0858    0.3133    0.1413    0.5099    1.0000
             |   0.5495    0.0252    0.3226    0.0001
             |
     college |  -0.3729    0.6400    0.6091    0.7234    0.5319    1.0000
             |   0.0070    0.0000    0.0000    0.0000    0.0001
             |
```

It is worth recalling here that if we drew many random samples from a population in which all variables really had 0 correlations, about 5% of the sample correlations would nonetheless test "statistically significant" at the .05 level. Analysts who review many individual hypothesis tests, such as those in a **pwcorr** matrix, to identify the handful that are significant at the .05 level, therefore run a much higher than .05 risk of making a Type I error. This problem is called the "multiple comparison fallacy." **pwcorr** offers two methods, Bonferroni and Šidák, for adjusting significance levels to take multiple comparisons into account. Of these, the Šidák method is more precise.

```
. pwcorr csat expense percent income high college, sidak sig

             |     csat   expense   percent    income      high   college
-------------+------------------------------------------------------------
        csat |   1.0000
             |
             |
     expense |  -0.4663    1.0000
             |   0.0084
             |
     percent |  -0.8758    0.6509    1.0000
             |   0.0000    0.0000
             |
      income |  -0.4713    0.6784    0.6733    1.0000
             |   0.0072    0.0000    0.0000
             |
        high |   0.0858    0.3133    0.1413    0.5099    1.0000
             |   1.0000    0.3180    0.9971    0.0020
             |
     college |  -0.3729    0.6400    0.6091    0.7234    0.5319    1.0000
             |   0.1004    0.0000    0.0000    0.0000    0.0009
```

Comparing the test probabilities in the table above with those of the previous **pwcorr** provides some idea of how much adjustment occurs. In general, the more variables we correlate, the more the adjusted probabilities will exceed their unadjusted counterparts. See the *Base Reference Manual*'s discussion of **oneway** for the formulas involved.

correlate itself offers several important options. Adding the **covariance** option produces a matrix of variances and covariances instead of correlations:

```
. correlate w x y z, covariance
```

Typing the following after a regression analysis displays the matrix of correlations between estimated coefficients, sometimes used to diagnose multicollinearity (see Chapter 7):

```
. correlate, _coef
```

The following command will display the estimated coefficients' variance–covariance matrix, from which standard errors are derived:

```
. correlate, _coef covariance
```

Pearson correlation coefficients measure how well an OLS regression line fits the data. They consequently share the assumptions and weaknesses of OLS, and like OLS, should generally not be interpreted without first reviewing the corresponding scatterplots. A scatterplot matrix provides a quick way to do this, using the same organization as the correlation matrix. Figure 6.5 shows a scatterplot matrix corresponding to the **pwcorr** matrix given earlier. Only the lower-triangular half of the matrix is drawn, and plus signs are used as plotting symbols. We suppress *y* and *x*-axis labeling here to keep the graph uncluttered.

```
. graph matrix csat expense percent income high college,
     half msymbol(+) maxis(ylabel(none) xlabel(none))
```

Figure 6.5

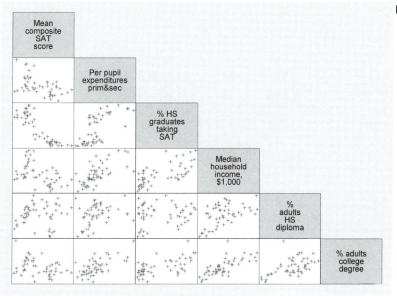

To obtain a scatterplot matrix corresponding to a **correlate** correlation matrix, from which all observations having missing values have been dropped, we would need to qualify the command. If all of the variables had some missing values, we could type a command such as

```
. graph matrix csat expense percent income high college if csat < .
     & expense < . & income < . & high < . & college < .
```

To reduce the likelihood of confusion and mistakes, it might make sense to create a new dataset keeping only those observations that have no missing values:

```
. keep if csat < . & expense < . & income < . & high < .
     & college < .
. save nmvstate
```

In this example, we immediately saved the reduced dataset with a new name, so as to avoid inadvertently writing over and losing the information in the old, more complete dataset. An alternative way to eliminate missing values uses **drop** instead of **keep** :

```
. drop if csat >= . | expense >= . | income >= . | high >= .
     | college >= .
. save nmvstate
```

In addition to Pearson correlations, Stata can also calculate several rank-based correlations. These can be employed to measure associations between ordinal variables, or as an outlier-resistant alternative to Pearson correlation for measurement variables. To obtain the Spearman rank correlation between *csat* and *expense*, equivalent to the Pearson correlation if these variables were transformed into ranks, type

```
. spearman csat expense

 Number of obs =       51
Spearman's rho =     -0.4282

Test of Ho: csat and expense are independent
     Prob > |t| =       0.0017
```

Kendall's τ_a (tau-a) and τ_b (tau-b) rank correlations can be found easily for these data, although with larger datasets their calculation becomes slow:

```
. ktau csat expense

  Number of obs =       51
Kendall's tau-a =     -0.2925
Kendall's tau-b =     -0.2932
Kendall's score =    -373
    SE of score =     123.095    (corrected for ties)

Test of Ho: csat and expense are independent
     Prob > |z| =       0.0025   (continuity corrected)
```

For comparison, here is the Pearson correlation with its (unadjusted) *P*-value:

```
. pwcorr csat expense, sig

             |     csat  expense
-------------+------------------
        csat |   1.0000
             |
             |
     expense |  -0.4663   1.0000
             |   0.0006
             |
```

In this example, both **spearman** (−.4282) and **pwcorr** (−.4663) yield higher correlations than **ktau** (−.2925 or −.2932). All three agree that null hypotheses of no association can be rejected.

Hypothesis Tests

Two types of hypothesis tests appear in **regress** output tables. As with other common hypothesis tests, they begin from the assumption that observations in the sample at hand were drawn randomly and independently from an infinitely large population.

1. Overall *F* test: The *F* statistic at the upper right in the regression table evaluates the null hypothesis that in the population, coefficients on all the model's *x* variables equal zero.

2. Individual *t* tests: The third and fourth columns of the regression table contain *t* tests for each individual regression coefficient. These evaluate the null hypotheses that in the population, the coefficient on each particular *x* variable equals zero.

The *t* test probabilities are two-sided. For one-sided tests, divide these *P*-values in half.

In addition to these standard *F* and *t* tests, Stata can perform *F* tests of user-specified hypotheses. The **test** command refers back to the most recent model-fitting command such as **anova** or **regress**. For example, individual *t* tests from the following regression report that neither the percent of adults with at least high school diplomas (*high*) nor the percent with college degrees (*college*) has a statistically significant individual effect on composite SAT scores.

```
. regress csat expense percent income high college
```

Conceptually, however, both predictors reflect the level of education attained by a state's population, and for some purposes we might want to test the null hypothesis that *both* have zero effect. To do this, we begin by repeating the multiple regression **quietly**, because we do not need to see its full output again. Then use the **test** command:

```
. quietly regress csat expense percent income high college
. test high college

 ( 1)  high = 0.0
 ( 2)  college = 0.0

       F(  2,    45) =    3.32
            Prob > F =    0.0451
```

Unlike the individual null hypotheses, the joint hypothesis that coefficients on *high* and *college* both equal zero can reasonably be rejected ($P = .0451$). Such tests on subsets of coefficients are useful when we have several conceptually related predictors or when individual coefficient estimates appear unreliable due to multicollinearity (Chapter 7).

test could duplicate the overall F test:

. test *expense percent income high college*

test could also duplicate the individual-coefficient tests:

. test *expense*
. test *percent*
. test *income*

and so forth. Applications of **test** more useful in advanced work include

1. Test whether a coefficient equals a specified constant. For example, to test the null hypothesis that the coefficient on *income* equals 1 ($H_0 : \beta_3 = 1$), instead of the usual null hypothesis that it equals 0 ($H_0 : \beta_3 = 0$), type

   . test *income = 1*

2. Test whether two coefficients are equal. For example, the following command evaluates the null hypothesis $H_0 : \beta_4 = \beta_5$.

   . test *high = college*

3. Finally, **test** understands some algebraic expressions. We could request something like the following, which would test $H_0 : \beta_3 = (\beta_4 + \beta_5) / 100$:

   . test *income = (high + college)/100*

Consult **help test** for more information and examples.

Dummy Variables

Categorical variables can become predictors in a regression when they are expressed as one or more {0,1} dichotomies called "dummy variables." For example, we have reason to suspect that regional differences exist in states' mean SAT scores. The **tabulate** command will generate one dummy variable for each category of the tabulated variable if we add a **gen** (generate) option. Below, we create four dummy variables from the four-category variable *region*. The dummies are named *reg1, reg2, reg3* and *reg4. reg1* equals 1 for Western states and 0 for others; *reg2* equals 1 for Northeastern states and 0 for others; and so forth.

. tabulate *region*, gen(*reg*)

```
Geographica |
  l region |       Freq.      Percent        Cum.
------------+-----------------------------------
      West  |          13        26.00       26.00
   N. East  |           9        18.00       44.00
     South  |          16        32.00       76.00
   Midwest  |          12        24.00      100.00
------------+-----------------------------------
     Total  |          50       100.00
```

```
. describe reg1-reg4

                storage  display    value
variable name   type     format     label        variable label
------------------------------------------------------------------------
reg1            byte     %8.0g                    region==West
reg2            byte     %8.0g                    region==N. East
reg3            byte     %8.0g                    region==South
reg4            byte     %8.0g                    region==Midwest
. tabulate reg1

region==Wes |
          t |       Freq.      Percent        Cum.
------------+-----------------------------------
          0 |          37        74.00       74.00
          1 |          13        26.00      100.00
------------+-----------------------------------
      Total |          50       100.00

. tabulate reg2

region==N.  |
       East |       Freq.      Percent        Cum.
------------+-----------------------------------
          0 |          41        82.00       82.00
          1 |           9        18.00      100.00
------------+-----------------------------------
      Total |          50       100.00
```

Regressing *csat* on one dummy variable, *reg2* (Northeast), is equivalent to performing a two-sample *t* test of whether mean *csat* is the same across categories of *reg2*. That is, is the mean *csat* the same in the Northeast as in other U.S. states?

```
. regress csat reg2

      Source |        SS        df        MS              Number of obs =        50
-------------+------------------------------            F(  1,    48) =      9.50
       Model |   35191.4017       1   35191.4017         Prob > F      =    0.0034
    Residual |   177769.978      48   3703.54121         R-squared     =    0.1652
-------------+------------------------------            Adj R-squared =    0.1479
       Total |   212961.38       49   4346.15061         Root MSE      =    60.857


------------------------------------------------------------------------------
        csat |      Coef.   Std. Err.       t     P>|t|     [95% Conf. Interval]
-------------+----------------------------------------------------------------
        reg2 |   -69.0542   22.40167     -3.08    0.003    -114.0958   -24.01262
       _cons |   958.6098   9.504224    100.86    0.000     939.5002    977.7193
------------------------------------------------------------------------------
```

The dummy variable coefficient's *t* statistic ($t = -3.08$, $P = .003$) indicates a significant difference. According to this regression, mean SAT scores are 69.0542 points lower (because $b = -69.0542$) among Northeastern states. We get exactly the same result ($t = 3.08$, $P = .003$) from a simple *t* test, which also shows the means as 889.5556 (Northeast) and 958.6098 (other states), a difference of 69.0542.

. **ttest** *csat*, **by**(*reg2*)

Two-sample t test with equal variances

```
-----------------------------------------------------------------------------
  Group |      Obs        Mean    Std. Err.    Std. Dev.   [95% Conf. Interval]
--------+--------------------------------------------------------------------
      0 |       41    958.6098    10.36563     66.37239       937.66    979.5595
      1 |        9    889.5556    4.652094     13.95628     878.8278    900.2833
--------+--------------------------------------------------------------------
combined|       50      946.18    9.323251     65.92534     927.4442    964.9158
--------+--------------------------------------------------------------------
   diff |              69.0542    22.40167                  24.01262    114.0958
-----------------------------------------------------------------------------
```
Degrees of freedom: 48

                        Ho: mean(0) - mean(1) = diff = 0

    Ha: diff < 0   Ha: diff != 0   Ha: diff > 0
      t =   3.0825   t =   3.0825   t =   3.0825
  P < t =   0.9983   P > |t| =   0.0034   P > t =   0.0017

This conclusion proves spurious, however, once we control for the percentage of students taking the test. We do so with a multiple regression of *csat* on both *reg2* and *percent*.

. **regress** *csat* *reg2* *percent*

```
      Source |       SS       df       MS              Number of obs =      50
-------------+------------------------------           F(  2,    47) =  107.18
       Model |  174664.983     2   87332.4916           Prob > F      =  0.0000
    Residual |  38296.3969    47   814.816955           R-squared     =  0.8202
-------------+------------------------------           Adj R-squared =  0.8125
       Total |   212961.38    49   4346.15061           Root MSE      =  28.545
```

```
-----------------------------------------------------------------------------
        csat |      Coef.   Std. Err.       t    P>|t|     [95% Conf. Interval]
-------------+---------------------------------------------------------------
        reg2 |   57.52437   14.28326      4.03    0.000     28.79016    86.25858
     percent |  -2.793009   .2134796    -13.08    0.000    -3.222475   -2.363544
       _cons |   1033.749   7.270285    142.19    0.000     1019.123    1048.374
-----------------------------------------------------------------------------
```

The Northeastern region variable *reg2* now has a statistically significant *positive* coefficient ($b = 57.52437$, $P < .0005$). The earlier negative relationship was misleading. Although mean SAT scores among Northeastern states really are lower, they are lower *because higher percentages of students take this test in the Northeast*. A smaller, more "elite" group of students, often less than 20% of high school seniors, take the SAT in many of the non-Northeast states. In all Northeastern states, however, large majorities (64% to 81%) do so. Once we adjust for differences in the percentages taking the test, SAT scores actually tend to be higher in the Northeast.

To understand dummy variable regression results, it can help to write out the regression equation, substituting zeroes and ones. For Northeastern states, the equation is approximately

$$\text{predicted } csat = 1033.7 + 57.5reg2 - 2.8percent$$
$$= 1033.7 + 57.5 \times 1 - 2.8percent$$
$$= 1091.2 - 2.8percent$$

For other states, the predicted *csat* is 57.5 points lower at any given level of *percent*:

$$\text{predicted } csat = 1033.7 + 57.5 \times 0 - 2.8percent$$
$$= 1033.7 - 2.8percent$$

Dummy variables in models such as this are termed "intercept dummy variables," because they describe a shift in the *y*-intercept or constant.

From a categorical variable with *k* categories we can define *k* dummy variables, but one of these will be redundant. Once we know a state's values on the West, Northeast, and Midwest dummy variables, for example, we can already guess its value on the South variable. For this reason, no more than *k* – 1 of the dummy variables — three, in the case of *region* — can be included in a regression. If we try to include all the possible dummies, Stata will automatically drop one because multicollinearity otherwise makes the calculation impossible.

```
. regress csat reg1 reg2 reg3 reg4 percent
```

Source	SS	df	MS		
Model	181378.099	4	45344.5247		
Residual	31583.2811	45	701.850691		
Total	212961.38	49	4346.15061		

Number of obs = 50
F(4, 45) = 64.61
Prob > F = 0.0000
R-squared = 0.8517
Adj R-squared = 0.8385
Root MSE = 26.492

csat	Coef.	Std. Err.	t	P>|t|	[95% Conf. Interval]
reg1	-23.77315	11.12578	-2.14	0.038	-46.18162 -1.364676
reg2	25.79985	16.96365	1.52	0.135	-8.366693 59.96639
reg3	-33.29951	10.85443	-3.07	0.004	-55.16146 -11.43757
reg4	(dropped)				
percent	-2.546058	.2140196	-11.90	0.000	-2.977116 -2.115001
_cons	1047.638	8.273625	126.62	0.000	1030.974 1064.302

The model's fit — including R^2, *F* tests, predictions, and residuals — remains essentially the same regardless of which dummy variable we (or Stata) choose to omit. Interpretation of the coefficients, however, occurs with reference to that omitted category. In this example, the Midwest dummy variable (*reg4*) was omitted. The regression coefficients on *reg1*, *reg2*, and *reg3* tell us that, at any given level of *percent*, the predicted mean SAT scores are approximately as follows:

23.8 points lower in the West (*reg1* = 1) than in the Midwest;

25.8 points higher in the Northeast (*reg2* = 1) than in the Midwest; and

33.3 points lower in the South (*reg3* = 1) than in the Midwest.

The West and South both differ significantly from the Midwest in this respect, but the Northeast does not.

An alternative command, **areg**, fits the same model without going through dummy variable creation. Instead, it "absorbs" the effect of a *k*-category variable such as *region*. The model's fit, *F* test on the absorbed variable, and other key aspects of the results are the same as those we could obtain through explicit dummy variables. Note that **areg** does not provide estimates of the coefficients on individual dummy variables, however.

```
. areg csat percent, absorb(region)
```

```
                                           Number of obs =        50
                                           F(  1,    45) =    141.52
                                           Prob > F      =    0.0000
                                           R-squared     =    0.8517
                                           Adj R-squared =    0.8385
                                           Root MSE      =    26.492

------------------------------------------------------------------------
      csat |     Coef.   Std. Err.      t     P>|t|    [95% Conf. Interval]
-----------+------------------------------------------------------------
   percent | -2.546058   .2140196   -11.90   0.000   -2.977116   -2.115001
     _cons |  1035.445    8.38689   123.46   0.000    1018.553    1052.337
-----------+------------------------------------------------------------
    region |              F(3, 45) =      9.465   0.000      (4 categories)
```

Although its output is less informative than regression with explicit dummy variables, **areg** does have two advantages. It speeds up exploratory work, providing quick feedback about whether a dummy variable approach is worthwhile. Secondly, when the variable of interest has many values, creating dummies for each of them could lead to too many variables or too large a model for our particular Stata configuration. **areg** thus works around the usual limitations on dataset and matrix size.

Explicit dummy variables have other advantages, however, including ways to model interaction effects. Interaction terms called "slope dummy variables" can be formed by multiplying a dummy times a measurement variable. For example, to model an interaction between Northeast/other region and *percent*, we create a slope dummy variable called *reg2perc*.

```
. generate reg2perc = reg2 * percent
(1 missing value generated)
```

The new variable, *reg2perc*, equals *percent* for Northeastern states and zero for all other states. We can include this interaction term among the regression predictors:

```
. regress csat reg2 percent reg2perc
```

```
    Source |       SS       df       MS              Number of obs =        50
-----------+------------------------------          F(  3,    46) =     82.27
     Model |  179506.19        3   59835.3968        Prob > F      =    0.0000
  Residual | 33455.1897       46   727.286733        R-squared     =    0.8429
-----------+------------------------------          Adj R-squared =    0.8327
     Total |  212961.38       49   4346.15061        Root MSE      =    26.968

------------------------------------------------------------------------
      csat |     Coef.   Std. Err.      t     P>|t|    [95% Conf. Interval]
-----------+------------------------------------------------------------
      reg2 | -241.3574   116.6278    -2.07   0.044    -476.117   -6.597821
   percent | -2.858829   .2032947   -14.06   0.000    -3.26804   -2.449618
  reg2perc |  4.179666   1.620009     2.58   0.013    .9187559    7.440576
     _cons |  1035.519   6.902898   150.01   0.000    1021.624    1049.414
------------------------------------------------------------------------
```

The interaction is statistically significant ($t = 2.58$, $P = .013$). Because this analysis includes both intercept (*reg2*) and slope (*reg2perc*) dummy variables, it is worthwhile to write out the equations. The regression equation for Northeastern states is approximately

$$\begin{aligned}
\text{predicted } csat &= 1035.5 - 241.4reg2 - 2.9percent + 4.2reg2perc \\
&= 1035.5 - 241.4 \times 1 - 2.9percent + 4.2 \times 1 \times percent \\
&= 794.1 + 1.3percent
\end{aligned}$$

For other states it is

$$\begin{aligned}
\text{predicted } csat &= 1035.5 - 241.4 \times 0 - 2.9percent + 4.2 \times 0 \times percent \\
&= 1035.5 - 2.9percent
\end{aligned}$$

An interaction implies that the effect of one variable changes, depending on the values of some other variable. From this regression, it appears that *percent* has a relatively weak and positive effect among Northeastern states, whereas its effect is stronger and negative among the rest.

To visualize the results from a slope-and-intercept dummy variable regression, we have several graphing possibilities. Without even fitting the model, we could ask **lfit** to do the work as follows, with the results seen in Figure 6.6.

```
. label define reg2 0 "other regions" 1 "Northeast"

. label values reg2 reg2

. graph twoway lfit csat percent
      ||   scatter csat percent
      ||   , by(reg2, legend(off) note(""))
      ytitle("Mean composite SAT score")
```

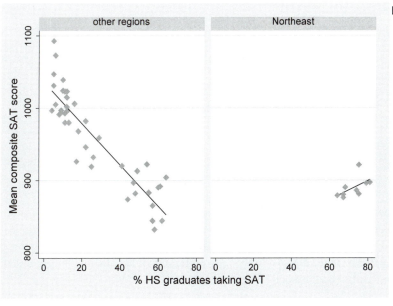

Figure 6.6

Alternatively, we could fit the regression model, calculate predicted values, and use those to make the a more refined plot such as Figure 6.7. The **bands(50)** options with both **mspline** commands specify median splines based on 50 vertical bands, which is more than enough to cover the range of the data.

```
. quietly regress csat reg2 percent reg2perc

. predict yhat1

. graph twoway scatter csat percent if reg2 == 0
        ||   mspline yhat1 percent if reg2 == 0, clpattern(solid)
     bands(50)
        ||   scatter csat percent if reg2 == 1, msymbol(Sh)
        ||   mspline yhat1 percent if reg2 == 1, clpattern(solid)
     bands(50)
        ||   , ytitle("Composite mean SAT score")
     legend(order(1 3) label(1 "other regions")
        label(3 "Northeast states") position(12) ring(0))
```

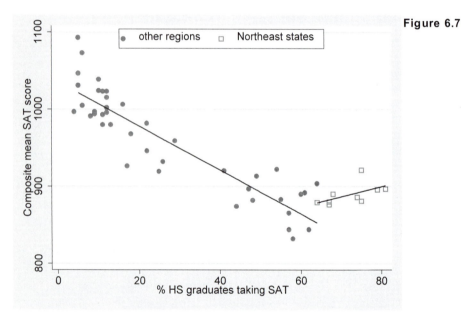

Figure 6.7

Figure 6.7 involves four overlays: two scatterplots (*csat* vs. *percent* for Northeast and other states) and two median-spline plots (connecting predicted values, *yhat1*, graphed against *percent* for Northeast and others). The Northeast states are plotted as hollow squares, **msymbol(Sh)**. **ytitle** and **legend** options simplify the *y*-axis title and the legend; in their default form, both would be crowded and unclear.

Figures 6.6 and 6.7 both show the striking difference, captured by our interaction effect, between Northeastern and other states. This raises the question of what other regional differences exist. Figure 6.8 explores this question by drawing a *csat–percent* scatterplot with different symbols for each of the four regions. In this plot, the Midwestern states, with one exception (Indiana), seem to have their own steeply negative regional pattern at the left side of the graph. Southern states are the most heterogeneous group.

```
. graph twoway scatter csat percent if reg1 ==1
     || scatter csat percent if reg2 ==1, msymbol(Sh)
     || scatter csat percent if reg3 == 1, msymbol(T)
     || scatter csat percent if reg4 == 1, msymbol(+)
     || , legend(position(1) ring(0) label(1 "West")
        label(2 "Northeast") label(3 "South") label(4 "Midwest"))
```

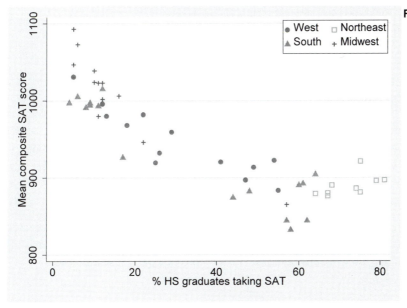

Figure 6.8

Automatic Categorical-Variable Indicators and Interactions

The **xi** (**ex**pand **i**nteractions) command simplifies the jobs of expanding multiple-category variables into sets of dummy and interaction variables, and including these as predictors in regression or other models. For example, in dataset *student2.dta* (introduced in Chapter 5) there is a four-category variable *year*, representing a student's year in college (freshman, sophomore, etc.). We could automatically create a set of three dummy variables by typing

```
. xi, prefix(ind) i.year
```

The three new dummy variables will be named *indyear_2*, *indyear_3*, and *indyear_4*. The **prefix()** option specified the prefix used in naming the new dummy variables. If we typed simply

```
. xi i.year
```

giving no **prefix()** option, the names *_Iyear_2*, *_Iyear_3*, and *_Iyear_4* would be assigned (and any previously calculated variables with those names would be overwritten by the new variables). Typing

```
. drop _I*
```

employs the wildcard * notation to drop all variables that have names beginning with *_I*.

By default, **xi** omits the lowest value of the categorical variable when creating dummies, but this can be controlled. Typing the command

```
. char _dta[omit] prevalent
```

will cause subsequent **xi** commands to automatically omit the most prevalent category (note the use of square brackets). **char _dta[]** preferences are saved with the data; to restore the default, type

```
. char _dta[omit]
```

Typing

```
. char year[omit] 3
```

would omit *year* 3. To restore the default, type

```
. char year[omit]
```

xi can also create interaction terms involving two categorical variables, or one categorical and one measurement variable. For example, we could create a set of interaction terms for *year* and *gender* by typing

```
. xi i.year*i.gender
```

From the four categories of *year* and the two categories of *gender*, this **xi** command creates seven new variables — four dummy variables and three interactions. Because their names all begin with *_I* , we can use the wildcard notation *_I** to **describe** these variables:

```
. describe _I*
```

variable name	storage type	display format	value label	variable label
_Iyear_2	byte	%8.0g		year==2
_Iyear_3	byte	%8.0g		year==3
_Iyear_4	byte	%8.0g		year==4
_Igender_1	byte	%8.0g		gender==1
_IyeaXgen_2_1	byte	%8.0g		year==2 & gender==1
_IyeaXgen_3_1	byte	%8.0g		year==3 & gender==1
_IyeaXgen_4_1	byte	%8.0g		year==4 & gender==1

To create interaction terms for categorical variable *year* and measurement variable *drink* (33-point drinking behavior scale), type

```
. xi i.year*drink
```

Six new variables result: three dummy variables for *year*, and three interaction terms representing each of the *year* dummies times *drink*. For example, for a sophomore student *_Iyear2* = 1 and *_IyeaXdrink_2* = 1×*drink* = *drink*. For a junior student, *_Iyear2* = 0 and *_IyearXdrink_2* = 0×*drink* =0; also *_Iyear3* = 1 and *_IyeaXdrink_3* = 1×*drink* = *drink*, and so forth.

```
. describe _Iyea*
```

variable name	storage type	display format	value label	variable label
_Iyear_2	byte	%8.0g		year==2
_Iyear_3	byte	%8.0g		year==3
_Iyear_4	byte	%8.0g		year==4

```
_IyeaXdrink_2    float    %9.0g              (year==2)*drink
_IyeaXdrink_3    float    %9.0g              (year==3)*drink
_IyeaXdrink_4    float    %9.0g              (year==4)*drink
```

The real convenience of **xi** comes from its ability to generate dummy variables and interactions automatically within a regression or other model-fitting command. For example, to regress variable *gpa* (student's college grade point average) on *drink* and a set of dummy variables for *year*, simply type

. **xi: regress** *gpa drink i.year*

This command automatically creates the necessary dummy variables, following the same rules described above. Similarly, to regress *gpa* on *drink*, *year*, and the interaction of *drink* and *year*, type

. **xi: regress** *gpa drink i.year*drink*

```
i.year              _Iyear_1-4           (naturally coded; _Iyear_1 omitted)
i.year*drink        _IyeaXdrink_#        (coded as above)

      Source |       SS          df        MS            Number of obs =      218
-------------+------------------------------            F(  7,    210) =     3.75
       Model |  5.08865901        7   .726951288         Prob > F      =   0.0007
    Residual |  40.6630801      210   .193633715         R-squared     =   0.1112
-------------+------------------------------            Adj R-squared =   0.0816
       Total |  45.7517391      217   .210837507         Root MSE      =  .44004


------------------------------------------------------------------------------
         gpa |      Coef.   Std. Err.       t    P>|t|     [95% Conf. Interval]
-------------+----------------------------------------------------------------
       drink |  -.0285369   .0140402    -2.03   0.043    -.0562146   -.0008591
    _Iyear_2 |  -.5839268    .314782    -1.86   0.065    -1.204464    .0366107
    _Iyear_3 |  -.2859424   .3044178    -0.94   0.349    -.8860487    .3141639
    _Iyear_4 |  -.2203783   .2939595    -0.75   0.454     -.799868    .3591114
       drink |  (dropped)
_IyeaXdrin~2 |   .0199977   .0164436     1.22   0.225    -.0124179    .0524133
_IyeaXdrin~3 |   .0108977    .016348     0.67   0.506    -.0213297     .043125
_IyeaXdrin~4 |   .0104239    .016369     0.64   0.525    -.0218446    .0426925
       _cons |   3.432132   .2523984    13.60   0.000     2.934572    3.929691
------------------------------------------------------------------------------
```

The **xi:** command can be applied in the same way before many other model-fitting procedures such as **logistic** (Chapter 10). In general, it allows us to include predictor (right-hand-side) variables such as the following, without first creating the actual dummy variable or interaction terms.

i. *catvar*	Creates $j-1$ dummy variables representing the j categories of *catvar*.
i. *catvar1*__*i.__ *catvar2*	Creates $j-1$ dummy variables representing the j categories of *catvar1*; $k-1$ dummy variables from the k categories of *catvar2*; and $(j-1)(k-1)$ interaction variables (dummy × dummy).
i. *catvar*measvar*	Creates $j-1$ dummy variables representing the j categories of *catvar*, and $j-1$ variables representing interactions with the measurement variable (dummy × *measvar*).

After any **xi** command, the new variables remain in the dataset.

Stepwise Regression

With the regional dummy variable terms we added earlier to the state-level data in *states.dta*, we have many possible predictors of *csat*. This results in an overly complicated model, with several coefficients statistically indistinguishable from zero.

```
. regress csat expense percent income college high reg1 reg2
    reg2perc reg3

      Source |       SS       df       MS              Number of obs =      50
-------------+------------------------------           F(  9,    40) =   49.51
       Model |  195420.517      9  21713.3908           Prob > F      =  0.0000
    Residual |   17540.863     40  438.521576           R-squared     =  0.9176
-------------+------------------------------           Adj R-squared =  0.8991
       Total |   212961.38     49  4346.15061           Root MSE      =  20.941


        csat |      Coef.   Std. Err.       t    P>|t|     [95% Conf. Interval]
-------------+----------------------------------------------------------------
     expense |  -.0022508   .0041333    -0.54   0.589    -.0106045     .006103
     percent |   -2.93786   .2302596   -12.76   0.000    -3.403232   -2.472488
      income |  -.0004919   .0010255    -0.48   0.634    -.0025645    .0015806
     college |   3.900087   1.719409     2.27   0.029     .4250318    7.375142
        high |   2.175542   1.171767     1.86   0.071    -.192688     4.543771
        reg1 |  -33.78456   9.302983    -3.63   0.001    -52.58659   -14.98253
        reg2 |  -143.5149   101.1244    -1.42   0.164    -347.8949    60.86509
    reg2perc |   2.506616   1.404483     1.78   0.082    -.3319506    5.345183
        reg3 |  -8.799205   12.54658    -0.70   0.487    -34.15679    16.55838
       _cons |   839.2209   76.35942    10.99   0.000     684.8927     993.549
```

We might now try to simplify this model, dropping first that predictor with the highest *t* probability (*income*, *P* = .634), then refitting the model and deciding whether to drop something further. Through this process of backward elimination, we seek a more parsimonious model; one that is simpler but fits almost equally well. Ideally, this strategy is pursued with attention both to the statistical results and to the substantive or theoretical implications of keeping or discarding certain variables.

For analysts in a hurry, stepwise methods provide ways to automate the process of model selection. They work either by subtracting predictors from a complicated model, or by adding predictors to a simpler one according to some pre-set statistical criteria. Stepwise methods cannot consider the substantive or theoretical implications of their choices, nor can they do much troubleshooting to evaluate possible weaknesses in the models produced at each step. Despite their drawbacks, stepwise methods meet certain practical needs and have been widely used.

For automatic backward elimination, we issue a **sw regress** command that includes all of our possible predictor variables, and a maximum *P* value required to retain them. Setting the *P*-to-retain criteria as **pr(.05)** ensures that only predictors having coefficients that are significantly different from zero at the .05 level will be kept in the model.

```
. sw regress csat expense percent income college high reg1 reg2
    reg2perc reg3, pr(.05)
```

```
                            begin with full model
p = 0.6341 >= 0.0500   removing income
p = 0.5273 >= 0.0500   removing reg3
p = 0.4215 >= 0.0500   removing expense
p = 0.2107 >= 0.0500   removing reg2
```

```
      Source |       SS          df       MS              Number of obs =       50
-------------+------------------------------             F(  5,    44) =    91.01
       Model |  194185.761        5    38837.1521         Prob > F      =   0.0000
    Residual |  18775.6194       44    426.718624         R-squared     =   0.9118
-------------+------------------------------             Adj R-squared =   0.9018
       Total |  212961.38        49    4346.15061         Root MSE      =   20.657
```

```
------------------------------------------------------------------------------
        csat |      Coef.   Std. Err.       t    P>|t|     [95% Conf. Interval]
-------------+----------------------------------------------------------------
        reg1 |  -30.59218   8.479395    -3.61   0.001    -47.68128   -13.50309
     percent |  -3.119155   .1804553   -17.28   0.000    -3.482839   -2.755471
    reg2perc |   .5833272   .1545969     3.77   0.000     .2717577    .8948967
     college |   3.995495   1.359331     2.94   0.005     1.255944    6.735046
        high |   2.231294   .8178968     2.73   0.009     .5829313    3.879657
       _cons |    806.672   49.98744    16.14   0.000     705.9289    907.4151
------------------------------------------------------------------------------
```

sw regress dropped first *income*, then *reg3*, *expense*, and finally *reg2* before settling on the final model. Although it has four fewer coefficients, this final model has almost the same R^2 (.9118 versus .9176) and a higher R^2_a (.9018 versus .8991) compared with the earlier version.

If, instead of a *P*-to-retain, **pr(.05)**, we specify a *P*-to-enter value such as **pe(.05)**, then **sw regress** performs forward inclusion (starting with an "empty" or constant-only model) instead of backward elimination. Other stepwise options include hierarchical selection and locking certain predictors into the model. For example, the following command specifies that the first term (*x1*) should be locked into the model and not subject to possible removal:

```
. sw regress y x1 x2 x3, pr(.05) lockterm1
```

The following command calls for forward inclusion of any predictors found significant at the .10 level, but with variables *x4*, *x5*, and *x6* treated as one unit — either entered or left out together:

```
. sw regress y x1 x2 x3 (x4 x5 x6), pe(.10)
```

The following command invokes hierarchical backward elimination with a *P* = .20 criterion:

```
. sw regress y x1 x2 x3 (x4 x5 x6) x7, pr(.20) hier
```

The **hier** option specifies that the terms are ordered: consider dropping the last term (*x7*) first, and stop if it is not dropped. If *x7* is dropped, next consider the second-to-last term *(x4 x5 x6)*, and so forth.

Many other Stata commands besides **regress** also have stepwise variants that work in a similar manner. Available stepwise procedures include the following:

sw clogit	Conditional (fixed-effects) logistic regression
sw cloglog	Maximum likelihood complementary log-log estimation

`sw cnreg`	Censored normal regression
`sw glm`	Generalized linear models
`sw logistic`	Logistic regression (odds)
`sw logit`	Logistic regression (coefficients)
`sw nbreg`	Negative binomial regression
`sw ologit`	Ordered logistic regression
`sw oprobit`	Ordered probit regression
`sw poisson`	Poisson regression
`sw probit`	Probit regression
`sw qreg`	Quantile regression
`sw regress`	OLS regression
`sw stcox`	Cox proportional hazard model regression
`sw streg`	Parametric survival-time model regression
`sw tobit`	Tobit regression

Type **help sw** for details about the stepwise options and logic.

Polynomial Regression

Earlier in this chapter, Figures 6.1 and 6.2 revealed an apparently curvilinear relationship between mean composite SAT scores (*csat*) and the percentage of high school seniors taking the test (*percent*). Figure 6.6 illustrated one way to model the upturn in SAT scores at high *percent* values: as a phenomenon peculiar to the Northeastern states. That interaction model fit reasonably well ($R^2_a = .8327$). But Figure 6.9 (next page), a residuals versus predicted values plot for the interaction model, still exhibits signs of trouble. Residuals appear to trend upwards at both high and low predicted values.

```
. quietly regress csat reg2 percent reg2perc
. rvfplot, yline(0)
```

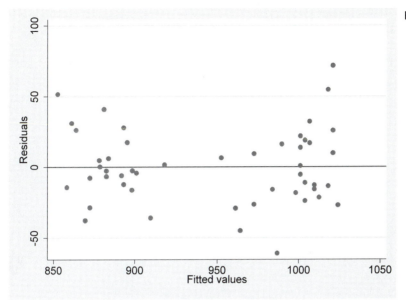

Figure 6.9

Chapter 8 presents a variety of techniques for curvilinear and nonlinear regression. "Curvilinear regression" here refers to intrinsically linear OLS regressions (for example, **regress**) that include nonlinear transformations of the original *y* or *x* variables. Although curvilinear regression fits a curved model with respect to the original data, this model remains linear in the transformed variables. (Nonlinear regression, also discussed in Chapter 8, applies non-OLS methods to fit models that cannot be linearized through transformation.)

One simple type of curvilinear regression, called polynomial regression, often succeeds in fitting U or inverted-U shaped curves. It includes as predictors both an independent variable and its square (and possibly higher powers if necessary). Because the *csat–percent* relationship appears somewhat U-shaped, we generate a new variable equal to *percent* squared, then include *percent* and *percent* ^{2} as predictors of *csat*. Figure 6.10 graphs the resulting curve.

```
. generate percent2 = percent^2

. regress csat percent percent2
```

```
      Source |       SS       df       MS              Number of obs =      51
-------------+------------------------------           F(  2,    48) =  153.48
       Model |  193721.829      2  96860.9146           Prob > F      =  0.0000
    Residual |  30292.6806     48  631.097513           R-squared     =  0.8648
-------------+------------------------------           Adj R-squared =  0.8591
       Total |   224014.51     50   4480.2902           Root MSE      =  25.122

------------------------------------------------------------------------------
        csat |      Coef.   Std. Err.      t    P>|t|     [95% Conf. Interval]
-------------+----------------------------------------------------------------
     percent |  -6.111993   .6715406    -9.10   0.000    -7.462216   -4.76177
    percent2 |   .0495819   .0084179     5.89   0.000     .0326566   .0665072
       _cons |   1065.921   9.285379   114.80   0.000     1047.252   1084.591
------------------------------------------------------------------------------
```

```
. predict yhat2
(option xb assumed; fitted values)

. graph twoway mspline yhat2 percent, bands(50)
       ||   scatter csat percent
       ||   , legend(off) ytitle("Mean composite SAT score")
```

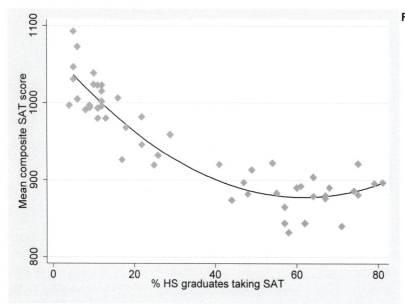

Figure 6.10

If we only wanted to see the graph, and did not need the regression analysis, there is a quicker way to achieve this. The command **graph twoway qfit** fits a quadratic (second-order polynomial) regression model; **qfitci** draws confidence bands as well. For example, a curve similar to Figure 6.10 could have been obtained by typing

```
. graph twoway qfit csat percent
       ||   scatter csat percent
```

The polynomial model in Figure 6.10 matches the data slightly better than our interaction model in Figure 6.6 (R^2_a = .8591 versus .8327). Because the curvilinear pattern is now less striking in a residual versus predicted values plot (Figure 6.11), the usual assumption of independent, identically distributed errors also appears more plausible with respect to this polynomial model.

```
. quietly regress csat percent percent2
. rvfplot, yline(0)
```

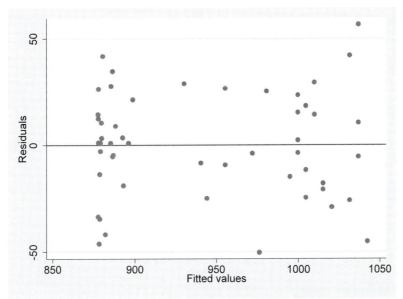

Figure 6.11

In Figures 6.7 and 6.10, we have two alternative models for the observed upturn in SAT scores at high levels of student participation. Statistical evidence seems to lean towards the polynomial model at this point. For serious research, however, we ought to choose between similar-fitting alternative models on substantive as well as statistical grounds. Which model seems more useful, or makes more sense? Which, if either, regression model suggests or corresponds to a good real-world explanation for the upturn in test scores at high levels of student participation?

Although it can closely fit sample data, polynomial regression also has important statistical weaknesses. The different powers of x might be highly correlated with each other, giving rise to multicollinearity. Furthermore, polynomial regression tends to track observations that have unusually large positive or negative x values, so a few data points can exert disproportionate influence on the results. For both reasons, polynomial regression results can sometimes be sample-specific, fitting one dataset well but generalizing poorly to other data. Chapter 7 takes a second look at this example, using tools that check for potential problems.

Panel Data

Panel data, also called cross-sectional time series, consist of observations on i analytical units or cases, repeated over t points in time. The *Longitudinal/Panel Data Reference Manual* describes a wide range of methods for analyzing such data. Most of the relevant Stata commands begin with the letters xt; type **help xt** for an overview. As mentioned in the documentation, some **xt** procedures require time series or **tsset** data; see Chapter 13, or type **help tsset**, for more about this step.

This section considers the relatively simple case of linear regression with panel data, accomplished by the command **xtreg** . Our example dataset, *newfdiv.dta* contains information about the 10 census divisions of the Canadian province of Newfoundland (Avalon Peninsula, Burin Peninsula, and 8 others), for the years 1992–96.

```
Contains data from C:\data\newfdiv.dta
  obs:              50                    Newfoundland Census divisions
                                            (source:  Statistics Canada)
  vars:              7                    18 Jul 2005 10:28
  size:          2,250 (99.9% of memory free)
-------------------------------------------------------------------------------
               storage  display   value
variable name   type    format    label       variable label
-------------------------------------------------------------------------------
cendiv         byte     %9.0g      cd          Census Division
divname        str20    %20s                   Census Division name
year           int      %9.0g                  Year
pop            double   %9.0g                  Population, 1000s
unemp          float    %9.0g                  Total unemployment, 1000s
outmig         int      %9.0g                  Out-migration
tcrime         float    %9.0g                  Total crimes reported, 1000s
-------------------------------------------------------------------------------
Sorted by:  cendiv   year
```

```
. list in 1/10

     +-------------------------------------------------------------------------+
     | cendiv           divname   year       pop    unemp    outmig    tcrime  |
     |-------------------------------------------------------------------------|
  1. | Avalon   Avalon Peninsula   1992   259.587    58.56      6556    26.211  |
  2. | Avalon   Avalon Peninsula   1993   261.083    52.23      6449    21.039  |
  3. | Avalon   Avalon Peninsula   1994   259.296    44.81      6907    20.201  |
  4. | Avalon   Avalon Peninsula   1995   257.546    39.35         .    19.536  |
  5. | Avalon   Avalon Peninsula   1996   255.723    38.68         .    21.268  |
     |-------------------------------------------------------------------------|
  6. | Burin     Burin Peninsula   1992    29.865      9.5       874     1.903  |
  7. | Burin     Burin Peninsula   1993    29.611     9.18       928     1.953  |
  8. | Burin     Burin Peninsula   1994    29.327     8.41       884      1.94  |
  9. | Burin     Burin Peninsula   1995    28.898     7.12         .     2.063  |
 10. | Burin     Burin Peninsula   1996    28.126     6.81         .     1.923  |
     +-------------------------------------------------------------------------+
```

Figure 6.12 visualizes the panel data, graphing variations in the number of crimes reported each year for 9 of the 10 census divisions. Census division 1, the Avalon Peninsula, is by far the largest in Newfoundland. Setting it temporarily aside by specifying **if *cendiv* != 1** makes the remaining 9 plots in Figure 6.12 more readable. The **imargin(l=3 r=3)** option in this example calls for left and right margins subplot margins equal to 3% of the graph width, giving more separation than the default.

```
. graph twoway connected tcrime year if cendiv != 1,
    by(cendiv, note("")) xtitle("") imargin(left=3 right=3)
```

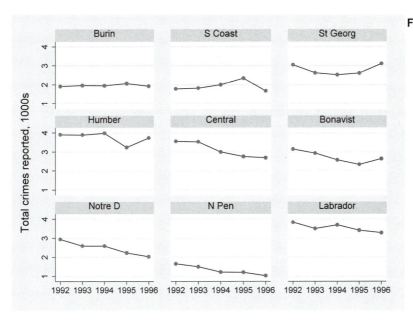

Figure 6.12

The dataset contains 50 observations total. Because the 50 observations represent only 10 individual cases, however, the usual assumptions of OLS and other common statistical methods do not apply. Instead, we need models with complex error specifications, allowing for both unit-specific and individual-observation disturbances.

Consider the regression of y on two predictors, x and w. OLS regression estimates the regression coefficients a, b, and c, and calculates the associated standard errors and tests, assuming a model of the form

$$y_i = a + bx_i + cw_i + e_i$$

where the residuals for each observation, e_i, are assumed to represent errors that have independent and identical distributions. The i.i.d. errors assumption appears unlikely with panel data, where the observations consist of the same units measured repeatedly.

A more plausible panel-data model includes two error terms. One is common to each of the i units, but differs between units (u_i). The second is unique to each of the i, t observations (e_{it}).

$$y_{it} = a + bx_{it} + cw_{it} + u_i + e_{it}$$

In order to fit such a model, Stata needs to know which variable identifies the i units, and which variable is the time index t. This can be done within an **xt** command, or more efficiently for the dataset as a whole. The commands **iis** ("i is") and **tis** ("t is") specify the i and t variables, respectively. For *newfdiv.dta*, the units are census divisions (*cendiv*) and the time index is *year*.

```
. iis cendiv

. tis year

. save, replace
```

Saving the dataset preserves the *i* and *t* specifications, so the **iis** and **tis** commands are not required in a future session. Having set these variables, we can now fit a random-effects (meaning that the common errors u_i are assumed to be variable, rather than fixed) model regressing *tcrime* on *unemp* and *pop*.

```
. xtreg tcrime unemp pop, re

Random-effects GLS regression            Number of obs      =          50
Group variable (i): cendiv               Number of groups   =          10

R-sq:  within  = 0.5265                   Obs per group: min =           5
       between = 0.9717                                  avg =         5.0
       overall = 0.9634                                  max =           5

Random effects u_i ~ Gaussian            Wald chi2(2)       =      705.54
corr(u_i, X)       = 0 (assumed)         Prob > chi2        =      0.0000

------------------------------------------------------------------------------
      tcrime |      Coef.   Std. Err.       z    P>|z|     [95% Conf. Interval]
-------------+----------------------------------------------------------------
       unemp |   .1645266   .0381813     4.31   0.000     .0896925    .2393607
         pop |   .0558997   .0073437     7.61   0.000     .0415062    .0702931
       _cons |  -.7264381    .301522    -2.41   0.016    -1.31741   -.1354659
-------------+----------------------------------------------------------------
     sigma_u |  .34458437
     sigma_e |  .42064667
         rho |  .40157462   (fraction of variance due to u_i)
------------------------------------------------------------------------------
```

The **xtreg** output table contains regression coefficients, standard errors, *t* tests, and confidence intervals that resemble those of an OLS regression. In this example we see that the coefficient on *unemp* (.1645) is positive and statistically significant. The predicted number of crimes increases by .1645 for each additional person unemployed, if population is held constant. Holding unemployment constant, predicted crimes increase by 5.59 with each 100-person increase in population. Echoing the individual-coefficient *z* tests, the Wald chi-square test at upper right ($\chi^2 = 705.54$, $df = 2$, $P < .00005$) allows us to reject the joint null hypothesis that the coefficients on *unemp* and *pop* are both zero.

This output table gives further information related to the two error terms. At lower left in the table we find

sigma_u   standard deviation of the common residuals u_i

sigma_e   standard deviation of the unique residuals e_i

rho   fraction of the unexplained variance due to differences among the units (i.e., differences among the 10 Newfoundland census divisions).

$$\text{Var}[u_i]/(\text{Var}[u_i] + \text{Var}[e_{it}])$$

At upper left the table gives three "R^2" statistics. The definitions for these differ from the true R^2 of OLS. In the case of **xtreg**, the "R^2" are based on fits between several kinds of observed and predicted *y* values.

R^2 within	Explained variation within units — defined as the squared correlation between deviations of y_{it} values from unit means ($y_{it} - \bar{y}_i$) and deviations of predicted values from unit mean predicted values ($\hat{y}_{it} - \hat{\bar{y}}_i$).
R^2 between	Explained variation between units — defined as the squared correlation between unit means ($\bar{y}_i$) and $\hat{\bar{y}}_i$ values predicted from unit means of the independent variables.
R^2 overall	Explained variation overall — defined as the squared correlation between observed (y_{it}) and predicted ($\hat{y}_{it}$) values.

Our example model does a very good job fitting the observed crimes overall ($R^2 = .96$), and also the variations among census division means ($R^2 = .97$). Variations around the means within census divisions are somewhat less predictable ($R^2 = .53$).

The random-effects option employed for this example is one of several possible choices.

re   Generalized least squares (GLS) random-effects estimator; default

be   between regression estimator

fe   fixed-effects (within) regression estimator

mle   maximum-likelihood random-effects estimator

pa   population-averaged estimator

Consult **help xtreg** for further options and syntax. The *Longitudinal/Panel Data Reference Manual* gives examples, references, and technical details.

7

Regression Diagnostics

Do the data give us any reason to distrust our regression results? Can we find better ways to specify the model, or to estimate its parameters? Careful diagnostic work, checking for potential problems and evaluating the plausibility of key assumptions, forms a crucial step in modern data analysis. We fit an initial model, but then look closely at our results for signs of trouble or ways in which the model needs improvement. Many of the general methods introduced in earlier chapters, such as scatterplots, box plots, normality tests, or just sorting and listing the data, prove useful for troubleshooting. Stata also provides a toolkit of specialized diagnostic techniques designed for this purpose.

Autocorrelation, a complication that often affects regression with time series data, is not covered in this chapter. Chapter 13, Time Series Analysis, introduces Stata's library of time series procedures including Durbin–Watson tests, autocorrelation graphs, lag operators, and time-series regression techniques.

Regression diagnostic procedures can be found under these menu selections:

Statistics – Linear regression and related – Regression diagnostics

Statistics – General post-estimation – Obtain predictions, residuals, etc., after estimation

Example Commands

The commands illustrated in this section all assume that you have just fit a model using either **anova** or **regress**. The commands' results refer back to that model. These followup commands are of three basic types:

1. **predict** options that generate new variables containing case statistics such as predicted values, residuals, standard errors, and influence statistics. Chapter 6 noted some key options; type **help regress** for a complete listing.

2. Diagnostic tests for statistical problems such as autocorrelation, heteroskedasticity, specification errors, or variance inflation (multicollinearity). Type **help regdiag** for a list.

3. Diagnostic plots such as added-variable or leverage plots, residual-versus-fitted plots, residual-versus-predictor plots, and component-versus-residual plots. Again, typing **help regdiag** obtains a full listing of regression and ANOVA diagnostic plots. General graphs for diagnosing distribution shape and normality were covered in Chapter 2; type **help diagplots** for a list of those.

`predict` Options

. `predict` *new*, `cooksd`

Generates a new variable equal to Cook's distance D, summarizing how much each observation influences the fitted model.

. `predict` *new*, `covratio`

Generates a new variable equal to Belsley, Kuh, and Welsch's *COVRATIO* statistic. *COVRATIO* measures the ith case's influence upon the variance–covariance matrix of the estimated coefficients.

. `predict` *DFx1*, `dfbeta(x1)`

Generates *DFBETA* case statistics measuring how much each observation affects the coefficient on predictor $x1$. The **dfbeta** command accomplishes the same thing more conveniently, and in this example will automatically name the resulting statistics *DFx1*:

. `dfbeta x1`

To create a complete set of *DFBETA*s for all predictors in the model, simply type the command **dfbeta** without arguments.

. `predict` *new*, `dfits`

Generates *DFITS* case statistics, summarizing the influence of each observation on the fitted model (similar in purpose to Cook's D and Welsch's W).

Diagnostic Tests

. `dwstat`

Calculates the Durbin–Watson test for first-order autocorrelation. Chapter 13 gives examples of this and other time series procedures. See also:

`help durbina`	Durbin–Watson h statistic
`help bgodfrey`	Breusch–Godfrey LM (Lagrange multiplier) statistic

. `hettest`

Performs Cook and Weisberg's test for heteroskedasticity. If we have reason to suspect that heteroskedasticity is a function of a particular predictor $x1$, we could focus on that predictor by typing **hettest x1** .

. `ovtest`, `rhs`

Performs the Ramsey regression specification error test (*RESET*) for omitted variables. The option **rhs** calls for using powers of the right-hand-side variables, instead of powers of predicted y (default).

. `vif`

Calculates variance inflation factors to check for multicollinearity.

Diagnostic Plots

. `acprplot` *x1*, `mspline msopts(bands(7))`

Constructs an augmented component-plus-residual plot (also known as an augmented partial residual plot), often better than **cprplot** in screening for nonlinearities. The options **mspline msopts(bands(7))** call for connecting with line segments the cross-medians of seven vertical bands. Alternatively, we might ask for a lowess-smoothed curve with bandwidth 0.5 by specifying the options **lowess lsopts(bwidth(.5))**.

. **avplot** *x1*

Constructs an added-variable plot (also called a partial-regression or leverage plot) showing the relationship between *y* and *x1*, both adjusted for other *x* variables. Such plots help to notice outliers and influence points.

. **avplots**

Draws and combines in one image all the added-variable plots from the recent **anova** or **regress**.

. **cprplot** *x1*

Constructs a component-plus-residual plot (also known as a partial-residual plot) showing the adjusted relationship between *y* and predictor *x1*. Such plots help detect nonlinearities in the data.

. **lvr2plot**

Constructs a leverage-versus-squared-residual plot (also known as an L–R plot).

. **rvfplot**

Graphs the residuals versus the fitted (predicted) values of *y*.

. **rvpplot** *x1*

Graphs the residuals against values of predictor *x1*.

SAT Score Regression, Revisited

Diagnostic techniques have been described as tools for "regression criticism," because they help us examine our regression models for possible flaws and for ways that the models could be improved. In this spirit, we return now to the state Scholastic Aptitude Test regressions of Chapter 6. A three-predictor model explains about 92% of the variance in mean state SAT scores. The predictors are *percent* (percent of high school graduates taking the test), *percent2* (*percent* squared), and *high* (percent of adults with a high school diploma).

. **generate** *percent2* = *percent*^2

. **regress** *csat percent percent2 high*

Source	SS	df	MS		Number of obs =	51
Model	207225.103	3	69075.0343		F(3, 47) =	193.37
Residual	16789.4069	47	357.221424		Prob > F =	0.0000
					R-squared =	0.9251
					Adj R-squared =	0.9203
Total	224014.51	50	4480.2902		Root MSE =	18.90

csat	Coef.	Std. Err.	t	P>|t|	[95% Conf. Interval]	
percent	-6.520312	.5095805	-12.80	0.000	-7.545455	-5.495168
percent2	.0536555	.0063678	8.43	0.000	.0408452	.0664658
high	2.986509	.4857502	6.15	0.000	2.009305	3.963712
_cons	844.8207	36.63387	23.06	0.000	771.1228	918.5185

The regression equation is

predicted *csat* = 844.82 – 6.52*percent* + .05*percent2* + 2.99*high*

The scatterplot matrix in Figure 7.1 depicts interrelations among these four variables. As noted in Chapter 6, the squared term *percent2* allows our regression model to fit the visibly curvilinear relationship between *csat* and *percent*.

```
. graph matrix percent percent2 high csat, half msymbol(+)
```

Figure 7.1

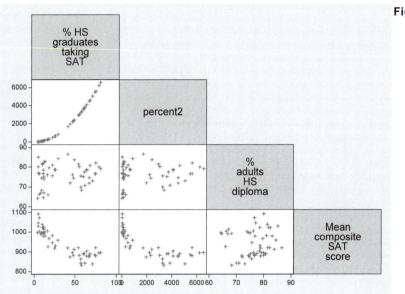

Several post-regression hypothesis tests perform checks on the model specification. The omitted-variables test **ovtest** essentially regresses *y* on the *x* variables, and also the second, third, and fourth powers of predicted *y* (after standardizing $\hat{y}$ to have mean 0 and variance 1). It then performs an *F* test of the null hypothesis that all three coefficients on those powers of $\hat{y}$ equal zero. If we reject this null hypothesis, further polynomial terms would improve the model. With the *csat* regression, we need not reject the null hypothesis.

```
. ovtest

Ramsey RESET test using powers of the fitted values of csat
      Ho:  model has no omitted variables
                 F(3, 44) =         1.48
                 Prob > F =       0.2319
```

A heteroskedasticity test, **hettest**, tests the assumption of constant error variance by examining whether squared standardized residuals are linearly related to $\hat{y}$ (see Cook and Weisberg 1994 for discussion and example). Results from the *csat* regression suggest that in this instance we should reject the null hypothesis of constant variance.

```
. hettest

Cook-Weisberg test for heteroskedasticity using fitted values of csat
      Ho: Constant variance
           chi2(1)      =         4.86
           Prob > chi2  =       0.0274
```

"Significant" heteroskedasticity implies that our standard errors and hypothesis tests might be invalid. Figure 7.2, in the next section, shows why this result occurs.

Diagnostic Plots

Chapter 6 demonstrated how **predict** can create new variables holding residual and predicted values after a **regress** command. To obtain these values from our regression of *csat* on *percent*, *percent2*, and *high*, we type the two commands:

```
. predict yhat3
. predict e3, resid
```

The new variables named *e3* (residuals) and *yhat3* (predicted values) could be displayed in a residual-versus-predicted graph by typing **graph twoway scatter e3 yhat, yline(0)**. The **rvfplot** (residual-versus-fitted) command obtains such graphs in a single step. The version in Figure 7.2 includes a horizontal line at 0 (the residual mean), which helps in reading such plots.

```
. rvfplot, yline(0)
```

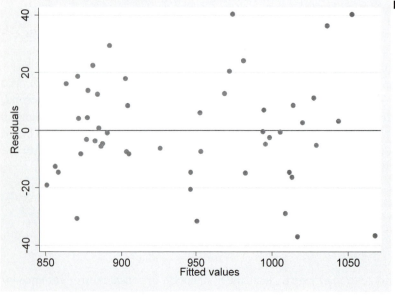

Figure 7.2

Figure 7.2 shows residuals symmetrically distributed around 0 (symmetry is consistent with the normal-errors assumption), and with no evidence of outliers or curvilinearity. The dispersion of the residuals appears somewhat greater for above-average predicted values of *y*, however, which is why **hettest** earlier rejected the constant-variance hypothesis.

Residual-versus-fitted plots provide a one-graph overview of the regression residuals. For more detailed study, we can plot residuals against each predictor variable separately through a series of "residual-versus-predictor" commands. To graph the residuals against predictor *high* (not shown), type

. rvpplot *high*

The one-variable graphs described in Chapter 3 can also be employed for residual analysis. For example, we could use box plots to check the residuals for outliers or skew, or quantile–normal plots to evaluate the assumption of normal errors.

Added-variable plots are valuable diagnostic tools, known by different names including partial-regression leverage plots, adjusted partial residual plots, or adjusted variable plots. They depict the relationship between *y* and one *x* variable, adjusting for the effects of other *x* variables. If we regressed *y* on *x2* and *x3*, and likewise regressed *x1* on *x2* and *x3*, then took the residuals from each regression and graphed these residuals in a scatterplot, we would obtain an added-variable plot for the relationship between *y* and *x1*, adjusted for *x2* and *x3*. An **avplot** command performs the necessary calculations automatically. We can draw the adjusted-variable plot for predictor *high*, for example, just by typing

. avplot *high*

Speeding the process further, we could type **avplots** to obtain a complete set of tiny added-variable plots with each of the predictor variables in the preceding regression. Figure 7.3 shows the results from the regression of *csat* on *percent*, *percent2*, and *high*. The lines drawn in added-variable plots have slopes equal to the corresponding partial regression coefficients. For example, the slope of the line at lower left in Figure 7.3 equals 2.99, which is the coefficient on *high*.

. avplots

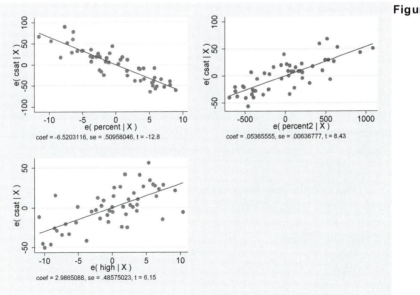

Figure 7.3

Added-variable plots help to uncover observations exerting a disproportionate influence on the regression model. In simple regression with one *x* variable, ordinary scatterplots suffice for this purpose. In multiple regression, however, the signs of influence become more subtle. An observation with an unusual combination of values on several *x* variables might have high leverage, or potential to influence the regression, even though none of its individual *x* values is unusual by itself. High-leverage observations show up in added-variable plots as points horizontally distant from the rest of the data. We see no such problems in Figure 7.3, however.

If outliers appear, we might identify which observations these are by including observation labels for the markers in an added-variable plot. This is done using the **mlabel( )** option, just as with scatterplots. Figure 7.4 illustrates using state names (values of the string variable *state*) as labels. Although such labels tend to overprint each other where the data are dense, individual outliers remain more readable.

```
. avplot high, mlabel(state)
```

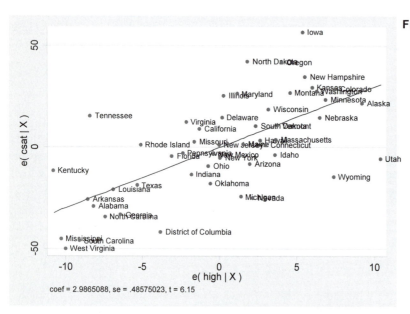

Figure 7.4

Component-plus-residual plots, produced by commands of the form **cprplot x1**, take a different approach to graphing multiple regression. The component-plus residual plot for variable *x1* graphs each observation's residual plus its component predicted from *x1*,

$$e_i + b_1 x1_i$$

against values of *x1*. Such plots might help diagnose nonlinearities and suggest alternative functional forms. An augmented component-plus-residual plot (Mallows 1986) works somewhat better, although both types often seem inconclusive. Figure 7.5 shows an augmented component-plus-residual plot from the regression of *csat* on *percent*, *percent2*, and *high*.

. acprplot *high*, lowess

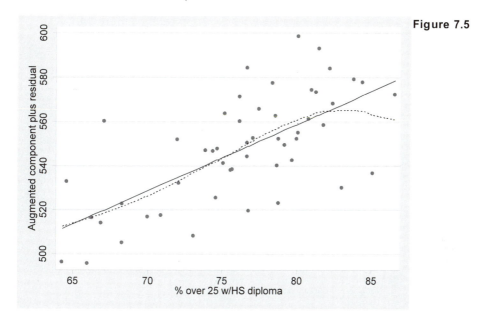

Figure 7.5

The straight line in Figure 7.5 corresponds to the regression model. The curved line reflects lowess smoothing based on the default bandwidth of .5, or half the data. The curve's downturn at far right can be disregarded as a lowess artifact, because only a few cases determine its location toward the extremes (see Chapter 8). If more central parts of the lowess curve showed a systematically curved pattern, departing from the linear regression model, we would have reason to doubt the model's adequacy. In Figure 7.5, however, the component-plus-residuals medians closely follow the regression model. This plot reinforces the conclusion we reached earlier from Figure 7.2, that the present regression model adequately accounts for all nonlinearity visible in the raw data (Figure 7.1), leaving none apparent in its residuals.

As its name implies, a leverage-versus-squared-residuals plot graphs leverage (hat matrix diagonals) against the residuals squared. Figure 7.6 shows such a plot for the *csat* regression. To identify individual outliers, we label the markers with the values of *state*. The option **mlabsize(medsmall)** calls for "medium small" marker labels, somewhat larger than the default size of "small." (See **help testsizestyle** for a list of other choices.) Most of the state names form a jumble at lower left in Figure 7.6, but a few outliers stand out.

```
. lvr2plot, mlabel(state) mlabsize(medsmall)
```

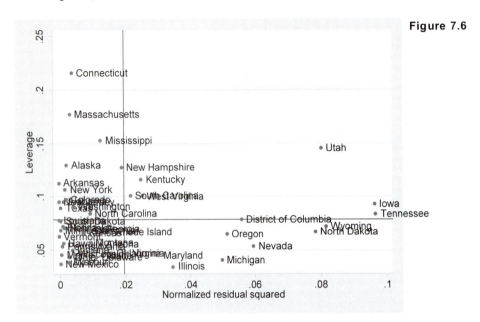

Figure 7.6

Lines in a leverage-versus-squared-residuals plot mark the means of leverage (horizontal line) and squared residuals (vertical line). Leverage tells us how much potential for influencing the regression an observation has, based on its particular combination of x values. Extreme x values or unusual combinations give an observation high leverage. A large squared residual indicates an observation with y value much different from that predicted by the regression model. Connecticut, Massachusetts, and Mississippi have the greatest potential leverage, but the model fits them relatively well. (This is not necessarily good. Sometimes, although not here, high-leverage observations exert so much influence that they control the regression, and it *must* fit them well.) Iowa and Tennessee are poorly fit, but have less potential influence. Utah stands out as one observation that is both ill fit and potentially influential. We can read its values by listing just this state. Because *state* is a string variable, we enclose the value "Utah" in double quotes.

```
. list csat yhat3 percent high e3 if state == "Utah"

     +------------------------------------------------+
     | csat      yhat3   percent   high          e3   |
     |------------------------------------------------|
  1. | 1031   1067.712         5   85.1   -36.71239   |
     +------------------------------------------------+
```

Only 5% of Utah students took the SAT, and 85.1% of the state's adults graduated from high school. This unusual combination of near-extreme values on both x variables is the source of the state's leverage, and leads our model to predict mean SAT scores 36.7 points higher than what Utah students actually achieved. To see exactly how much difference this one observation makes, we could repeat the regression using Stata's "not equal to" qualifier `!=` to set Utah aside.

```
. regress csat percent percent2 high if state != "Utah"

      Source |       SS       df       MS              Number of obs =      50
-------------+------------------------------           F(  3,    46) =  202.67
       Model |  201097.423     3  67032.4744           Prob > F      =  0.0000
    Residual |  15214.0968    46  330.741235           R-squared     =  0.9297
-------------+------------------------------           Adj R-squared =  0.9251
       Total |   216311.52    49  4414.52082           Root MSE      =  18.186


------------------------------------------------------------------------------
        csat |      Coef.   Std. Err.      t    P>|t|     [95% Conf. Interval]
-------------+----------------------------------------------------------------
     percent |  -6.778706   .5044217   -13.44   0.000    -7.794054   -5.763357
    percent2 |   .0563562   .0062509     9.02   0.000     .0437738    .0689387
        high |   3.281765   .4865854     6.74   0.000     2.302319     4.26121
       _cons |   827.1159   36.17138    22.87   0.000     754.3067    899.9252
------------------------------------------------------------------------------
```

In the $n = 50$ (instead of $n = 51$) regression, all three coefficients strengthened a bit because we deleted an ill-fit observation. The general conclusions remain unchanged, however.

Chambers et al. (1983) and Cook and Weisberg (1994) provide more detailed examples and explanations of diagnostic plots and other graphical methods for data analysis.

Diagnostic Case Statistics

After using **regress** or **anova**, we can obtain a variety of diagnostic statistics through the **predict** command (see Chapter 6 or type **help regress**). The variables created by **predict** are case statistics, meaning that they have values for each observation in the data. Diagnostic work usually begins by calculating the predicted values and residuals.

There is some overlap in purpose among other **predict** statistics. Many attempt to measure how much each observation influences regression results. "Influencing regression results," however, could refer to several different things — effects on the *y*-intercept, on a particular slope coefficient, on all the slope coefficients, or on the estimated standard errors, for example. Consequently, we have a variety of alternative case statistics designed to measure influence.

Standardized and studentized residuals (**rstandard** and **rstudent**) help to identify outliers among the residuals — observations that particularly contradict the regression model. Studentized residuals have the most straightforward interpretation. They correspond to the *t* statistic we would obtain by including in the regression a dummy predictor coded 1 for that observation and 0 for all others. Thus, they test whether a particular observation significantly shifts the *y*-intercept.

Hat matrix diagonals (**hat**) measure leverage, meaning the potential to influence regression coefficients. Observations possess high leverage when their *x* values (or their combination of *x* values) are unusual.

Several other statistics measure actual influence on coefficients. *DFBETAs* indicate by how many standard errors the coefficient on *x1* would change if observation *i* were dropped from the regression. These can be obtained for a single predictor, *x1*, in either of two ways: through the **predict** option **dfbeta(x1)** or through the command **dfbeta**.

Cook's *D* (**cooksd**), Welsch's distance (**welsch**), and *DFITS* (**dfits**), unlike *DFBETA*, all summarize how much observation *i* influences the regression model as a whole — or equivalently, how much observation *i* influences the set of predicted values. *COVRATIO* measures the influence of the *i*th observation on the estimated standard errors. Below we generate a full set of diagnostic statistics including *DFBETA*s for all three predictors. Note that **predict** supplies variable labels automatically for the variables it creates, but **dfbeta** does not. We begin by repeating our original regression to ensure that these post-regression diagnostics refer to the proper (*n* = 51) model.

```
. quietly regress csat percent percent2 high

. predict standard, rstandard

. predict student, rstudent

. predict h, hat

. predict D, cooksd

. predict DFITS, dfits

. predict W, welsch

. predict COVRATIO, covratio

. dfbeta
              DFpercent:   DFbeta(percent)
             DFpercent2:   DFbeta(percent2)
                DFhigh:    DFbeta(high)

. describe standard - DFhigh

              storage   display    value
variable name   type    format     label      variable label
-------------------------------------------------------------------------------
standard        float   %9.0g                 Standardized residuals
student         float   %9.0g                 Studentized residuals
h               float   %9.0g                 Leverage
D               float   %9.0g                 Cook's D
DFITS           float   %9.0g                 Dfits
W               float   %9.0g                 Welsch distance
COVRATIO        float   %9.0g                 Covratio
DFpercent       float   %9.0g
DFpercent2      float   %9.0g
DFhigh          float   %9.0g

. summarize standard - DFhigh

    Variable |     Obs        Mean     Std. Dev.        Min         Max
-------------+-------------------------------------------------------------
    standard |      51   -.0031359     1.010579   -2.099976    2.233379
     student |      51     -.00162     1.032723   -2.182423    2.336977
           h |      51    .0784314     .0373011    .0336437    .2151227
           D |      51    .0219941     .0364003    .0000135    .1860992
       DFITS |      51   -.0107348     .3064762    -.896658    .7444486
-------------+-------------------------------------------------------------
           W |      51    -.089723     2.278704   -6.854601     5.52468
    COVRATIO |      51    1.092452     .1316834    .7607449    1.360136
   DFpercent |      51     .000938     .1498813   -.5067295    .5269799
  DFpercent2 |      51   -.0010659     .1370372    -.440771    .4253958
      DFhigh |      51   -.0012204     .1747835   -.6316988    .3414851
```

summarize shows us the minimum and maximum values of each statistic, so we can quickly check whether any are large enough to cause concern. For example, special tables could be used to determine whether the observation with the largest absolute studentized residual (*student*) constitutes a significant outlier. Alternatively, we could apply the Bonferroni inequality and *t* distribution table: max| *student* | is significant at level α if | *t* | is significant at α/n. In this example, we have max| *student* | = 2.337 (Iowa) and n = 51. For Iowa to be a significant outlier (cause a significant shift in intercept) at α = .05, t = 2.337 must be significant at .05 / 51:

```
. display .05/51
.00098039
```

Stata's **ttail( )** function can approximate the probability of | *t* | > 2.337, given $df = n - K - 1 = 51 - 3 - 1 = 47$:

```
. display 2*ttail(47, 2.337)
.02375138
```

The obtained *P*-value (P = .0238) is not below α/n = .00098, so Iowa is not a significant outlier at α = .05.

Studentized residuals measure the *i*th observation's influence on the *y*-intercept. Cook's *D*, *DFITS*, and Welsch's distance all measure the *i*th observation's influence on all coefficients in the model (or, equivalently, on all n predicted *y* values). To list the 5 most influential observations as measured by Cook's *D*, type

```
. sort D
. list state yhat3 D DFITS W in -5/1
```

	state	yhat3	D	DFITS	W
47.	North Dakota	1036.696	.0705921	.5493086	4.020527
48.	Wyoming	1017.005	.0789454	-.5820746	-4.270465
49.	Tennessee	974.6981	.111718	.6992343	5.162398
50.	Iowa	1052.78	.1265392	.7444486	5.52468
51.	Utah	1067.712	.1860992	-.896658	-6.854601

The **in -5/1** qualifier tells Stata to list only the fifth-from-last (–5) through last (lowercase letter "l") observations. Figure 7.7 shows one way to display influence graphically: symbols in a residual-versus-predicted plot are given sizes proportional to values of Cook's *D*, through the "analytical weight" option **[aweight = D]**. Five influential observations stand out, with large positive or negative residuals and high predicted *csat* values.

```
. graph twoway scatter e3 yhat3 [aweight = D], msymbol(oh) yline(0)
```

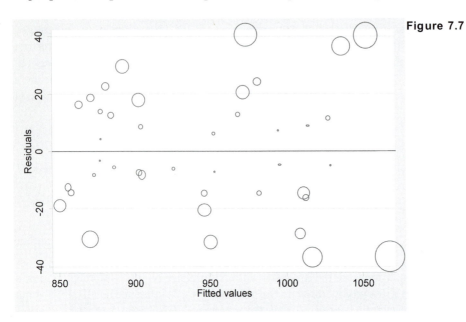

Figure 7.7

Although they have different statistical rationales, Cook's *D*, Welsch's distance, and *DFITS* are closely related. In practice they tend to flag the same observations as influential. Figure 7.8 shows their similarity in the example at hand.

```
. graph matrix D W DFITS, half
```

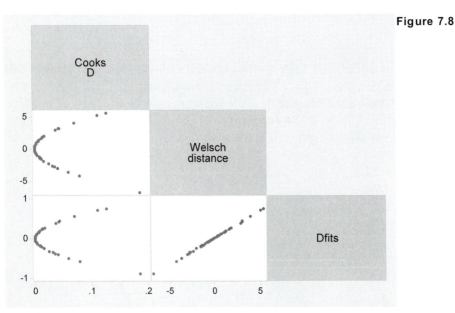

Figure 7.8

*DFBETA*s indicate how much each observation influences each regression coefficient. Typing **dfbeta** after a regression automatically generates *DFBETA*s for each predictor. In this example, they received the names *DFpercent* (*DFBETA* for predictor *percent*), *DFpercent2*, and *DFhigh*. Figure 7.9 graphs their distributions as box plots.

```
. graph box DFpercent DFpercent2 DFhigh, legend(cols(3))
```

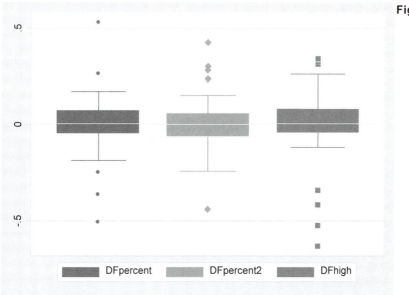

Figure 7.9

From left to right, Figure 7.9 shows the distributions of *DFBETAs* for *percent*, *percent2*, and *high*. (We could more easily distinguish them in color.) The extreme values in each plot belong to Iowa and Utah, which also have the two highest Cook's *D* values. For example, Utah's *DFhigh* = –.63. This tells us that Utah causes the coefficient on *high* to be .63 standard errors lower than it would be if Utah were set aside. Similarly, *DFpercent* = .53 indicates that with Utah present, the coefficient on *percent* is .53 standard errors higher (because the *percent* regression coefficient is negative, "higher" means closer to 0) than it otherwise would be. Thus, Utah weakens the apparent effects of both *high* and *percent*.

The most direct way to learn how particular observations affect a regression is to repeat the regression with those observations set aside. For example, we could set aside all states that move any coefficient by half a standard error (that is, have absolute *DFBETAs* of .5 or more):

```
. regress csat percent percent2 high if abs(DFpercent) < .5 &
     abs(DFpercent2) < .5 & abs(DFhigh) < .5
```

Source	SS	df	MS		Number of obs =	48
					F(3, 44) =	215.47
Model	175366.782	3	58455.5939		Prob > F =	0.0000
Residual	11937.1351	44	271.298525		R-squared =	0.9363
					Adj R-squared =	0.9319
Total	187303.917	47	3985.18972		Root MSE =	16.471

csat	Coef.	Std. Err.	t	P>|t|	[95% Conf.	Interval]
percent	-6.510868	.4700719	-13.85	0.000	-7.458235	-5.5635
percent2	.0538131	.005779	9.31	0.000	.0421664	.0654599
high	3.35664	.4577103	7.33	0.000	2.434186	4.279095
_cons	815.0279	33.93199	24.02	0.000	746.6424	883.4133

Careful inspection will reveal the details in which this regression table (based on $n = 48$) differs from its $n = 51$ or $n = 50$ counterparts seen earlier. Our central conclusion — that mean state SAT scores are well predicted by the percent of adults with high school diplomas and, curvilinearly, by the percent of students taking the test — remains unchanged, however.

Although diagnostic statistics draw attention to influential observations, they do not answer the question of whether we should set those observations aside. That requires a substantive decision based on careful evaluation of the data and research context. In this example, we have no substantive reason to discard any states, and even the most influential of them do not fundamentally change our conclusions.

Using any fixed definition of what constitutes an "outlier," we are liable to see more of them in larger samples. For this reason, sample-size-adjusted cutoffs are sometimes recommended for identifying unusual observations. After fitting a regression model with K coefficients (including the constant) based on n observations, we might look more closely at those observations for which any of the following are true:

> leverage $h > 2K/n$
>
> Cook's $D > 4/n$
>
> $DFITS > 2\sqrt{K/n}$
>
> Welsch's $W > 3\sqrt{K}$
>
> $DFBETA > 2/\sqrt{n}$
>
> $|COVRATIO - 1| \geq 3K/n$

The reasoning behind these cutoffs, and the diagnostic statistics more generally, can be found in Cook and Weisberg (1982, 1994); Belsley, Kuh, and Welsch (1980); or Fox (1991).

Multicollinearity

If perfect multicollinearity (linear relationship) exists among the predictors, regression equations become unsolvable. Stata handles this by warning the user and then automatically dropping one of the offending predictors. High but not perfect multicollinearity causes more subtle problems. When we add a new x variable that is strongly related to x variables already in the model, symptoms of possible trouble include the following:

1. Substantially higher standard errors, with correspondingly lower t statistics.

2. Unexpected changes in coefficient magnitudes or signs.

3. Nonsignificant coefficients despite a high R^2.

Multiple regression attempts to estimate the independent effects of each x variable. There is little information for doing so, however, if one or more of the x variables does not have much independent variation. The symptoms listed above warn that coefficient estimates have become unreliable, and might shift drastically with small changes in the sample or model. Further troubleshooting is needed to determine whether multicollinearity really is at fault and, if so, what should be done about it.

Multicollinearity cannot necessarily be detected, or ruled out, by examining a matrix of correlations between variables. A better assessment comes from regressing each x on all of the other x variables. Then we calculate $1 - R^2$ from this regression to see what fraction of the first *x* variable's variance is independent of the other *x* variables. For example, about 97% of *high*'s variance is independent of *percent* and *percent2*:

```
. quietly regress high percent percent2
. display 1 - e(r2)
.96942331
```

After regression, e(r2) holds the value of R^2. Similar commands reveal that only 4% of *percent*'s variance is independent of the other two predictor variables:

```
. quietly regress percent high percent2
. display 1 - e(r2)
.04010307
```

This finding about *percent* and *percent2* is not surprising. In polynomial regression or regression with interaction terms, some *x* variables are calculated directly from other *x* variables. Although strictly speaking their relationship is nonlinear, it often is close enough to linear to raise problems of multicollinearity.

The post-regression command **vif**, for **v**ariance **i**nflation **f**actor, performs similar calculations automatically. This gives a quick and straightforward check for multicollinearity.

```
. quietly regress csat percent percent2 high

. vif
```

Variable	VIF	1/VIF
percent	24.94	0.040103
percent2	24.78	0.040354
high	1.03	0.969423
Mean VIF	16.92	

The 1/VIF column at right in a **vif** table gives values equal to $1 - R^2$ from the regression of each *x* on the other *x* variables, as can be seen by comparing the values for *high* (.969423) or *percent* (.040103) with our earlier **display** calculations. That is, 1/VIF (or $1 - R^2$) tells us what proportion of an *x* variable's variance is independent of all the other *x* variables. A low proportion, such as the .04 (4% independent variation) of *percent* and *percent2*, indicates potential trouble. Some analysts set a minimum level, called *tolerance*, for the 1/VIF value, and automatically exclude predictors that fall below their tolerance criterion.

The VIF column at center in a **vif** table reflects the degree to which other coefficients' variances (and standard errors) are increased due to the inclusion of that predictor. We see that *high* has virtually no impact on other variances, but *percent* and *percent2* affect the variances substantially. VIF values provide guidance but not direct measurements of the increase in coefficient variances. The following commands show the impact directly by displaying standard error estimates for the coefficient on *percent*, when *percent2* is and is not included in the model.

```
. quietly regress csat percent percent2 high
. display _se[percent]
.50958046
. quietly regress csat percent high
. display _se[percent]
.16162193
```

With *percent2* included in the model, the standard error for *percent* is three times higher:

.50958046 /.16162193 = 3.1529166

This corresponds to a tenfold increase in the coefficient's variance.

How much variance inflation is too much? Chatterjee, Hadi, and Price (2000) suggest the following as guidelines for the presence of multicollinearity:

1. The largest VIF is greater than 10; or
2. the mean VIF is larger than 1.

With our largest VIFs close to 25, and the mean almost 17, the *csat* regression clearly meets both criteria. How troublesome the problem is, and what, if anything, should be done about it, are the next questions to consider.

Because *percent* and *percent2* are closely related, we cannot estimate their separate effects with nearly as much precision as we could the effect of either predictor alone. That is why the standard error for the coefficient on *percent* increases threefold when we compare the regression of *csat* on *percent* and *high* to a polynomial regression of *csat* on *percent*, *percent2*, and *high*. Despite this loss of precision, however, we can still distinguish all the coefficients from zero. Moreover, the polynomial regression obtains a better prediction model. For these reasons, the multicollinearity in this regression does not necessarily pose a great problem, or require a solution. We could simply live with it as one feature of an otherwise acceptable model.

When solutions are needed, a simple trick called "centering" often succeeds in reducing multicollinearity in polynomial or interaction-effect models. Centering involves subtracting the mean from *x* variable values before generating polynomial or product terms. Subtracting the mean creates a new variable centered on zero and much less correlated with its own squared values. The resulting regression fits the same as an uncentered version. By reducing multicollinearity, centering often (*but not always*) yields more precise coefficient estimates with lower standard errors. The commands below generate a centered version of *percent* named *Cpercent*, and then obtain squared values of *Cpercent* named *Cpercent2*.

```
. summarize percent

    Variable |       Obs        Mean    Std. Dev.       Min        Max
-------------+--------------------------------------------------------
     percent |        51    35.76471    26.19281          4         81

. generate Cpercent = percent - r(mean)
. generate Cpercent2 = Cpercent ^2
. correlate Cpercent Cpercent2 percent percent2 high csat
(obs=51)

             | Cpercent Cperce~2  percent percent2     high     csat
-------------+------------------------------------------------------
   Cpercent |   1.0000
  Cpercent2 |   0.3791   1.0000
    percent |   1.0000   0.3791   1.0000
   percent2 |   0.9794   0.5582   0.9794   1.0000
       high |   0.1413  -0.0417   0.1413   0.1176   1.0000
       csat |  -0.8758  -0.0428  -0.8758  -0.7946   0.0858   1.0000
```

Whereas *percent* and *percent2* have a near-perfect correlation with each other ($r = .9794$), the centered versions *Cpercent* and *Cpercent2* are just moderately correlated ($r = .3791$). Otherwise, correlations involving *percent* and *Cpercent* are identical because centering is a linear transformation. Correlations involving *Cpercent2* are different from those with *percent2*, however. Figure 7.10 shows scatterplots that help to visualize these correlations, and the transformation's effects.

```
. graph matrix Cpercent Cpercent2 percent percent2 high csat,
       half msymbol(+)
```

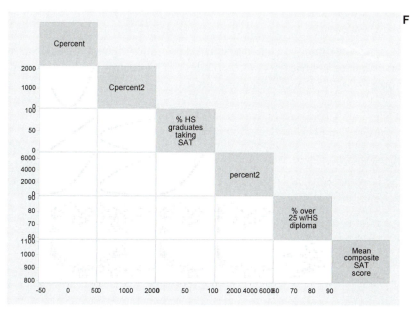

Figure 7.10

The R^2, overall F test, predictions, and many other aspects of a model should be unchanged after centering. Differences will be most noticeable in the centered variable's coefficient and standard error.

```
. regress csat Cpercent Cpercent2 high
```

Source	SS	df	MS		Number of obs	=	51
					F(3, 47)	=	193.37
Model	207225.103	3	69075.0343		Prob > F	=	0.0000
Residual	16789.407	47	357.221426		R-squared	=	0.9251
					Adj R-squared	=	0.9203
Total	224014.51	50	4480.2902		Root MSE	=	18.90

csat	Coef.	Std. Err.	t	P>|t|	[95% Conf. Interval]
Cpercent	-2.682362	.1119085	-23.97	0.000	-2.907493 -2.457231
Cpercent2	.0536555	.0063678	8.43	0.000	.0408452 .0664659
high	2.986509	.4857502	6.15	0.000	2.009305 3.963712
_cons	680.2552	37.82329	17.99	0.000	604.1646 756.3458

In this example, the standard error of the coefficient on *Cpercent* is actually lower (.1119085 compared with .16162193) when *Cpercent2* is included in the model. The *t* statistic is correspondingly larger. Thus, it appears that centering did improve that coefficient estimate's precision. The VIF table now gives less cause for concern: each of the three predictors has more than 80% independent variation, compared with 4% for *percent* and *percent2* in the uncentered regression.

```
. vif

    Variable |       VIF       1/VIF
-------------+----------------------
    Cpercent |      1.20     0.831528
   Cpercent2 |      1.18     0.846991
        high |      1.03     0.969423
-------------+----------------------
    Mean VIF |      1.14
```

Another diagnostic table sometimes consulted to check for multicollinearity is the matrix of correlations between estimated coefficients (*not* variables). This matrix can be displayed after **regress**, **anova**, or other model-fitting procedures by typing

```
. correlate, _coef

             | Cpercent Cperce~2     high    _cons
-------------+------------------------------------
    Cpercent |   1.0000
   Cpercent2 |  -0.3893   1.0000
        high |  -0.1700   0.1040   1.0000
       _cons |   0.2105  -0.2151  -0.9912   1.0000
```

High correlations between pairs of coefficients indicate possible collinearity problems.

By adding the option **covariance**, we can see the coefficients' variance–covariance matrix, from which standard errors are derived.

```
. correlate, _coef covariance

             | Cpercent Cperce~2     high    _cons
-------------+------------------------------------
    Cpercent |  .012524
   Cpercent2 | -.000277  .000041
        high | -.009239  .000322  .235953
       _cons |  .891126 -.051817 -18.2105   1430.6
```

8

Fitting Curves

Basic regression and correlation methods assume linear relationships. Linear models provide reasonable and simple approximations for many real phenomena, over a limited range of values. But analysts also encounter phenomena where linear approximations are too simple; these call for nonlinear alternatives. This chapter describes three broad approaches to modeling nonlinear or curvilinear relationships:

1. Nonparametric methods, including band regression and lowess smoothing.
2. Linear regression with transformed variables ("curvilinear regression"), including Box–Cox methods.
3. Nonlinear regression.

Nonparametric regression serves as an exploratory tool because it can summarize data patterns visually without requiring the analyst to specify a particular model in advance. Transformed variables extend the usefulness of linear parametric methods, such as OLS regression (**regress**), to encompass curvilinear relationships as well. Nonlinear regression, on the other hand, requires a different class of methods that can estimate parameters of intrinsically nonlinear models.

The following menu groups cover many of the operations discussed in this chapter. The final topic, nonlinear regression, requires a command-based approach.

Graphics – Twoway

Statistics – Nonparametric analysis – Lowess smoothing

Data – Create or change variables – Create new variable

Statistics – Linear regression and related

Example Commands

. **boxcox y x1 x2 x3, model(lhs)**

Finds maximum-likelihood estimates of the parameter λ (lambda) for a Box–Cox transformation of y, assuming that $y^{(\lambda)}$ is a linear function of $x1$, $x2$, and $x3$ plus Gaussian constant-variance errors. The **model(lhs)** option restricts transformation to the left-hand-side variable y. Other options could transform right-hand-side (x) variables by the same or different parameters, and control further details of the model. Type **help boxcox** for the syntax and a complete list of options. The *Base Reference Manual* gives technical details.

. `graph twoway mband y x, bands(10)  ||  scatter y x`

Produces a y versus x scatterplot with line segments connecting the cross-medians (median x, median y points) within 10 equal-width vertical bands. This is one form of "band regression." Typing **mspline** in place of **mband** in this command would result in the cross-medians being connected by a smooth cubic spline curve instead of by line segments.

. `graph twoway lowess y x, bwidth(.4)  ||  scatter y x`

Draws a lowess-smoothed curve with a scatterplot of y versus x. Lowess calculations use a bandwidth of .4 (40% of the data). In order to calculate and keep the smoothed values as a new variable, use the related command **lowess**.

. `lowess y x, bwidth(.3) gen(newvar)`

Draws a lowess-smoothed curve on a scatterplot of y versus x, using a bandwidth of .3 (30% of the data). Predicted values for this curve are saved as a variable named *newvar*. The **lowess** command offers more options than **graph twoway lowess**, including fitting methods and the ability to save predicted values. See **help lowess** for details.

. `nl exp2 y x`

Uses iterative nonlinear least squares to fit a 2-parameter exponential growth model,

$$\text{predicted } y = b_1 b_2{}^x$$

The term **exp2** refers to a separate program that specifies the model itself. You can write a program to define your own model, or use one of the common models (including exponential, logistic, and Gompertz) supplied with Stata. After **nl**, use **predict** to generate predicted values or residuals.

. `nl log4 y x, init(B0=5, B1=25, B2=.1, B3=50)`

Fits a 4-parameter logistic growth model (**log4**) of the form

$$\text{predicted } y = b_0 + b_1 /(1 + \exp(-b_2 (x - b_3)))$$

Sets initial parameter values for the iterative estimation process at $b_0 = 5, b_1 = 25, b_2 = .1$, and $b_3 = 50$.

. `regress lny x1 sqrtx2 invx3`

Performs curvilinear regression using the variables *lny*, *x1*, *sqrtx2*, and *invx3*. These variables were previously generated by nonlinear transformations of the raw variables y, *x2*, and *x3* through commands such as the following:

. `generate lny = ln(y)`

. `generate sqrtx2 = sqrt(x2)`

. `generate invx3 = 1/x3`

When, as in this example, the y variable was transformed, the predicted values generated by **predict yhat**, or residuals generated by **predict e, resid**, will be also in transformed units. For graphing or other purposes, we might want to return predicted values or residuals to raw-data units, using inverse transformations such as

. `replace yhat = exp(yhat)`

Band Regression

Nonparametric regression methods generally do not yield an explicit regression equation. They are primarily graphic tools for displaying the relationship, possibly nonlinear, between *y* and *x*. Stata can draw a simple kind of nonparametric regression, band regression, onto any scatterplot or scatterplot matrix. For illustration, consider these sobering Cold War data (*missile.dta*) from MacKenzie (1990). The observations are 48 types of long-range nuclear missiles, deployed by the U.S. and Soviet Union during their arms race, 1958 to 1990:

```
Contains data from C:\data\missile.dta
  obs:            48                          Missiles (MacKenzie 1990)
  vars:            6                          16 Jul 2005 14:57
  size:        1,392 (99.9% of memory free)
-------------------------------------------------------------------------
              storage  display    value
variable name   type   format     label    variable label
-------------------------------------------------------------------------
missile         str15  %15s                 Missile
country         byte   %8.0g      soviet    US or Soviet missile?
year            int    %8.0g                Year of first deployment
type            byte   %8.0g      type      ICBM or submarine-launched?
range           int    %8.0g                Range in nautical miles
CEP             float  %9.0g                Circular Error Probable (miles)
-------------------------------------------------------------------------
Sorted by:  country  year
```

Variables in *missile.dta* include an accuracy measure called the "Circular Error Probable" (*CEP*). *CEP* represents the radius of a bulls eye within which 50% of the missile's warheads should land. Year by year, scientists on both sides worked to improve accuracy (Figure 8.1).

```
. graph twoway mband CEP year, bands(8)
      ||   scatter CEP year
      ||   , ytitle("Circular Error Probable, miles") legend(off)
```

Figure 8.1

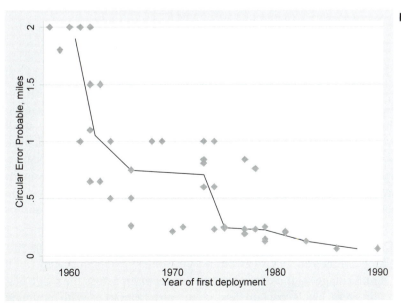

Figure 8.1 shows *CEP* declining (accuracy increasing) over time. The option **bands(8)** instructs **graph twoway mband** to divide the scatterplot into 8 equal-width vertical bands and draw line segments connecting the points (median *x*, median *y*) within each band. This curve traces how the median of *CEP* changes with *year*.

Nonparametric regression does not require the analyst to specify a relationship's functional form in advance. Instead, it allows us to explore the data with an "open mind." This process often uncovers interesting results, such as when we view trends in U.S. and Soviet missile accuracy separately (Figure 8.2). The **by(country)** option in the following command produces separate plots for each country, each with overlaid band-regression curve and scatterplot. Within the **by( )** option are suboptions controlling the legend and note.

```
. graph twoway mband CEP year, bands(8)
       ||   scatter CEP year
       ||   , ytitle("Circular Error Probable, miles")
          by(country, legend(off) note(""))
```

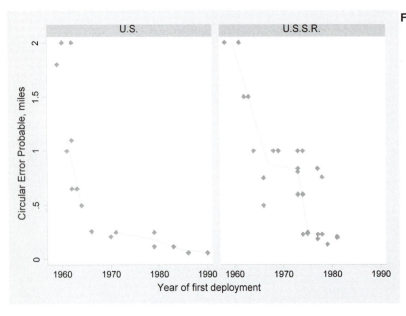

Figure 8.2

The shapes of the two curves in Figure 8.2 differ substantially. U.S. missiles became much more accurate in the 1960s, permitting a shift to smaller warheads. Three or more small warheads would fit on the same size missile that formerly carried one large warhead. The accuracy of Soviet missiles improved more slowly, apparently stalling during the late 1960s to early 1970s, and remained a decade or so behind their American counterparts. To make up for this accuracy disadvantage, Soviet strategy emphasized larger rockets carrying high-yield warheads. Nonparametric regression can assist with a qualitative description of this sort or serve as a preliminary to fitting parametric models such as those described later.

We can add band regression curves to any scatterplot by overlaying an **mband** (or **mspline**) plot. Band regression's simplicity makes it a convenient exploratory tool, but it possesses one notable disadvantage — the bands have the same width across the range of *x* values, although some of these bands contain few or no observations. With normally distributed variables, for example, data density decreases toward the extremes. Consequently, the left and right endpoints of the band regression curve (which tend to dominate its appearance) often reflect just a few data points. The next section describes a more sophisticated, computation-intensive approach.

Lowess Smoothing

The **lowess** and **graph twoway lowess** commands accomplish a form of nonparametric regression called lowess smoothing (for **lo**cally **we**ighted scatterplot smoothing). In general the **lowess** command is more specialized and more powerful, with options that control details of the fitting process. **graph twoway lowess** has advantages of simplicity, and follows the familiar syntax of the **graph twoway** family. The following example uses **graph twoway lowess** to plot *CEP* against *year* for U.S. missiles only (*country* == 0).

```
. graph twoway lowess CEP year if country == 0, bwidth(.4)
       ||   scatter CEP year
       ||   , legend(off) ytitle("Circular Error Probable, miles")
```

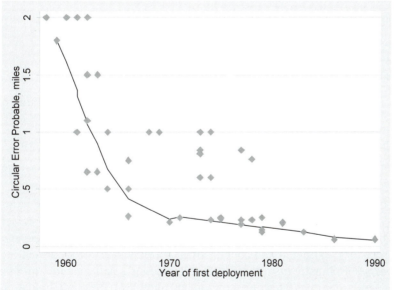

Figure 8.3

A graph very similar to Figure 8.2 would result if we had typed instead

```
. lowess CEP year if country == 0, bwidth(.4)
```

Like Figure 8.2, Figure 8.3 (next page) shows U.S. missile accuracy improving rapidly during the 1960s and progressing at a more gradual rate in the 1970s and 1980s. Lowess-smoothed values of *CEP* are generated here with the name *lsCEP*. The **bwidth(.4)** option specifies the lowess bandwidth: the fraction of the sample used in smoothing each point. The default is **bwidth(.8)**. The closer bandwidth is to 1, the greater the degree of smoothing.

Lowess predicted (smoothed) *y* values for *n* observations result from *n* weighted regressions. Let *k* represent the half-bandwidth, truncated to an integer. For each y_i, a smoothed value y_i^s is obtained by weighted regression involving only those observations within the interval from $i = \max(1, i - k)$ through $i = \min(i + k, n)$. The jth observation within this interval receives weight w_j according to a tricube function:

$$w_j = (1 - |u_j|^3)^3$$

where

$$u_j = (x_i - x_j) / \Delta$$

Δ stands for the distance between x_i and its furthest neighbor within the interval. Weights equal 1 for $x_i = x_j$, but fall off to zero at the interval's boundaries. See Chambers et al. (1983) or Cleveland (1993) for more discussion and examples of lowess methods.

   lowess options include the following.

mean   For running-mean smoothing. The default is running-line least squares smoothing.

noweight   Unweighted smoothing. The default is Cleveland's tricube weighting function.

bwidth( )   Specifies the bandwidth. Centered subsets of approximately bwidth $\times n$ observations are used for smoothing, except towards the endpoints where smaller, uncentered bands are used. The default is **bwidth(.8)**.

logit   Transforms smoothed values to logits.

adjust   Adjusts the mean of smoothed values to equal the mean of the original y variable; like **logit**, **adjust** is useful with dichotomous y.

gen(*newvar*)   Creates *newvar* containing smoothed values of y.

nograph   Suppresses displaying the graph.

plot( )   Provides a way to add other plots to the generated graph; see **help** **plot_option**.

rlopts()   Affects the rendition of the reference line; see **help cline_options**.

Because it requires n weighted regressions, lowess smoothing proceeds slowly with large samples.

   In addition to smoothing scatterplots, **lowess** can be used for exploratory time series smoothing. The file *ice.dta* contains results from the Greenland Ice Sheet 2 (GISP2) project described in Mayewski, Holdsworth, and colleagues (1993) and Mayewski, Meeker, and colleagues (1993). Researchers extracted and chemically analyzed an ice core representing more than 100,000 years of climate history. *ice.dta* includes a small fraction of this information: measured non-sea salt sulfate concentration and an index of "Polar Circulation Intensity" since AD 1500.

```
Contains data from C:\data\ice.dta
  obs:            271                      Greenland ice (Mayewski 1995)
  vars:             3                      14 Jul 2005 14:57
  size:         5,962 (99.9% of memory free)
-----------------------------------------------------------------------------
              storage  display    value
variable name   type   format     label     variable label
-----------------------------------------------------------------------------
year            int    %ty                  Year
sulfate         double %10.0g               SO4 ion concentration, ppb
PCI             double %6.0g                Polar Circulation Intensity
-----------------------------------------------------------------------------
Sorted by:  year
```

To retain more detail from this 271-point time series, we smooth with a relatively narrow bandwidth, only 5% of the sample. Figure 8.4 graphs the results. The smoothed curve has been drawn with "thick" width, to visually distinguish it from the raw data. (Type **help linewidthstyle** for other choices of line width.)

```
. graph twoway lowess sulfate year, bwidth(.05) clwidth(thick)
      ||   line sulfate year, clpattern(solid)
      ||   , ytitle("SO4 ion concentration, ppb")
      legend(label(1 "lowess smoothed") label(2 "raw data"))
```

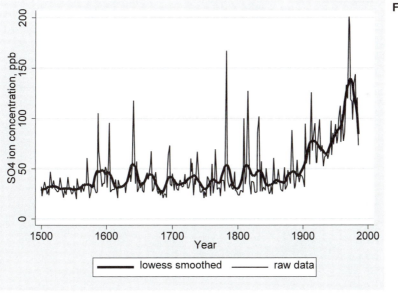

Figure 8.4

Non-sea salt sulfate (SO_4) reached the Greenland ice after being injected into the atmosphere, chiefly by volcanoes or the burning of fossil fuels such as coal and oil. Both the smoothed and raw curves in Figure 8.4 convey information. The smoothed curve shows oscillations around a slightly rising mean from 1500 through the early 1800s. After 1900, fossil fuels drive the smoothed curve upward, with temporary setbacks after 1929 (the Great Depression) and the early 1970s (combined effects of the U.S. Clean Air Act, 1970; the Arab oil embargo, 1973; and subsequent oil price hikes). Most of the sharp peaks of the raw data have been identified with known volcanic eruptions such as Iceland's Hekla (1970) or Alaska's Katmai (1912).

After smoothing time series data, it is often useful to study the smooth and rough (residual) series separately. The following commands create two new variables: lowess-smoothed values of sulfate (*smooth*) and the residuals or rough values (*rough*) calculated by subtracting the smoothed values from the raw data.

```
. lowess sulfate year, bwidth(.05) gen(smooth)
. label variable smooth "SO4 ion concentration (smoothed)"
. gen rough = sulfate - smooth
. label variable rough "SO4 ion concentration (rough)"
```

Figure 8.5 compares the *smooth* and *rough* time series in a pair of graphs annotated using the `text( )` option, then combined.

```
. graph twoway line smooth year, ylabel(0(50)150)  xtitle("")
     ytitle("Smoothed") text(20 1540 "Renaissance")
     text(20 1900 "Industrialization")
     text(90 1860 "Great Depression 1929")
     text(150 1935 "Oil Embargo 1973") saving(fig08_05a, replace)
. graph twoway line rough year, ylabel(0(50)150) xtitle("")
     ytitle("Rough") text(75 1630 "Awu 1640", orientation(vertical))
      text(120 1770 "Laki 1783", orientation(vertical))
     text(90 1805 "Tambora 1815", orientation(vertical))
     text(65 1902 "Katmai 1912", orientation(vertical))
     text(80 1960 "Hekla 1970", orientation(vertical))
     yline(0) saving(fig08_05b, replace)
. graph combine fig08_05a.gph fig08_05b.gph, rows(2)
```

Figure 8.5

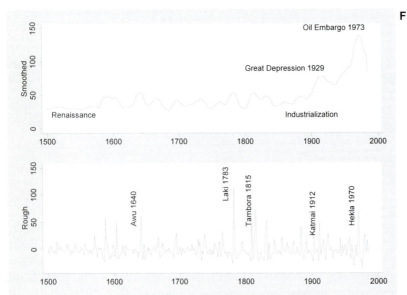

Regression with Transformed Variables — 1

By subjecting one or more variables to nonlinear transformation, and then including the transformed variable(s) in a linear regression, we implicitly fit a curvilinear model to the underlying data. Chapters 6 and 7 gave one example of this approach, polynomial regression, which incorporates second (and perhaps higher) powers of at least one *x* variable among the predictors. Logarithms also are used routinely in many fields. Other common transformations include those of the ladder of powers and Box–Cox transformations, introduced in Chapter 4.

Dataset *tornado.dta* provides a simple illustration involving U.S. tornados from 1916 to 1986 (from the Council on Environmental Quality, 1988).

```
Contains data from C:\data\tornado.dta
  obs:            71                   U.S. tornados 1916-1986
                                       (Council on Env. Quality 1988)
  vars:            4                   16 Jul 2005 14:57
  size:          994 (99.9% of memory free)
-------------------------------------------------------------------------
              storage  display   value
variable name   type   format    label      variable label
-------------------------------------------------------------------------
year            int    %8.0g                 Year
tornado         int    %8.0g                 Number of tornados
lives           int    %8.0g                 Number of lives lost
avlost          float  %9.0g                 Average lives lost/tornado
-------------------------------------------------------------------------
Sorted by:  year
```

The number of fatalities decreased over this period, while the number of recognized tornados increased, because of improvements in warnings and our ability to detect more tornados, even those that do little damage. Consequently, the average lives lost per tornado (*avlost*) declined with time, but a linear regression (Figure 8.6, following page) does not well describe this trend. The scatter descends more steeply than the regression line at first, then levels off in the mid-1950s. The regression line actually predicts negative numbers of deaths in later years. Furthermore, average tornado deaths exhibit more variation in early years than later — evidence of heteroskedasticity.

```
. graph twoway scatter avlost year
        || lfit avlost year, clpattern(solid)
        || , ytitle("Average number of lives lost") xlabel(1920(10)1990)
        xtitle("") legend(off) ylabel(0(1)7) yline(0)
```

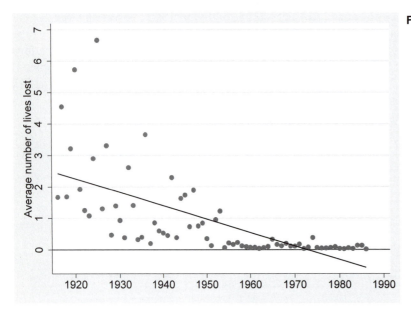

Figure 8.6

The relationship becomes linear, and heteroskedasticity vanishes if we work instead with logarithms of the average number of lives lost (Figure 8.7):

```
. generate loglost = ln(avlost)
```

```
. label variable loglost "ln(avlost)"
```

```
. regress loglost year
```

```
      Source |       SS       df       MS              Number of obs =      71
-------------+------------------------------           F(  1,    69) =  182.24
       Model |  115.895325        1  115.895325        Prob > F      =  0.0000
    Residual |  43.8807356       69   .63595269        R-squared     =  0.7254
-------------+------------------------------           Adj R-squared =  0.7214
       Total |   159.77606       70  2.28251515        Root MSE      =  .79747

-------------------------------------------------------------------------------
     loglost |      Coef.   Std. Err.      t    P>|t|     [95% Conf. Interval]
-------------+-----------------------------------------------------------------
        year |  -.0623418    .004618   -13.50   0.000    -.0715545   -.053129
       _cons |   120.5645   9.010312    13.38   0.000     102.5894   138.5395
-------------------------------------------------------------------------------
```

```
. predict yhat2
(option xb assumed; fitted values)
```

```
. label variable yhat2 "ln(avlost) = 120.56 - .06year"
```

```
. label variable loglost "ln(avlost)"
```

```
. graph twoway scatter loglost year
      ||   mspline yhat2 year, clpattern(solid) bands(50)
      ||   , ytitle("Natural log(average lives lost)")
      xlabel(1920(10)1990) xtitle("") legend(off) ylabel(-4(1)2)
      yline(0)
```

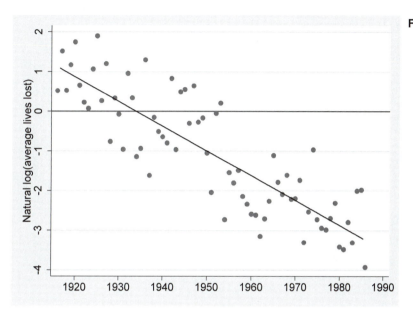

Figure 8.7

The regression model is approximately

predicted ln(*avlost*) = 120.56 − .06*year*

Because we regressed logarithms of lives lost on *year*, the model's predicted values are also measured in logarithmic units. Return these predicted values to their natural units (lives lost) by inverse transformation, in this case exponentiating (*e* to power) *yhat2*:

```
. replace yhat2 = exp(yhat2)
(71 real changes made)
```

Graphing these inverse-transformed predicted values reveals the curvilinear regression model, which we obtained by linear regression with a transformed *y* variable (Figure 8.8). Contrast Figures 8.7 and 8.8 with Figure 8.6 to see how transformation made the analysis both simpler and more realistic.

```
. graph twoway scatter avlost year
      ||  mspline yhat2 year, clpattern(solid) bands(50)
      ||  , ytitle("Average number of lives lost") xlabel(1920(10)1990)
      xtitle("") legend(off) ylabel(0(1)7) yline(0)
```

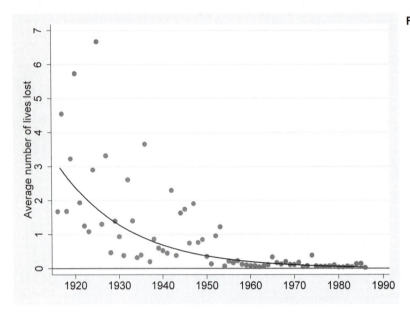

Figure 8.8

The **boxcox** command employs maximum-likelihood methods to fit curvilinear models involving Box–Cox transformations (introduced in Chapter 4). Fitting a model with Box–Cox transformation of the dependent variable (**model(lhs)** specifies left-hand side) to the tornado data, we obtain results quite similar to the model of Figures 8.7 and 8.8. The **nolog** option in the following command does not affect the model, but suppresses display of log likelihood after each iteration of the fitting process.

```
. boxcox avlost year, model(lhs) nolog
```

```
                                      Number of obs   =          71
                                      LR chi2(1)      =       92.28
Log likelihood = -7.7185533           Prob > chi2     =       0.000


------------------------------------------------------------------------------
     avlost |      Coef.   Std. Err.      z    P>|z|     [95% Conf. Interval]
------------+-----------------------------------------------------------------
     /theta |  -.0560959   .0646726    -0.87   0.386    -.1828519      .07066
------------------------------------------------------------------------------

Estimates of scale-variant parameters
----------------------------
            |        Coef.
------------+---------------
Notrans     |
       year |   -.0661891
      _cons |    127.9713
------------+---------------
     /sigma |    .8301177
----------------------------
```

```
------------------------------------------------------------------
    Test            Restricted      LR statistic       P-Value
    H0:            log likelihood      chi2          Prob > chi2
------------------------------------------------------------------
theta = -1         -84.928791         154.42            0.000
theta =  0         -8.0941678           0.75            0.386
theta =  1        -101.50385          187.57            0.000
------------------------------------------------------------------
```

The **boxcox** output shows theta = –.056 as the optimal Box–Cox parameter for transforming *avlost*, in order to linearize its relationship with *year*. Therefore, the left-hand-side transformation is

$$avlost^{(-.056)} = (avlost^{-.056} - 1)/-.056$$

Box–Cox transformation by a parameter close to zero, such as –.056, produces results similar to the natural-logarithm transformation we applied earlier to this variable "by hand." It is therefore not surprising that the **boxcox** regression model

$$\text{predicted } avlost^{(-.056)} = 127.97 - .07year$$

resembles the earlier model (predicted $\ln(avlost) = 120.56 - .06year$) drawn in Figures 8.7 and 8.8. The **boxcox** procedure assumes normal, independent, and identically distributed errors. It does not select transformations with the aim of normalizing residuals, however.

 boxcox can fit several types of models, including multiple regressions in which some or all of the right-hand-side variables are transformed by a parameter different from the *y*-variable transformation. It cannot apply different transformations to each separate right-hand-side predictor. To do that, we return to a "by hand" curvilinear-regression approach, as illustrated in the next section.

Regression with Transformed Variables — 2

For a multiple-regression example, we will use data on living conditions in 109 countries found in dataset *nations.dta* (from World Bank 1987; World Resources Institute 1993).

```
Contains data from C:\data\nations.dta
  obs:           109                        Data on 109 nations, ca. 1985
  vars:           15                        16 Jul 2005 14:57
  size:         4,033 (99.9% of memory free)
------------------------------------------------------------------------
              storage  display    value
variable name  type    format     label      variable label
------------------------------------------------------------------------
country        str8    %9s                    Country
pop            float   %9.0g                  1985 population in millions
birth          byte    %8.0g                  Crude birth rate/1000 people
death          byte    %8.0g                  Crude death rate/1000 people
chldmort       byte    %8.0g                  Child (1-4 yr) mortality 1985
infmort        int     %8.0g                  Infant (<1 yr) mortality 1985
life           byte    %8.0g                  Life expectancy at birth 1985
food           int     %8.0g                  Per capita daily calories 1985
energy         int     %8.0g                  Per cap energy consumed, kg oil
gnpcap         int     %8.0g                  Per capita GNP 1985
gnpgro         float   %9.0g                  Annual GNP growth % 65-85
urban          byte    %8.0g                  % population urban 1985
```

```
school1          int     %8.0g              Primary enrollment % age-group
school2          byte    %8.0g              Secondary enroll % age-group
school3          byte    %8.0g              Higher ed. enroll % age-group
-------------------------------------------------------------------------------
```

Relationships among birth rate, per capita gross national product (GNP), and child mortality are not linear, as can be seen clearly in the scatterplot matrix of Figure 8.9. The skewed *gnpcap* and *chldmort* distributions also present potential leverage and influence problems.

```
. graph matrix gnpcap chldmort birth, half
```

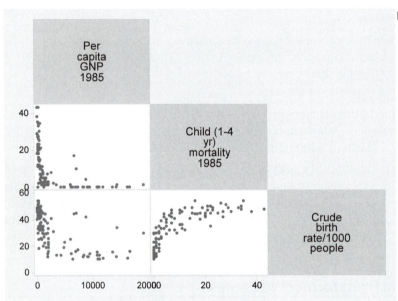

Figure 8.9

Experimenting with ladder-of-powers transformations reveals that the log of *gnpcap* and the square root of *chldmort* have distributions more symmetrical, with fewer outliers or potential leverage points, than the raw variables. More importantly, these transformations largely eliminate the nonlinearities: compare the raw-data scatterplots in Figure 8.9 with their transformed-variables counterparts in Figure 8.10, on the following page,

```
. generate loggnp = log10(gnpcap)
. label variable loggnp "Log-10 of per cap GNP"
. generate srmort = sqrt(chldmort)
. label variable srmort "Square root child mortality"
. graph matrix loggnp srmort birth, half
```

Figure 8.10

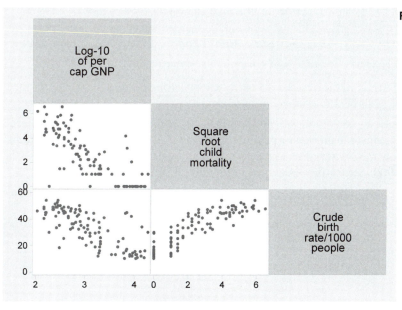

We can now apply linear regression using the transformed variables:

```
. regress birth loggnp srmort
```

Source	SS	df	MS
Model	15837.9603	2	7918.98016
Residual	4238.18646	106	39.9828911
Total	20076.1468	108	185.890248

Number of obs =	109
F(2, 106) =	198.06
Prob > F =	0.0000
R-squared =	0.7889
Adj R-squared =	0.7849
Root MSE =	6.3232

birth	Coef.	Std. Err.	t	P>|t|	[95% Conf. Interval]	
loggnp	-2.353738	1.686255	-1.40	0.166	-5.696903	.9894259
srmort	5.577359	.533567	10.45	0.000	4.51951	6.635207
_cons	26.19488	6.362687	4.12	0.000	13.58024	38.80953

Unlike the raw-data regression (not shown), this transformed-variables version finds that per capita gross national product does not significantly affect birth rate once we control for child mortality. The transformed-variables regression fits slightly better: $R^2_a = .7849$ instead of .6715. (We can compare R^2_a across models here only because both have the same untransformed *y* variable.) Leverage plots would confirm that transformations have much reduced the curvilinearity of the raw-data regression.

Conditional Effect Plots

Conditional effect plots trace the predicted values of *y* as a function of one *x* variable, with other *x* variables held constant at arbitrary values such as their means, medians, quartiles, or extremes. Such plots help with interpreting results from transformed-variables regression.

    Continuing with the previous example, we can calculate predicted birth rates as a function of *loggnp*, with *srmort* held at its mean (2.49):

```
. generate yhat1 = _b[_cons] + _b[loggnp]*loggnp + _b[srmort]*2.49
. label variable yhat1 "birth = f(gnpcap | srmort = 2.49)
```

The _b[*varname*] terms refer to the regression coefficient on *varname* from this session's most recent regression. _b[_cons] is the *y*-intercept or constant.

    For a conditional effect plot, graph *yhat1* (after inverse transformation if needed, although it is not needed here) against the untransformed *x* variable (Figure 8.11). Because conditional effect plots do not show the scatter of data, it can be useful to add reference lines such as the *x* variable's 10th and 90th percentiles, as shown in Figure 8.11.

```
. graph twoway line yhat1 gnpcap, sort xlabel(,grid) xline(230 10890)
```

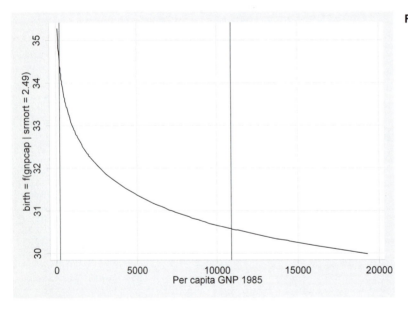

Figure 8.11

    Similarly, Figure 8.12 depicts predicted birth rates as a function of *srmort*, with *loggnp* held at its mean (3.09):

```
. generate yhat2 = _b[_cons] + _b[loggnp]*3.09 + _b[srmort]*srmort
. label variable yhat2 "birth = f(chldmort | loggnp = 3.09)"
. graph twoway line yhat2 chldmort, sort xlabel(,grid) xline(0 27)
```

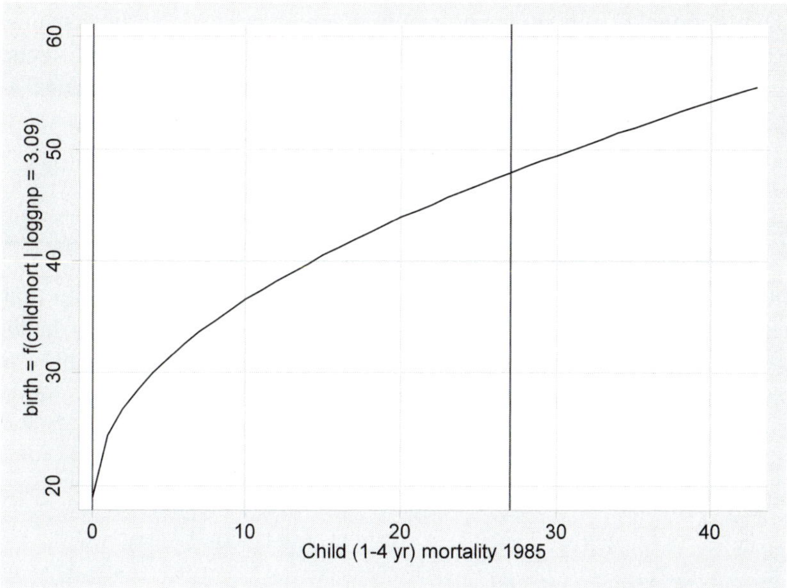

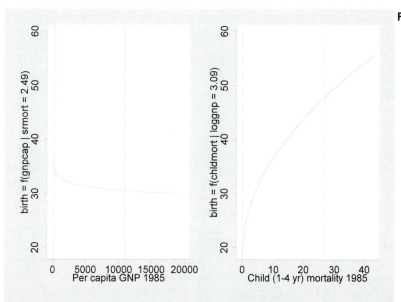

Figure 8.12

How can we compare the strength of different *x* variables' effects? Standardized regression coefficients (beta weights) are sometimes used for this purpose, but they imply a specialized definition of "strength" and can easily be misleading. A more substantively meaningful comparison might come from looking at conditional effect plots drawn with identical *y* scales. This can be accomplished easily by using **graph combine**, and specifying common *y*-axis scales, as done in Figure 8.13. The vertical distances traveled by the predicted values curve, particularly over the middle 80% of the *x* values (between 10th and 90th percentile lines), provide a visual comparison of effect magnitude.

```
. graph combine fig08_11.gph fig08_12.gph, ycommon cols(2) scale(1.25)
```

Figure 8.13

Combining several conditional effects plots into one image with common vertical scales, as done in Figure 8.13, allows quick visual comparison of the strength of different effects. Figure 8.13 makes obvious how much stronger is the effect of child mortality on birth rates — as separate plots (Figures 8.11 and 8.12) did not.

Nonlinear Regression — 1

Variable transformations allow fitting some curvilinear relationships using the familiar techniques of intrinsically linear models. Intrinsically nonlinear models, on the other hand, require a different class of fitting techniques. The **nl** command performs nonlinear regression by iterative least squares. This section introduces it using a dataset of simple examples, *nonlin.dta*:

```
Contains data from C:\data\nonlin.dta
  obs:             100                          Nonlinear model examples
                                                  (artificial data)
  vars:              5                          16 Jul 2005 14:57
  size:          2,100  (99.9% of memory free)
-------------------------------------------------------------------------------
              storage  display    value
variable name   type   format     label        variable label
-------------------------------------------------------------------------------
x              byte    %9.0g                    Independent variable
y1             float   %9.0g                    y1 = 10 * 1.03^x + e
y2             float   %9.0g                    y2 = 10 * (1 - .95^x) + e
y3             float   %9.0g                    y3 = 5 + 25/(1+exp(-.1*(x-50)))
                                                  + e
y4             float   %9.0g                    y4 = 5 +
                                                  25*exp(-exp(-.1*(x-50))) + e
-------------------------------------------------------------------------------
Sorted by:  x
```

The *nonlin.dta* data are manufactured, with *y* variables defined as various nonlinear functions of *x*, plus random Gaussian errors. *y1*, for example, represents the exponential growth process $y1 = 10 \times 1.03^x$. Estimating these parameters from the data, **nl** obtains $y1 = 11.20 \times 1.03^x$, which is reasonably close to the true model.

```
. nl exp2 y1 x

(obs = 100)

Iteration 0:  residual SS =  27625.96
Iteration 1:  residual SS =  26547.42
Iteration 2:  residual SS =  26138.3
Iteration 3:  residual SS =  26138.29
```

Source	SS	df	MS	
Model	667018.255	2	333509.128
Residual	26138.2933	98	266.717278
Total	693156.549	100	6931.56549

```
Number of obs =       100
F(  2,    98) =  1250.42
Prob > F      =   0.0000
R-squared     =   0.9623
Adj R-squared =   0.9615
Root MSE      =  16.33148
Res. dev.     = 840.3864
```

```
2-param. exp. growth curve, y1=b1*b2^x
------------------------------------------------------------------------
     y1 |      Coef.    Std. Err.        t    P>|t|    [95% Conf. Interval]
--------+---------------------------------------------------------------
     b1 |   11.20416    1.146682      9.77    0.000     8.928602    13.47971
     b2 |   1.028838    .0012404    829.41    0.000     1.026376    1.031299
------------------------------------------------------------------------
 (SE's, P values, CI's, and correlations are asymptotic approximations)
```

The **predict** command obtains predicted values and residuals for a nonlinear model estimated by **nl** . Figure 8.14 graphs predicted values from the previous example, showing the close fit ($R^2 = .96$) between model and data.

```
. predict yhat1
(option yhat assumed; fitted values)

. graph twoway scatter y1 x
      || line yhat1 x, sort
      || , legend(off) ytitle("y1 = 10 * 1.03^x + e") xtitle("x")
```

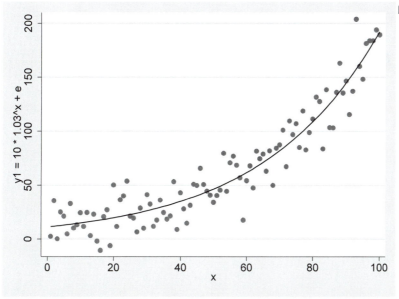

Figure 8.14

The **exp2** part of our **nl exp2 y1 x** command specified a particular exponential growth function by calling a brief program named *nlexp2.ad*o. Stata includes several such programs, defining the following functions:

exp3 3-parameter exponential: $y = b_0 + b_1 b_2^x$

exp2 2-parameter exponential: $y = b_1 b_2^x$

exp2a 2-parameter negative exponential: $y = b_1 (1 - b_2^x)$

log4 4-parameter logistic; b_0 starting level and ($b_0 + b_1$) asymptotic upper limit:
$$y = b_0 + b_1 /(1 + \exp(-b_2 (x - b_3)))$$

log3 3-parameter logistic; 0 starting level and b_1 asymptotic upper limit:
$$y = b_1 /(1 + \exp(-b_2 (x - b_3)))$$

      gom4   4-parameter Gompertz; b_0 starting level and $(b_0 + b_1)$ asymptotic upper limit:
$$y = b_0 + b_1 \exp(-\exp(-b_2 (x - b_3)))$$

      gom3   3-parameter Gompertz; 0 starting level and b_1 asymptotic upper limit:
$$y = b_1 \exp(-\exp(-b_2 (x - b_3)))$$

nonlin.dta contains examples corresponding to **exp2** (*y1*), **exp2a** (*y2*), **log4** (*y3*), and **gom4** (*y4*) functions. Figure 8.15 shows curves fit by **nl** to *y2*, *y3*, and *y4*.

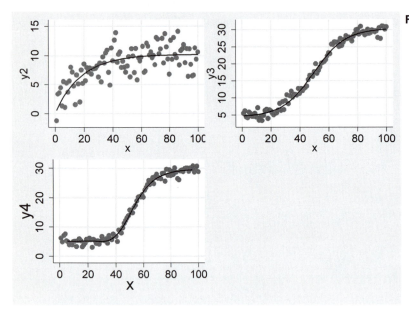

Figure 8.15

Users can write further nl*function* programs of their own. Here is the code for the nlexp2.ado program defining a 2-parameter exponential growth model:

```
*! version 1.1.3   12jun1998
program define nlexp2
    version 6
    if "`1'"=="?" {
        global S_2 "2-param. exp. growth curve, $S_E_depv=b1*b2^`2'"
        global S_1 "b1 b2"
/*
    Approximate initial values by regression of log Y on X.
*/
        local exp "[`e(wtype)' `e(wexp)']"
        tempvar Y
        quietly {
            gen `Y' = log(`e(depvar)') if e(sample)
            reg `Y' `2' `exp' if e(sample)
        }
        global b1 = exp(_b[_cons])
        global b2 = exp(_b[`2'])
        exit
    }
    replace `1'=$b1*($b2)^`2'
end
```

This program finds some approximate initial values of the parameters to be estimated, storing these as "global macros" named b1 and b2 . It then calculates an initial set of predicted values, as a "local macro" named 1 , employing the initial parameter estimates and the model equation:

```
replace `1' = $b1 * ($b2)^`2'
```

Subsequent iterations of **nl** will return to this line, calculating new predicted values (replacing the contents of macro 1) as they refine the parameter estimates b1 and b2 . In Stata programs, the notation $b1 means "the contents of global macro b1 ." Similarly, the notation `1' means "the contents of local macro 1 ."

Before attempting to write your own nonlinear function, examine nllog4.ado , nlgom4.ado , and others as examples, and consult the manual or **help nl** for explanations. Chapter 14 contains further discussion of macros and other aspects of Stata programming.

Nonlinear Regression — 2

Our second example involves real data, and illustrates some steps that can help in research. Dataset *lichen.dta* concerns measurements of lichen growth observed on the Norwegian arctic island of Svalbard (from Werner 1990). These slow-growing symbionts are often used to date rock monuments and other deposits, so their growth rates interest scientists in several fields.

```
Contains data from C:\data\lichen.dta
  obs:            11                          Lichen growth (Werner 1990)
  vars:            8                          14 Jul 2005 14:57
  size:          572 (99.9% of memory free)
-------------------------------------------------------------------------
              storage  display    value
variable name  type    format     label     variable label
-------------------------------------------------------------------------
locale        str31   %31s                   Locality and feature
point         str1    %9s                    Control point
date          int     %8.0g                  Date
age           int     %8.0g                  Age in years
rshort        float   %9.0g                  Rhizocarpon short axis mm
rlong         float   %9.0g                  Rhizocarpon long axis mm
pshort        int     %8.0g                  P.minuscula short axis mm
plong         int     %8.0g                  P.minuscula long axis mm
-------------------------------------------------------------------------
Sorted by:
```

Lichens characteristically exhibit a period of relatively fast early growth, gradually slowing, as suggested by the lowess-smoothed curve in Figure 8.16.

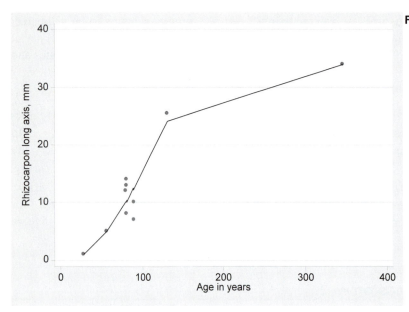

Figure 8.16

Lichenometricians seek to summarize and compare such patterns by drawing growth curves. Their growth curves might not employ an explicit mathematical model, but we can fit one here to illustrate the process of nonlinear regression. Gompertz curves are asymmetrical S-curves, which have been widely used to model biological growth:

$$y = b_1 \exp(-\exp(-b_2(x - b_3)))$$

They might provide a reasonable model for lichen growth.

If we intend to graph a nonlinear model, the data should contain a good range of closely spaced x values. Actual ages of the 11 lichen samples in *lichen.dta* range from 28 to 346 years. We can create 89 additional artificial observations, with "ages" from 0 to 352 in 4-year increments, by the following commands:

```
. range newage 0 396 100
obs was 11, now 100

. replace age = newage[_n-11] if age >= .
(89 real changes made)
```

The first command created a new variable, *newage*, with 100 values ranging from 0 to 396 in 4-year increments. In so doing, we also created 89 new artificial observations, with missing values on all variables except *newage*. The **replace** command substitutes the missing artificial-case *age* values with *newage* values, starting at 0. The first 15 observations in our data now look like this:

```
. list rlong age newage in 1/15

     +----------------------+
     | rlong   age   newage |
     |----------------------|
  1. |     1    28        0 |
  2. |     5    56        4 |
  3. |    12    79        8 |
  4. |    14    80       12 |
```

```
  5. |    13      80         16 |
     |---------------------------|
  6. |     8      80         20 |
  7. |     7      89         24 |
  8. |    10      89         28 |
  9. |    34     346         32 |
 10. |    34     346         36 |
     |---------------------------|
 11. |  25.5     131         40 |
 12. |     .       0         44 |
 13. |     .       4         48 |
 14. |     .       8         52 |
 15. |     .      12         56 |
     +---------------------------+
```

```
. summarize rlong age newage

    Variable |       Obs        Mean    Std. Dev.        Min         Max
-------------+--------------------------------------------------------------
       rlong |        11    14.86364    11.31391          1          34
         age |       100      170.68    104.7042          0         352
      newage |       100         198     116.046          0         396
```

We now could **drop newage**. Only the original 11 observations have nonmissing *rlong* values, so only they will enter into model estimation. Stata calculates predicted values for any observation with nonmissing *x* values, however. We can therefore obtain such predictions for both the 11 real observations and the 89 artificial ones, which will allow us to graph the regression curve accurately.

Lichen growth starts with a size close to zero, so we chose the **gom3** Gompertz function rather than **gom4** (which incorporates a nonzero takeoff level, the parameter b_0). Figure 8.16 suggests an asymptotic upper limit somewhere near 34, suggesting that 34 should be a good guess or starting value of the parameter b_1. Estimation of this model is accomplished by

```
. nl gom3 rlong age, init(B1=34) nolog

(obs = 11)

      Source |       SS       df       MS            Number of obs =        11
-------------+------------------------------          F( 3,     8) =    125.68
       Model |  3633.16112      3  1211.05371          Prob > F      =    0.0000
    Residual |  77.0888815      8  9.63611018          R-squared     =    0.9792
-------------+------------------------------          Adj R-squared =    0.9714
       Total |     3710.25     11  337.295455          Root MSE      =  3.104208
                                                       Res. dev.     =  52.63435
3-parameter Gompertz function, rlong=b1*exp(-exp(-b2*(age-b3)))
-----------------------------------------------------------------------------
       rlong |     Coef.   Std. Err.      t    P>|t|     [95% Conf. Interval]
-------------+---------------------------------------------------------------
          b1 |  34.36637   2.267186    15.16   0.000     29.13823    39.59451
          b2 |  .0217685   .0060806     3.58   0.007     .0077465    .0357904
          b3 |  88.79701   5.632545    15.76   0.000     75.80834    101.7857
-----------------------------------------------------------------------------
 (SE's, P values, CI's, and correlations are asymptotic approximations)
```

A **nolog** option suppresses displaying a log of iterations with the output. All three parameter estimates differ significantly from 1.

We obtain predicted values using **predict**, and graph these to see the regression curve. The **yline** option is used to display the lower and estimated upper limits (0 and 34.366) of this curve in Figure 8.17.

```
. predict yhat
(option yhat assumed; fitted values)

. graph twoway scatter rlong age
        || mspline yhat age, clpattern(solid) bands(50)
        || , legend(off) yline(0 34.366)
        ytitle("Rhizocarpon long axis, mm") xlabel(0(100)400, grid)
```

Figure 8.17

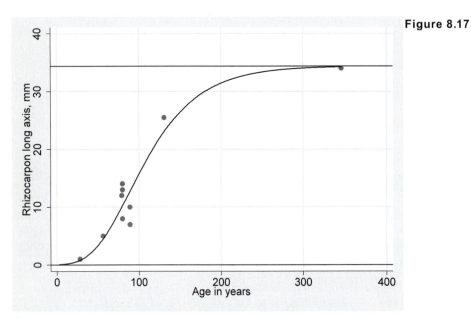

Especially when working with sparse data or a relatively complex model, nonlinear regression programs can be quite sensitive to their initial parameter estimates. The **init** option with **nl** permits researchers to suggest their own initial values if the default values supplied by an nl*function* program do not seem to work. Previous experience with similar data, or publications by other researchers, could help supply suitable initial values. Alternatively, we could estimate through trial and error by employing **generate** to calculate predicted values based on arbitrarily-chosen sets of parameter values and **graph** to compare the resulting predictions with the data.

9

Robust Regression

Stata's basic **regress** and **anova** commands perform ordinary least squares (OLS) regression. The popularity of OLS derives in part from its theoretical advantages given "ideal" data. If errors are normally, independently, and identically distributed (normal i.i.d.), then OLS is more efficient than any other unbiased estimator. The flip side of this statement often gets overlooked: if errors are not normal, or not i.i.d., then other unbiased estimators might outperform OLS. In fact, the efficiency of OLS degrades quickly in the face of heavy-tailed (outlier-prone) error distributions. Yet such distributions are common in many fields.

OLS tends to track outliers, fitting them at the expense of the rest of the sample. Over the long run, this leads to greater sample-to-sample variation or inefficiency when samples often contain outliers. Robust regression methods aim to achieve almost the efficiency of OLS with ideal data and substantially better-than-OLS efficiency in non-ideal (for example, nonnormal errors) situations. "Robust regression" encompasses a variety of different techniques, each with advantages and drawbacks for dealing with problematic data. This chapter introduces two varieties of robust regression, **rreg** and **qreg**, and briefly compares their results with those of OLS (**regress**).

rreg and **qreg** resist the pull of outliers, giving them better-than-OLS efficiency in the face of nonnormal, heavy-tailed error distributions. They share the OLS assumption that errors are independent and identically distributed, however. As a result, their standard errors, tests, and confidence intervals are not trustworthy in the presence of heteroskedasticity or correlated errors. To relax the assumption of independent, identically distributed errors when using **regress** or certain other modeling commands (although not **rreg** or **qreg**), Stata offers options that estimate robust standard errors.

For clarity, this chapter focuses mostly on two-variable examples, but robust multiple regression or *N*-way ANOVA are straightforward using the same commands. Chapter 14 returns to the topic of robustness, showing how we can use Monte Carlo experiments to evaluate competing statistical techniques.

Several of the techniques described in this chapter are available through menu selections:

Statistics – Nonparametric analysis – Quantile regression

Statistics – Linear regression and related – Linear regression – Robust SE

Example Commands

. `rreg y x1 x2 x3`

Performs robust regression of *y* on three predictors, using iteratively reweighted least squares with Huber and biweight functions tuned for 95% Gaussian efficiency. Given appropriately configured data, **rreg** can also obtain robust means, confidence intervals, difference of means tests, and ANOVA or ANCOVA.

. `rreg y x1 x2 x3, nolog tune(6) genwt(rweight) iterate(10)`

Performs robust regression of *y* on three predictors. The options shown above tell Stata not to print the iteration log, to use a tuning constant of 6 (which downweights outliers more steeply than the default 7), to generate a new variable (arbitrarily named *rweight*) holding the final-iteration robust weights for each observation, and to limit the maximum number of iterations to 10.

. `qreg y x1 x2 x3`

Performs quantile regression, also known as least absolute value (LAV) or minimum *L1*-norm regression, of *y* on three predictors. By default, **qreg** models the conditional .5 quantile (approximate median) of *y* as a linear function of the predictor variables, and thus provides "median regression."

. `qreg y x1 x2 x3, quantile(.25)`

Performs quantile regression modeling the conditional .25 quantile (first quartile) of *y* as a linear function of *x1*, *x2*, and *x3*.

. `bsqreg y x1 x2 x3, rep(100)`

Performs quantile regression, with standard errors estimated by bootstrap data resampling with 100 repetitions (default is **rep(20)**).

. `predict e, resid`

Calculates residual values (arbitrarily named *e*) after any **regress**, **rreg**, **qreg**, or **bsqreg** command. Similarly, **predict yhat** calculates the predicted values of *y*. Other **predict** options apply, with some restrictions.

. `regress y x1 x2 x3, robust`

Performs OLS regression of *y* on three predictors. Coefficient variances, and hence standard errors, are estimated by a robust method (Huber/White or sandwich) that does not assume identically distributed errors. With the **cluster()** option, one source of correlation among the errors can be accommodated as well. The *User's Guide* describes the reasoning behind these methods.

Regression with Ideal Data

To clarify the issue of robustness, we will explore the small (*n* = 20) contrived dataset *robust1.dta*:

```
Contains data from C:\data\robust1.dta
  obs:            20                     Robust regression examples 1
                                              (artificial data)
  vars:           10                     17 Jul 2005 09:35
  size:          880 (99.9% of memory free)
```

```
                  storage   display    value
variable name     type      format     label         variable label
------------------------------------------------------------------------
x                 float     %9.0g                     Normal X
e1                float     %9.0g                     Normal errors
y1                float     %9.0g                     y1 = 10 + 2*x + e1
e2                float     %9.0g                     Normal errors with 1 outlier
y2                float     %9.0g                     y2 = 10 + 2*x + e2
x3                float     %9.0g                     Normal X with 1 leverage obs.
e3                float     %9.0g                     Normal errors with 1 extreme
y3                float     %9.0g                     y3 = 10 + 2*x3 + e3
e4                float     %9.0g                     Skewed errors
y4                float     %9.0g                     y4 = 10 + 2*x + e4
------------------------------------------------------------------------
Sorted by:
```

The variables *x* and *e1* each contain 20 random values from independent standard normal distributions. *y1* contains 20 values produced by the regression model:

$$y1 = 10 + 2x + e1$$

The commands that manufactured these first three variables are

```
. clear
. set obs 20
. generate x = invnorm(uniform())
. generate e1 = invnorm(uniform())
. generate y1 = 10 + 2*x + e1
```

With real data, coding mistakes and measurement errors sometimes create wildly incorrect values. To simulate this, we might shift the second observation's error from –0.89 to 19.89:

```
. generate e2 = e1
. replace e2 = 19.89 in 2
. generate y2 = 10 + 2*x + e2
```

Similar manipulations produce the other variables in *robust1.dta*.

y1 and *x* present an ideal regression problem: the expected value of *y1* really is a linear function of *x*, and errors come from normal, independent, and identical distributions — because we defined them that way. OLS does a good job of estimating the true intercept (10) and slope (2), obtaining the line shown in Figure 9.1.

```
. regress y1 x

      Source |       SS       df       MS              Number of obs =      20
-------------+------------------------------           F(  1,     18) =  108.25
       Model |  134.059351     1   134.059351           Prob > F      =  0.0000
    Residual |   22.29157    18   1.23842055           R-squared     =  0.8574
-------------+------------------------------           Adj R-squared =  0.8495
       Total |  156.350921    19   8.22899586           Root MSE      =  1.1128

------------------------------------------------------------------------------
          y1 |      Coef.   Std. Err.       t    P>|t|     [95% Conf. Interval]
-------------+----------------------------------------------------------------
           x |   2.048057   .1968465     10.40   0.000     1.634498    2.461616
       _cons |   9.963161   .2499861     39.85   0.000      9.43796    10.48836
------------------------------------------------------------------------------

. predict yhat1o
```

```
. graph twoway scatter y1 x
     ||   line yhat1o x, clpattern(solid) sort
     ||   , ytitle("y1 = 10 + 2*x + e1") legend(order(2)
        label(2 "OLS line") position(11) ring(0) cols(1))
```

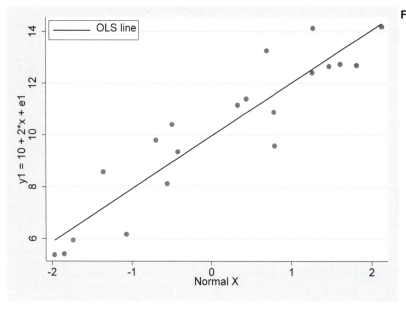

Figure 9.1

An iteratively reweighted least squares (IRLS) procedure, **rreg**, obtains robust regression estimates. The first **rreg** iteration begins with OLS. Any observations so influential as to have Cook's *D* values greater than 1 are automatically set aside after this first step. Next, weights are calculated for each observation using a Huber function, which downweights observations that have larger residuals, and weighted least squares is performed. After several WLS iterations, the weight function shifts to a Tukey biweight (as suggested by Li 1985), tuned for 95% Gaussian efficiency (see Hamilton 1992a for details). **rreg** estimates standard errors and tests hypotheses using a pseudovalues method (Street, Carroll and Ruppert 1988) that does not assume normality.

```
. rreg y1 x

   Huber iteration 1:  maximum difference in weights = .35774407
   Huber iteration 2:  maximum difference in weights = .02181578
Biweight iteration 3:  maximum difference in weights = .14421371
Biweight iteration 4:  maximum difference in weights = .01320276
Biweight iteration 5:  maximum difference in weights = .00265408

Robust regression estimates                      Number of obs =       20
                                                 F(  1,    18) =    79.96
                                                 Prob > F      =   0.0000

------------------------------------------------------------------------------
        y1 |      Coef.   Std. Err.       t     P>|t|     [95% Conf. Interval]
-----------+------------------------------------------------------------------
         x |   2.047813   .2290049      8.94    0.000     1.566692    2.528935
     _cons |   9.936163   .2908259     34.17    0.000     9.325161    10.54717
------------------------------------------------------------------------------
```

This "ideal data" example includes no serious outliers, so here **rreg** is unneeded. The **rreg** intercept and slope estimates resemble those obtained by **regress** (and are not far from the true values 10 and 2), but they have slightly larger estimated standard errors. Given normal i.i.d. errors, as in this example, **rreg** theoretically possesses about 95% of the efficiency of OLS.

rreg and **regress** both belong to the family of *M*-estimators (for maximum-likelihood). An alternative order-statistic strategy called *L*-estimation fits quantiles of y, rather than its expectation or mean. For example, we could model how the median (.5 quantile) of y changes with x. **qreg**, an *L1*-type estimator, accomplishes such quantile regression and provides another method with good resistance to outliers:

. **qreg y1 x**

```
Iteration  1:  WLS sum of weighted deviations =  17.711531

Iteration  1: sum of abs. weighted deviations =  17.130001
Iteration  2: sum of abs. weighted deviations =  16.858602

Median regression                          Number of obs =        20
  Raw sum of deviations    46.84 (about 10.4)
  Min sum of deviations  16.8586            Pseudo R2     =    0.6401

------------------------------------------------------------------------
        y1 |     Coef.   Std. Err.      t    P>|t|    [95% Conf. Interval]
-----------+------------------------------------------------------------
         x |   2.139896  .2590447     8.26   0.000    1.595664   2.684129
     _cons |    9.65342  .3564108    27.09   0.000    8.904628   10.40221
------------------------------------------------------------------------
```

Although **qreg** obtains reasonable parameter estimates, its standard errors here exceed those of **regress** (OLS) and **rreg**. Given ideal data, **qreg** is the least efficient of these three estimators. The following sections view their performance with less ideal data.

Y Outliers

The variable *y2* is identical to *y1*, but with one outlier caused by the "wild" error of observation #2. OLS has little resistance to outliers, so this shift in observation #2 (at upper left in Figure 9.2) substantially changes the **regress** results:

. **regress y2 x**

```
    Source |       SS       df       MS              Number of obs =        20
-----------+------------------------------          F(  1,    18) =      0.97
     Model |  18.764271       1   18.764271          Prob > F      =    0.3378
  Residual | 348.233471      18  19.3463039          R-squared     =    0.0511
-----------+------------------------------          Adj R-squared =   -0.0016
     Total | 366.997742      19  19.3156706          Root MSE      =    4.3984

------------------------------------------------------------------------
        y2 |     Coef.   Std. Err.      t    P>|t|    [95% Conf. Interval]
-----------+------------------------------------------------------------
         x |   .7662304   .7780232    0.98   0.338   -.8683356   2.400796
     _cons |   11.1579    .9880542   11.29   0.000    9.082078   13.23373
------------------------------------------------------------------------
```

```
. predict yhat2o
(option xb assumed; fitted values)

. label variable yhat2o "OLS line (regress)"
```

The outlier raises the OLS intercept (from 9.936 to 11.1579) and lessens the slope (from 2.048 to 0.766). R^2 has dropped from .8574 to .0511. Standard errors quadrupled, and the OLS slope (solid line in Figure 9.2) no longer significantly differs from zero.

The outlier has little impact on **rreg**, however, as shown by the dashed line in Figure 9.2. The robust coefficients barely change, and remain close to the true parameters 10 and 2; nor do the robust standard errors increase much.

```
. rreg y2 x, nolog genwt(rweight2)
```

```
Robust regression estimates                         Number of obs =        19
                                                    F(  1,     17) =     63.01
                                                    Prob > F       =    0.0000


-----------------------------------------------------------------------------
        y2 |      Coef.   Std. Err.       t    P>|t|     [95% Conf. Interval]
-----------+-----------------------------------------------------------------
         x |   1.979015   .2493146     7.94   0.000     1.453007    2.505023
     _cons |   10.00897   .3071265    32.59   0.000     9.360986    10.65695
-----------------------------------------------------------------------------
```

```
. predict yhat2r
(option xb assumed; fitted values)

. label variable yhat2r "robust regression (rreg)"

. graph twoway scatter y2 x
        ||   line yhat2o x, clpattern(solid) sort
        ||   line yhat2r x, clpattern(longdash) sort
        ||   , ytitle("y2 = 10 + 2*x + e2")
        legend(order(2 3) position(1) ring(0) cols(1) margin(sides))
```

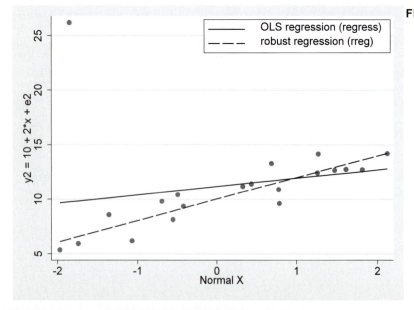

Figure 9.2

The **nolog** option above caused Stata not to print the iteration log. The **genwt(*rweight2*)** option saved robust weights as a variable named *rweight2*.

```
. predict resid2r, resid

. list y2 x resid2r rweight2
```

```
     +------------------------------------------+
     |   y2       x     resid2r    rweight2 |
     |------------------------------------------|
 1.  |  5.37   -1.97   -.7403071   .94644465 |
 2.  | 26.19   -1.85   19.84221           . |
 3.  |  5.93   -1.74   -.6354806   .96037073 |
 4.  |  8.58   -1.36    1.262494    .8493384 |
 5.  |  6.16   -1.07   -1.731421   .7257631 |
     |------------------------------------------|
 6.  |  9.80   -0.69    1.156554   .87273631 |
 7.  |  8.12   -0.55   -.8005085   .93758391 |
 8.  | 10.40   -0.49    1.36075    .82606386 |
 9.  |  9.35   -0.42     .17222    .99712388 |
10.  | 11.16    0.33    .4979582   .97581674 |
     |------------------------------------------|
11.  | 11.40    0.44    .5202664   .97360863 |
12.  | 13.26    0.69    1.885513   .68048066 |
13.  | 10.88    0.78   -.6725982   .95572833 |
14.  |  9.58    0.79   -1.992389   .64644918 |
15.  | 12.41    1.26   -.0925257   .99913568 |
     |------------------------------------------|
16.  | 14.14    1.27    1.617685   .75887073 |
17.  | 12.66    1.47   -.2581189   .99338589 |
18.  | 12.74    1.61   -.4551811   .97957817 |
19.  | 12.70    1.81   -.8909839   .92307041 |
20.  | 14.19    2.12   -.0144787   .99997651 |
     +------------------------------------------+
```

Residuals near zero produce weights near one; farther-out residuals get progressively lower weights. Observation #2 has been automatically set aside as too influential because of Cook's $D > 1$. **rreg** assigns its *rweight2* as "missing," so this observation has no effect on the final estimates. The same final estimates, although not the correct standard errors or tests, could be obtained using **regress** with analytical weights (results not shown):

```
. regress y2 x [aweight = rweight2]
```

Applied to the regression of *y2* on *x*, **qreg** also resists the outlier's influence and performs better than **regress**, but not as well as **rreg**. **qreg** appears less efficient than **rreg**, and in this sample its coefficient estimates are slightly farther from the true values of 10 and 2.

```
. qreg y2 x, nolog
```

```
Median regression                          Number of obs =        20
  Raw sum of deviations    56.68 (about 10.88)
  Min sum of deviations 36.20036            Pseudo R2     =    0.3613


--------------------------------------------------------------------------
        y2 |     Coef.   Std. Err.       t    P>|t|    [95% Conf. Interval]
-----------+--------------------------------------------------------------
         x |   1.821428   .4105944     4.44   0.000    .9588014    2.684055
     _cons |     10.115   .5088526    19.88   0.000    9.045941    11.18406
--------------------------------------------------------------------------
```

Monte Carlo researchers have also noticed that the standard errors calculated by **qreg** sometimes underestimate the true sample-to-sample variation, particularly with smaller samples. As an alternative, Stata provides the command **bsqreg**, which performs the same median or quantile regression as **qreg**, but employs bootstrapping (data resampling) to estimate the standard errors. The option **rep( )** controls the number of repetitions. Its default is **rep(20)**, which is enough for exploratory work. Before reaching "final" conclusions, we might take the time to draw 200 or more bootstrap samples. Both **qreg** and **bsqreg** fit identical models. In the example below, **bsqreg** also obtains similar standard errors. Chapter 14 returns to the topic of bootstrapping.

```
. bsqreg y2 x, rep(50)

(fitting base model)
(bootstrapping .................................................)

Median regression, bootstrap(50) SEs            Number of obs =         20
  Raw sum of deviations      56.68 (about 10.88)
  Min sum of deviations 36.20036                 Pseudo R2     =     0.3613


------------------------------------------------------------------------------
         y2 |      Coef.   Std. Err.       t    P>|t|     [95% Conf. Interval]
------------+-----------------------------------------------------------------
          x |   1.821428   .4084728     4.46   0.000     .9632587    2.679598
      _cons |     10.115   .4774718    21.18   0.000     9.111869    11.11813
------------------------------------------------------------------------------
```

X Outliers (Leverage)

rreg, **qreg**, and **bsqreg** deal comfortably with *y*-outliers, unless the observations with unusual *y* values have unusual *x* values (leverage) too. The *y3* and *x3* variables in *robust.dta* present an extreme example of leverage. Apart from the leverage observation (#2), these variables equal *y1* and *x*.

The high leverage of observation #2, combined with its exceptional *y3* value, make it influential: **regress** and **qreg** both track this outlier, reporting that the "best-fitting" line has a negative slope (Figure 9.3).

```
. regress y3 x3

      Source |       SS       df       MS              Number of obs =       20
-------------+------------------------------           F(  1,    18) =    11.01
       Model |  139.306724     1   139.306724          Prob > F      =   0.0038
    Residual |  227.691018    18   12.649501           R-squared     =   0.3796
-------------+------------------------------           Adj R-squared =   0.3451
       Total |  366.997742    19   19.3156706          Root MSE      =   3.5566


------------------------------------------------------------------------------
         y3 |      Coef.   Std. Err.       t    P>|t|     [95% Conf. Interval]
------------+-----------------------------------------------------------------
         x3 |  -.6212248   .1871973    -3.32   0.004    -1.014512    -.227938
      _cons |   10.80931   .8063436    13.41   0.000     9.115244    12.50337
------------------------------------------------------------------------------

. predict yhat3o

. label variable yhat3o "OLS regression (regress)"
```

```
. qreg y3 x3, nolog
```

```
Median regression                              Number of obs =        20
  Raw sum of deviations      56.68 (about 10.88)
  Min sum of deviations 56.19466                Pseudo R2       =    0.0086
```

```
------------------------------------------------------------------------------
         y3 |      Coef.   Std. Err.       t    P>|t|     [95% Conf. Interval]
------------+-----------------------------------------------------------------
         x3 |  -.6222217    .347103     -1.79   0.090    -1.351458    .1070146
      _cons |   11.36533   1.419214      8.01   0.000     8.383676    14.34699
------------------------------------------------------------------------------
```

```
. predict yhat3q
```

```
. label variable yhat3q "median regression (qreg)"
```

```
. rreg y3 x3, nolog
```

```
Robust regression estimates                    Number of obs =        19
                                               F(  1,    17) =     63.01
                                               Prob > F       =    0.0000
```

```
------------------------------------------------------------------------------
         y3 |      Coef.   Std. Err.       t    P>|t|     [95% Conf. Interval]
------------+-----------------------------------------------------------------
         x3 |   1.979015   .2493146      7.94   0.000     1.453007    2.505023
      _cons |   10.00897   .3071265     32.59   0.000     9.360986    10.65695
------------------------------------------------------------------------------
```

```
. predict yhat3r
```

```
. label variable yhat3r "robust regression (rreg)"
```

```
. graph twoway scatter y3 x3
        || line yhat3o x3, clpattern(solid) sort
        || line yhat3r x3, clpattern(longdash) sort
        || line yhat3q x3, clpattern(shortdash) sort ,
     ytitle("y3 = 10 + 2*x + e3") legend(order(4 3 2) position(5)
        ring(0) cols(1) margin(sides)) ylabel(-30(10)30)
```

Figure 9.3

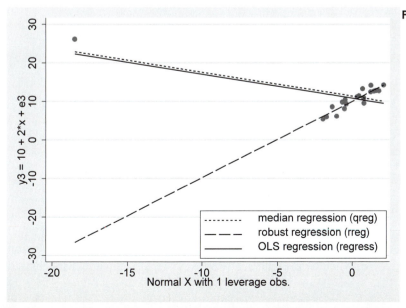

Figure 9.3 illustrates that **regress** and **qreg** are not robust against leverage (*x*-outliers). The **rreg** program, however, not only downweights large-residual observations (which by itself gives little protection against leverage), but also automatically sets aside observations with Cook's *D* (influence) statistics greater than 1. This happened when we regressed *y3* on *x3*; **rreg** ignored the one influential observation and produced a more reasonable regression line with a positive slope, based on the remaining 19 observations.

Setting aside high-influence observations, as done by **rreg**, provides a simple but not foolproof way to deal with leverage. More comprehensive methods, termed bounded-influence regression, also exist and could be implemented in a Stata program.

The examples in Figures 9.2 and 9.3 involve single outliers, but robust procedures can handle more. Too many severe outliers, or a cluster of similar outliers, might cause them to break down. But in such situations, which are often noticeable in diagnostic plots, the analyst must question whether fitting a linear model makes sense. It might be worthwhile to seek an explicit model for what is causing the outliers to be different.

Monte Carlo experiments (illustrated in Chapter 14) confirm that estimators like **rreg** and **qreg** generally remain unbiased, with better-than-OLS efficiency, when applied to heavy-tailed (outlier-prone) but symmetrical error distributions. The next section illustrates what can happen when errors have asymmetrical distributions.

Asymmetrical Error Distributions

The variable *e4* in *robust1.dta* has a skewed and outlier-filled distribution: *e4* equals *e1* (a standard normal variable) raised to the fourth power, and then adjusted to have 0 mean. These skewed errors, plus the linear relationship with *x*, define the variable $y4 = 10 + 2x + e4$. Regardless of an error distribution's shape, OLS remains an unbiased estimator. Over the long run, its estimates should center on the true parameter values.

```
. regress y4 x
```

Source	SS	df	MS		Number of obs =	20
					F(1, 18) =	6.97
Model	155.870383	1	155.870383		Prob > F =	0.0166
Residual	402.341909	18	22.3523283		R-squared =	0.2792
					Adj R-squared =	0.2392
Total	558.212291	19	29.3795943		Root MSE =	4.7278

y4	Coef.	Std. Err.	t	P>|t|	[95% Conf. Interval]
x	2.208388	.8362862	2.64	0.017	.4514157 3.96536
_cons	9.975681	1.062046	9.39	0.000	7.744406 12.20696

The same is not true for most robust estimators. Unless errors are symmetrical, the median line fit by **qreg**, or the biweight line fit by **rreg**, does not theoretically coincide with the expected-*y* line estimated by **regress**. So long as the errors' skew reflects only a small fraction of their distribution, **rreg** might exhibit little bias. But when the entire distribution is skewed, as with *e4*, **rreg** will downweight mostly one side, resulting in noticeably biased *y*-intercept estimates.

```
. rreg y4 x, nolog
```

```
Robust regression estimates                     Number of obs =      20
                                                F(  1,    18) = 1319.29
                                                Prob > F      =  0.0000


------------------------------------------------------------------------
      y4 |     Coef.   Std. Err.      t    P>|t|    [95% Conf. Interval]
---------+--------------------------------------------------------------
       x |  1.952073   .0537435    36.32   0.000    1.839163    2.064984
   _cons |  7.476669   .0682518   109.55   0.000    7.333278    7.620061
------------------------------------------------------------------------
```

Although the **rreg** *y*-intercept in Figure 9.4 is too low, the slope remains parallel to the OLS line and the true model. In fact, being less affected by outliers, the **rreg** slope (1.95) is closer to the true slope (2) and has a much smaller standard error than that of **regress** . This illustrates the tradeoff of using **rreg** or similar estimators with skewed errors: we risk getting biased estimates of the *y*-intercept, but can still expect unbiased and relatively precise estimates of other regression coefficients. In many applications, such coefficients are substantively more interesting than the *y*-intercept, making the tradeoff worthwhile. Moreover, the robust *t* and *F* tests, unlike those of OLS, do not assume normal errors.

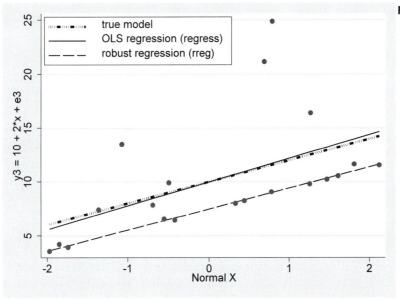

Figure 9.4

Robust Analysis of Variance

rreg can also perform robust analysis of variance or covariance once the model is recast in regression form. For illustration, consider the data on college faculty salaries in *faculty.dta*.

```
Contains data from C:\data\faculty.dta
  obs:          226                      College faculty salaries
  vars:           6                      17 Jul 2005 09:32
  size:       2,938 (99.9% of memory free)
```

```
------------------------------------------------------------------------
              storage  display     value
variable name  type    format      label       variable label
------------------------------------------------------------------------
rank           byte    %8.0g       rank        Academic rank
gender         byte    %8.0g       sex         Gender (dummy variable)
female         byte    %8.0g                   Gender (effect coded)
assoc          byte    %8.0g                   Assoc Professor (effect coded)
full           byte    %8.0g                   Full Professor (effect coded)
pay            float   %9.0g                   Annual salary
------------------------------------------------------------------------
Sorted by:
```

Faculty salaries increase with rank. In this sample, men have higher average salaries:

`. table gender rank, contents(mean pay)`

```
----------------------------------------
Gender     |
(dummy     |        Academic rank
variable)  |   Assist      Assoc      Full
-----------+----------------------------
     Male  |    29280   38622.22   52084.9
   Female  | 28711.04   38019.05     47190
----------------------------------------
```

An ordinary (OLS) analysis of variance indicates that both *rank* and *gender* significantly affect salary. Their interaction is not significant.

`. anova pay rank gender rank*gender`

```
                        Number of obs =      226    R-squared     =  0.7305
                        Root MSE      = 5108.21    Adj R-squared =  0.7244

            Source |  Partial SS    df       MS            F      Prob > F
        -----------+----------------------------------    ------------------
             Model |  1.5560e+10     5   3.1120e+09       119.26    0.0000
                   |
              rank |  7.6124e+09     2   3.8062e+09       145.87    0.0000
            gender |   127361829     1    127361829         4.88    0.0282
       rank*gender |  87997720.1     2   43998860.1         1.69    0.1876
                   |
          Residual |  5.7406e+09   220   26093824.5
        -----------+----------------------------------    ------------------
             Total |  2.1300e+10   225   94668810.3
```

But salary is not normally distributed, and the senior-rank averages reflect the influence of a few highly paid outliers. Suppose we want to check these results by performing a robust analysis of variance. We need effect-coded versions of the *rank* and *gender* variables, which this dataset also contains.

`. tabulate gender female`

```
    Gender |
   (dummy  | Gender (effect coded)
 variable) |        -1          1 |     Total
-----------+----------------------+----------
     Male  |       149          0 |       149
   Female  |         0         77 |        77
-----------+----------------------+----------
     Total |       149         77 |       226
```

```
. tabulate rank assoc

Academic |    Assoc Professor (effect coded)
    rank |       -1           0           1 |     Total
---------+--------------------------------+----------
  Assist |       64           0           0 |        64
   Assoc |        0           0         105 |       105
    Full |        0          57           0 |        57
---------+--------------------------------+----------
   Total |       64          57         105 |       226

. tab rank full

Academic |    Full Professor (effect coded)
    rank |       -1           0           1 |     Total
---------+--------------------------------+----------
  Assist |       64           0           0 |        64
   Assoc |        0         105           0 |       105
    Full |        0           0          57 |        57
---------+--------------------------------+----------
   Total |       64         105          57 |       226
```

If *faculty.dta* did not already have these effect-coded variables (*female, assoc,* and *full*), we could create them from *gender* and *rank* using a series of **generate** and **replace** statements. We also need two interaction terms representing female associate professors and female full professors:

```
. generate femassoc = female*assoc
. generate femfull = female*full
```

Males and assistant professors are "omitted categories" in this example. Now we can duplicate the previous ANOVA using regression:

```
. regress pay assoc full female femassoc femfull

      Source |       SS       df       MS              Number of obs =     226
-------------+------------------------------           F(  5,   220) =  119.26
       Model | 1.5560e+10        5  3.1120e+09         Prob > F      =  0.0000
    Residual | 5.7406e+09      220  26093824.5         R-squared     =  0.7305
-------------+------------------------------           Adj R-squared =  0.7244
       Total | 2.1300e+10      225  94668810.3         Root MSE      =  5108.2

-------------------------------------------------------------------------------
         pay |      Coef.   Std. Err.       t    P>|t|    [95% Conf. Interval]
-------------+-----------------------------------------------------------------
       assoc |  -663.8995   543.8499     -1.22   0.223    -1735.722    407.9229
        full |   10652.92   783.9227     13.59   0.000     9107.957    12197.88
      female |  -1011.174   457.6938     -2.21   0.028    -1913.199   -109.1483
    femassoc |   709.5864   543.8499      1.30   0.193    -362.2359    1781.409
     femfull |  -1436.277   783.9227     -1.83   0.068    -2981.236    108.6819
       _cons |   38984.53   457.6938     85.18   0.000     38082.51    39886.56
-------------------------------------------------------------------------------

. test assoc full

 ( 1)  assoc = 0.0
 ( 2)  full = 0.0

       F(  2,   220) =  145.87
            Prob > F =    0.0000
```

```
.  test female

( 1)   female = 0.0

        F(  1,   220) =     4.88
               Prob > F =    0.0282

.  test femassoc femfull

( 1)   femassoc = 0.0
( 2)   femfull = 0.0

        F(  2,   220) =     1.69
               Prob > F =    0.1876
```

regress followed by the appropriate **test** commands obtains exactly the same R^2 and F test results that we found earlier using **anova**. Predicted values from this regression equal the mean salaries.

```
.  predict predpay1
(option xb assumed; fitted values)

.  label variable predpay1 "OLS predicted salary"

.  table gender rank, contents(mean predpay1)
```

```
----------------------------------------
Gender    |
(dummy    |          Academic rank
variable) |   Assist      Assoc       Full
----------+-----------------------------
    Male  |    29280    38622.22    52084.9
  Female  | 28711.04    38019.05      47190
----------------------------------------
```

Predicted values (means), R^2, and F tests would also be the same regardless of which categories we chose to omit from the regression. Our "omitted categories," males and assistant professors, are not really absent. Their information is implied by the included categories: if a faculty member is not female, he must be male, and so forth.

To perform a robust analysis of variance, apply **rreg** to this model:

```
.  rreg pay assoc full female femassoc femfull, nolog
```

```
Robust regression estimates                    Number of obs =      226
                                               F(  5,   220) =   138.25
                                               Prob > F      =   0.0000

------------------------------------------------------------------------------
        pay |     Coef.    Std. Err.       t     P>|t|    [95% Conf. Interval]
------------+-----------------------------------------------------------------
       assoc | -315.6463    458.1588    -0.69    0.492   -1218.588     587.2956
        full |  9765.296    660.4048    14.79    0.000    8463.767     11066.83
      female | -749.4949    385.5778    -1.94    0.053   -1509.394     10.40395
    femassoc |  197.7833    458.1588     0.43    0.666   -705.1587     1100.725
     femfull |  -913.348    660.4048    -1.38    0.168   -2214.878     388.1815
       _cons |  38331.87    385.5778    99.41    0.000    37571.97     39091.77
------------------------------------------------------------------------------
```

```
.  test assoc full

( 1)   assoc = 0.0
( 2)   full = 0.0

       F(  2,    220) =   182.67
             Prob > F =    0.0000

.  test female

( 1)   female = 0.0

       F(  1,    220) =     3.78
             Prob > F =    0.0532

.  test femassoc femfull

( 1)   femassoc = 0.0
( 2)   femfull = 0.0

       F(  2,    220) =     1.16
             Prob > F =    0.3144
```

rreg downweights several outliers, mainly highly-paid male full professors. To see the robust means, again use predicted values:

```
.  predict predpay2
(option xb assumed; fitted values)

.  label variable predpay2 "Robust predicted salary"

.  table gender rank, contents(mean predpay2)

----------------------------------------
Gender    |
(dummy    |         Academic rank
variable) |   Assist    Assoc      Full
----------+-----------------------------
     Male | 28916.15  38567.93  49760.01
   Female | 28848.29  37464.51  46434.32
----------------------------------------
```

The male–female salary gap among assistant and full professors appears smaller if we use robust means. It does not entirely vanish, however, and the gender gap among associate professors slightly widens.

With effect coding and suitable interaction terms, **regress** can duplicate ANOVA exactly. **rreg** can do parallel analyses, testing for differences among robust means instead of ordinary means (as **regress** and **anova** do). Used in similar fashion, **qreg** opens the third possibility of testing for differences among medians. For comparison, here is a quantile regression version of the faculty pay analysis:

```
.  qreg pay assoc full female femassoc femfull, nolog
```

```
Median regression                              Number of obs  =        226
  Raw sum of deviations   1738010 (about 37360)
  Min sum of deviations    798870              Pseudo R2      =     0.5404

-------------------------------------------------------------------------------
     pay |      Coef.   Std. Err.       t    P>|t|     [95% Conf. Interval]
---------+---------------------------------------------------------------------
   assoc |       -760   440.1693    -1.73   0.086    -1627.488    107.4881
    full |      10335   615.7735    16.78   0.000      9121.43    11548.57
  female |  -623.3333   365.1262    -1.71   0.089    -1342.926     96.2594
femassoc |  -156.6667   440.1693    -0.36   0.722    -1024.155    710.8214
 femfull |  -691.6667   615.7735    -1.12   0.263    -1905.236    521.9031
   _cons |      38300   365.1262   104.90   0.000     37580.41    39019.59
-------------------------------------------------------------------------------
```

```
. test assoc full
```

```
 ( 1)   assoc = 0.0
 ( 2)   full = 0.0

       F(  2,   220) =  208.94
            Prob > F =    0.0000
```

```
. test female
```

```
 ( 1)   female = 0.0

       F(  1,   220) =    2.91
            Prob > F =    0.0892
```

```
. test femassoc femfull
```

```
 ( 1)   femassoc = 0.0
 ( 2)   femfull = 0.0

       F(  2,   220) =    1.60
            Prob > F =    0.2039
```

```
. predict predpay3
(option xb assumed; fitted values)
```

```
. label variable predpay3 "Median predicted salary"
```

```
. table gender rank, contents(mean predpay3)
```

```
----------------------------------
Gender     |
(dummy     |      Academic rank
variable)  | Assist    Assoc     Full
-----------+----------------------
    Male   |  28500    38320    49950
  Female   |  28950    36760    47320
----------------------------------
```

Predicted values from this quantile regression closely resemble the median salaries in each subgroup, as we can verify directly:

```
. table gender rank, contents(median pay)
```

```
----------------------------------
Gender     |
(dummy     |       Academic rank
variable)  | Assist    Assoc     Full
-----------+----------------------
     Male  |  28500    38320    49950
   Female  |  28950    36590    46530
----------------------------------
```

qreg thus allows us to fit models analogous to *N*-way ANOVA or ANCOVA, but involving .5 quantiles or approximate medians instead of the usual means. In theory, .5 quantiles and medians are the same. In practice, quantiles are approximated from actual sample data values, whereas the median is calculated by averaging the two central values, if a subgroup contains an even number of observations. The sample median and .5 quantile approximations then can be different, but in a way that does not much affect model interpretation.

Further **rreg** and **qreg** Applications

Diagnostic statistics and plots (Chapter 7) and nonlinear transformations (Chapter 8) extend the usefulness of robust procedures as they do in ordinary regression. With transformed variables, **rreg** or **qreg** fit curvilinear regression models. **rreg** can also robustly perform simpler types of analysis. To obtain a 90% confidence interval for the mean of a single variable, *y*, we could type either the usual confidence-interval command **ci** :

```
. ci y, level(90)
```

Or, we could get exactly the same mean and interval through a regression with no *x* variables:

```
. regress y, level(90)
```

Similarly, we can obtain robust mean with 90% confidence interval by typing

```
. rreg y, level(90)
```

qreg could be used in the same way, but keep in mind the previous section's note about how a .5 quantile found by **qreg** might differ from a sample median. In any of these commands, the **level( )** option specifies the desired degree of confidence. If we omit this option, Stata automatically displays a 95% confidence interval.

To compare two means, analysts typically employ a two-sample *t* test (**ttest**) or one-way analysis of variance (**oneway** or **anova**). As seen earlier, we can perform equivalent tests (yielding identical *t* and *F* statistics) with regression, for example, by regressing the measurement variable on a dummy variable (here called *group*) representing the two categories:

```
. regress y group
```

A robust version of this test results from typing the following command:

```
. rreg y group
```

qreg performs median regression by default, but it is actually a more general tool. It can fit linear models for any quantile of *y*, not just the median (.5 quantile). For example, commands such as the following analyze how the first quartile (.25 quantile) of *y* changes with *x*.

```
. qreg y x, quant(.25)
```

Assuming constant error variance, the slopes of the .25 and .75 quantile lines should be roughly the same. **qreg** thus could perform a check for heteroskedasticity or subtle kinds of nonlinearity.

Robust Estimates of Variance — 1

Both **rreg** and **qreg** tend to perform better than OLS (**regress** or **anova**) in the presence of outlier-prone, nonnormal errors. All of these procedures share the common assumption that errors follow independent and identical distributions, however. If the distributions of errors vary across *x* values or observations, then the standard errors calculated by **anova**, **regress**, **rreg**, or **qreg** probably will understate the true sample-to-sample variation, and yield unrealistically narrow confidence intervals.

regress and some other model fitting commands (although not **rreg** or **qreg**) have an option that estimates standard errors without relying on the strong and sometimes implausible assumptions of independent, identically distributed errors. This option uses an approach derived independently by Huber, White, and others that is sometimes referred to as a sandwich estimator of variance. The artificial dataset (*robust2.dta*) provides a first example.

```
Contains data from C:\data\robust2.dta
  obs:           500                      Robust regression examples 2
                                            (artificial data)
  vars:           12                      17 Jul 2005 09:03
  size:        24,500 (99.9% of memory free)
-------------------------------------------------------------------------------
              storage  display    value
variable name  type    format     label      variable label
-------------------------------------------------------------------------------
x             float    %9.0g                 Standard normal x
e5            float    %9.0g                 Standard normal errors
y5            float    %9.0g                 y5 = 10 + 2*x + e5 (normal
                                               i.i.d. errors)
e6            float    %9.0g                 Contaminated normal errors:
                                               95% N(0,1), 5%(N(0,10)
y6            float    %9.0g                 y6 = 10 + 2*x + e6
                                               (Contaminated normal errors)
e7            float    %9.0g                 Centered chi-square(1) errors
y7            float    %9.0g                 y7 = 10 + 2*x + e7 (skewed
                                               errors)
e8            float    %9.0g                 Normal errors, variance
                                               increases with x
y8            float    %9.0g                 y8 = 10 + 2*x + e8
                                               (heteroskedasticity)
group         byte     %9.0g
e9            float    %9.0g                 Normal errors, variance
                                               increases with x, mean &
                                               variance increase with cluster
y9            float    %9.0g                 y9 = 10 + 2*x + e9
                                               (heteroskedasticity &
                                               correlated errors)
-------------------------------------------------------------------------------
Sorted by:
```

When we regress *y8* on *x*, we obtain a significant positive slope. A scatterplot shows strong heteroskedasticity, however (Figure 9.5). Variation around the regression line increases with *x*. Because errors do not appear to be identically distributed at all values of *x*, the standard errors, confidence intervals, and tests printed by **regress** are untrustworthy.   **rreg** or **qreg** would face the same problem.

```
. regress y8 x

      Source |       SS       df       MS              Number of obs =     500
-------------+------------------------------           F(  1,   498) =  133.96
       Model | 1607.35658        1  1607.35658         Prob > F      =  0.0000
    Residual | 5975.19162      498  11.9983767         R-squared     =  0.2120
-------------+------------------------------           Adj R-squared =  0.2104
       Total | 7582.5482       499  15.1954874         Root MSE      =  3.4639

------------------------------------------------------------------------------
          y8 |      Coef.   Std. Err.       t    P>|t|     [95% Conf. Interval]
-------------+----------------------------------------------------------------
           x |   1.819032   .1571612     11.57   0.000     1.510251    2.127813
       _cons |   10.06642    .154919     64.98   0.000     9.762047     10.3708
------------------------------------------------------------------------------
```

Figure 9.5

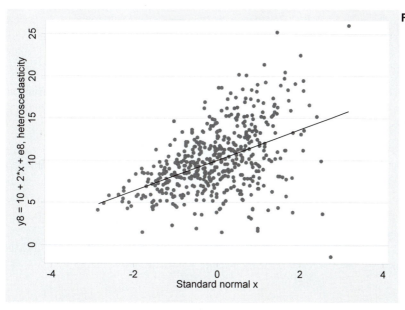

More credible standard errors and confidence intervals for this OLS regression can be obtained by using the **robust** option:

```
. regress y8 x, robust
```

```
Regression with robust standard errors                Number of obs =      500
                                                      F(  1,    498) =    83.80
                                                      Prob > F       =   0.0000
                                                      R-squared      =   0.2120
                                                      Root MSE       =   3.4639

-----------------------------------------------------------------------------
             |               Robust
         y8  |    Coef.    Std. Err.       t     P>|t|     [95% Conf. Interval]
-------------+---------------------------------------------------------------
          x  |  1.819032   .1987122      9.15   0.000     1.428614    2.209449
       _cons |  10.06642   .1561846     64.45   0.000     9.759561    10.37328
-----------------------------------------------------------------------------
```

Although the fitted model remains unchanged, the robust standard error for the slope is 27% larger (.199 vs. .157) than its nonrobust counterpart. With the **robust** option, the regression output does not show the usual ANOVA sums of squares because these no longer have their customary interpretation.

The rationale underlying these robust standard-error estimates is explained in the *User's Guide*. Briefly, we give up on the classical goal of estimating true population parameters (β's) for a model such as

$$y_i = \beta_0 + \beta_1 x_i + \epsilon_i$$

Instead, we pursue the less ambitious goal of simply estimating the sample-to-sample variation that our *b* coefficients might have, if we drew many random samples and applied OLS repeatedly to calculate *b* values for a model such as

$$y_i = b_0 + b_1 x_i + e_i$$

We do not assume that these *b* estimates will converge on some "true" population parameter. Confidence intervals formed using the robust standard errors therefore lack the classical interpretation of having a certain likelihood (across repeated sampling) of containing the true value of β. Rather, the robust confidence intervals have a certain likelihood (across repeated sampling) of containing *b*, defined as the value upon which sample *b* estimates converge. Thus, we pay for relaxing the identically-distributed-errors assumption by settling for a less impressive conclusion.

Robust Estimates of Variance — 2

Another robust-variance option, **cluster**, allows us to relax the independent-errors assumption in a limited way, when errors are correlated within subgroups or clusters of the data. The data in *attract.dta* describe an undergraduate social experiment that can be used for illustration. In this experiment, 51 college students were asked to individually rate the attractiveness, on a scale from 1 to 10, of photographs of unknown men and women. The rating exercise was repeated by each participant, given the same photos shuffled in random order, on four occasions during evening social events. Variable *ratemale* is the mean rating each participant gave to all the male photos in one sitting, and *ratefem* is the mean rating given to female photos. *gender* records the participant's (rater's) own gender, and *bac* his or her blood alcohol content at the time, measured by Breathalyzer.

```
Contains data from C:\data\attract.dta
  obs:            204                      Perceived attractiveness and
                                             drinking (D. C. Hamilton 2003)
  vars:             8                      18 Jul 2005 17:27
  size:         5,508 (99.9% of memory free)
-------------------------------------------------------------------------------
              storage  display   value
variable name   type   format    label    variable label
-------------------------------------------------------------------------------
id             byte    %9.0g              Participant number
gender         byte    %9.0g     sex      Participant gender (female)
bac            float   %9.0g              Blood alchohol content
genbac         float   %9.0g              gender*bac interaction
relstat        byte    %9.0g     rel      Relationship status (single)
drinkfrq       float   %9.0g              Days drinking in previous week
ratefem        float   %9.0g              Rated attractiveness of females
ratemale       float   %9.0g              Rated attractiveness of males
-------------------------------------------------------------------------------
Sorted by:  id
```

Although the data contain 204 observations, these represent only 51 individual participants. It seems reasonable to assume that disturbances (unmeasured influences on the ratings) were correlated across the repetitions by each individual. Viewing each participant's four rating sessions as a cluster should yield more realistic standard error estimates. Adding the option **cluster(id)** to a regression command, as seen below, obtains robust standard errors across clusters defined by *id* (individual participant).

```
. regress ratefem bac gender genbac, cluster(id)

Regression with robust standard errors          Number of obs =     204
                                                F( 3,    50) =      7.75
                                                Prob > F      =   0.0002
                                                R-squared     =   0.1264
Number of clusters (id) = 51                    Root MSE      =   1.1219

-------------------------------------------------------------------------------
             |              Robust
   ratefem   |    Coef.    Std. Err.     t     P>|t|    [95% Conf. Interval]
-------------+-----------------------------------------------------------------
         bac |  2.896741   .8543378    3.39   0.001    1.180753    4.612729
      gender | -.7299888   .3383096   -2.16   0.036   -1.409504   -.0504741
      genbac |  .2080538   1.708146    0.12   0.904   -3.222859    3.638967
       _cons |  6.486767   .229689    28.24   0.000    6.025423    6.94811
-------------------------------------------------------------------------------
```

Blood alcohol content (*bac*) has a significant positive effect: as *bac* goes up, predicted attractiveness rating of female photos increases as well. Gender (female) has a negative effect: female participants tended to rate female photos as somewhat less attractive (about .73 lower) than male participants did. The interaction of *gender* and *bac* is weak (.21). The intercept- and slope-dummy variable regression model, approximately

predicted *ratefem* $= 6.49 + 2.90bac - .73gender + .21genbac$

can be reduced for male participants (*gender* = 0) to

$$\text{predicted } ratefem = 6.49 + 2.90bac - (.73 \times 0) + (.21 \times 0 \times bac)$$

$$= 6.49 + 2.90bac$$

and for female participants (*gender* = 1) to

$$\text{predicted } ratefem = 6.49 + 2.90bac - (.73 \times 1) + (.21 \times 1 \times bac)$$

$$= 6.49 + 2.90bac - .73 + .21bac$$

$$= 5.76 + 3.11bac$$

The slight difference between the effects of alcohol on males (2.90) and females (3.11) equals the interaction coefficient, .21.

Attractiveness ratings for photographs of males were likewise positively affected by blood alcohol content. Gender has a stronger effect on the ratings of male photos: female participants tended to give male photos much higher ratings than male participants did. For male-photo ratings, the *gender* × *bac* interaction is substantial (−4.36), although it falls short of the .05 significance level.

```
. regress ratemal bac gender genbac, cluster(id)

Regression with robust standard errors              Number of obs =      201
                                                    F(  3,     50) =    10.96
                                                    Prob > F       =   0.0000
                                                    R-squared      =   0.3516
Number of clusters (id) = 51                        Root MSE       =   1.3931

------------------------------------------------------------------------------
             |             Robust
    ratemale |      Coef.   Std. Err.      t    P>|t|     [95% Conf. Interval]
-------------+----------------------------------------------------------------
         bac |   4.246042   2.261792     1.88   0.066    -.2969004    8.788985
      gender |   2.443216   .4529047     5.39   0.000     1.53353     3.352902
      genbac |  -4.364301   3.573689    -1.22   0.228    -11.54227    2.813663
       _cons |   3.628043   .2504253    14.49   0.000     3.125049    4.131037
------------------------------------------------------------------------------
```

The regression equation for ratings of male photos by male participants is approximately

$$\text{predicted } ratemale = 3.63 + 4.25bac + (2.44 \times 0) - (4.36 \times 0 \times bac)$$

$$= 3.63 + 4.25bac$$

and for rating of male photos by female participants,

$$\text{predicted } ratemale = 3.63 + 4.25bac + (2.44 \times 1) - (4.36 \times 1 \times bac)$$

$$= 6.07 - 0.11bac$$

The difference between the substantial alcohol effect on male participants (4.25) and the near-zero alcohol effect on females (−0.11) equals the interaction coefficient, −4.36. In this sample, males' ratings of male photos increase steeply, and females' ratings of male photos remain virtually steady, as the rater's *bac* increases.

Figure 9.6 visualizes these results in a graph. We see positive *rating–bac* relationships across all subplots except for females rating males. The graphs also show other gender differences, including higher *bac* values among male participants.

Figure 9.6

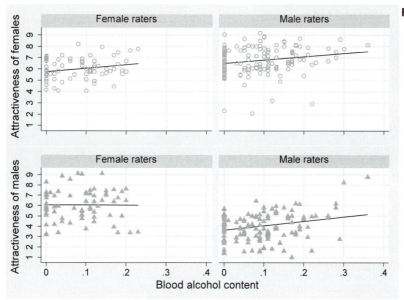

OLS regression with robust standard errors, estimated by **regress** with the **robust** option, should not be confused with the robust regression estimated by **rreg**. Despite similar-sounding names, the two procedures are unrelated, and solve different problems.

10

Logistic Regression

The regression and ANOVA methods described in Chapters 5 through 9 require measured dependent or y variables. Stata also offers a full range of techniques for modeling categorical, ordinal, and censored dependent variables. A list of some relevant commands follows. For more details on any of these, type **help** *command* .

binreg Binomial regression (generalized linear models).

blogit Logit estimation with grouped (blocked) data.

bprobit Probit estimation with grouped (blocked) data.

clogit Conditional fixed-effects logistic regression.

cloglog Complementary log-log estimation.

cnreg Censored-normal regression, assuming that y follows a Gaussian distribution but is censored at a point that might vary from observation to observation.

constraint Defines, lists, and drops linear constraints.

dprobit Probit regression giving changes in probabilities instead of coefficients.

glm Generalized linear models. Includes option to model logistic, probit, or complementary log-log links. Allows response variable to be binary or proportional for grouped data.

glogit Logit regression for grouped data.

gprobit Probit regression for grouped data.

heckprob Probit estimation with selection.

hetprob Heteroskedastic probit estimation.

intreg Interval regression, where y is either point data, interval data, left-censored data, or right-censored data.

logistic Logistic regression, giving odds ratios.

logit Logistic regression — similar to **logistic** , but giving coefficients instead of odds ratios.

mlogit Multinomial logistic regression, with polytomous y variable.

nlogit Nested logit estimation.

ologit Logistic regression with ordinal y variable.

oprobit Probit regression with ordinal y variable.

probit Probit regression, with dichotomous y variable.

`rologit`   Rank-ordered logit model for rankings (also known as the Plackett–Luce model, exploded logit model, or choice-based conjoint analysis).

`scobit`   Skewed probit estimation.

`svy: logit`   Logistic regression with complex survey data. Survey (**svy**) versions of many other categorical-variables modeling commands also exist.

`tobit`   Tobit regression, assuming *y* follows a Gaussian distribution but is censored at a known, fixed point (see **cnreg** for a more general version).

`xtcloglog` Random-effects and population-averaged cloglog models. Panel (**xt**) versions of **logit** , **probit** , and population-averaged generalized linear models (see **help xtgee**) also exist.

After most model-fitting commands, **predict** can calculate predicted values or probabilities. **predict** also obtains appropriate diagnostic statistics, such as those described for logistic regression in Hosmer and Lemeshow (2000). Specific **predict** options depend on the type of model just fitted. A different post-fitting command, **predictnl** , obtains nonlinear predictions and their confidence intervals (see **help predictnl**).

Examples of several of these commands appear in the next section. Most of the methods for modeling categorical dependent variables can be found under the following menus:

Statistics – Binary outcomes

Statistics – Ordinal outcomes

Statistics – Categorical outcomes

Statistics – Generalized linear models (GLM)

Statistics – Cross-sectional time series

Statistics – Linear regression and related – Censored regression

After the Example Commands section below, the remainder of this chapter concentrates on an important family of methods called logit or logistic regression. We review basic logit methods for dichotomous, ordinal, and polytomous dependent variables.

Example Commands

```
. logistic y x1 x2 x3
```
Performs logistic regression of {0,1} variable *y* on predictors *x1*, *x2*, and *x3*. Predictor variable effects are reported as odds ratios. A closely related command,
```
. logit y x1 x2 x3
```
performs essentially the same analysis, but reports effects as logit regression coefficients. The underlying models fit by **logistic** and **logit** are the same, so subsequent predictions or diagnostic tests will be identical.

. `lfit`

Presents a Pearson chi-squared goodness-of-fit test for the fitted logistic model: observed versus expected frequencies of $y = 1$, using cells defined by the covariate (x-variable) patterns. When a large number of x patterns exist, we might want to group them according to estimated probabilities. `lfit, group(10)` would perform the test with 10 approximately equal-size groups.

. `lstat`

Presents classification statistics and classification table. `lstat`, `lroc`, and `lsens` (see below) are particularly useful when the point of analysis is classification. These commands all refer to the previously-fit `logistic` model.

. `lroc`

Graphs the receiver operating characteristic (ROC) curve, and calculates area under the curve.

. `lsens`

Graphs both sensitivity and specificity versus the probability cutoff.

. `predict phat`

Generates a new variable (here arbitrarily named *phat*) equal to predicted probabilities that $y = 1$ based on the most recent `logistic` model.

. `predict dX2, dx2`

Generates a new variable (arbitrarily named *dX2*), the diagnostic statistic measuring change in Pearson chi-squared, from the most recent `logistic` analysis.

. `mlogit y x1 x2 x3, base(3) rrr nolog`

Performs multinomial logistic regression of multiple-category variable y on three x variables. Option `base(3)` specifies $y = 3$ as the base category for comparison; `rrr` calls for relative risk ratios instead of regression coefficients; and `nolog` suppresses display of the log likelihood on each iteration.

. `predict P2, outcome(2)`

Generates a new variable (arbitrarily named *P2*) representing the predicted probability that $y = 2$, based on the most recent `mlogit` analysis.

. `glm success x1 x2 x3, family(binomial trials) eform`

Performs a logistic regression via generalized linear modeling using tabulated rather than individual-observation data. The variable *success* gives the number of times that the outcome of interest occurred, and *trials* gives the number of times it could have occurred for each combination of the predictors *x1*, *x2*, and *x3*. That is, *success / trials* would equal the proportion of times that an outcome such as "patient recovers" occurred. The `eform` option asks for results in the form of odds ratios ("exponentiated form") rather than logit coefficients.

. `cnreg y x1 x2 x3, censored(cen)`

Performs censored-normal regression of measurement variable y on three predictors *x1*, *x2*, and *x3*. If an observation's true y value is unknown due to left or right censoring, it is replaced for this regression by the nearest y value at which censoring occurs. The censoring variable *cen* is a {–1,0,1} indicator of whether each observation's value of y has been left censored, not censored, or right censored.

Space Shuttle Data

Our main example for this chapter, *shuttle.dta*, involves data covering the first 25 flights of the U.S. space shuttle. These data contain evidence that, if properly analyzed, might have persuaded NASA officials not to launch *Challenger* on its last, fatal flight in 1985 (that was 25th shuttle flight, designated STS 51-L). The data are drawn from the *Report of the Presidential Commission on the Space Shuttle Challenger Accident* (1986) and from Tufte (1997). Tufte's book contains an excellent discussion about data and analytical issues. His comments regarding specific shuttle flights are included as a string variable in these data.

```
Contains data from C:\data\shuttle.dta
  obs:            25                          First 25 space shuttle flights
  vars:            8                          20 Jul 2005 10:40
  size:         1,675 (99.9% of memory free)
-------------------------------------------------------------------------------
              storage   display    value
variable name   type    format     label     variable label
-------------------------------------------------------------------------------
flight         byte     %8.0g      flbl      Flight
month          byte     %8.0g                Month of launch
day            byte     %8.0g                Day of launch
year           int      %8.0g                Year of launch
distress       byte     %8.0g      dlbl      Thermal distress incidents
temp           byte     %8.0g                Joint temperature, degrees F
damage         byte     %9.0g                Damage severity index (Tufte
                                               1997)
comments       str55    %55s                 Comments (Tufte 1997)
-------------------------------------------------------------------------------
Sorted by:
```

```
. list flight-temp, sepby(year)
```

```
     +--------------------------------------------------------------+
     |  flight    month    day    year    date    distress    temp  |
     |--------------------------------------------------------------|
  1. |   STS-1        4     12    1981    7772        none      66   |
  2. |   STS-2       11     12    1981    7986     1 or 2       70   |
     |--------------------------------------------------------------|
  3. |   STS-3        3     22    1982    8116        none      69   |
  4. |   STS-4        6     27    1982    8213           .      80   |
  5. |   STS-5       11     11    1982    8350        none      68   |
     |--------------------------------------------------------------|
  6. |   STS-6        4      4    1983    8494     1 or 2       67   |
  7. |   STS-7        6     18    1983    8569        none      72   |
  8. |   STS-8        8     30    1983    8642        none      73   |
  9. |   STS-9       11     28    1983    8732        none      70   |
     |--------------------------------------------------------------|
 10. | STS_41-B       2      3    1984    8799     1 or 2       57   |
 11. | STS_41-C       4      6    1984    8862     3 plus       63   |
 12. | STS_41-D       8     30    1984    9008     3 plus       70   |
 13. | STS_41-G      10      5    1984    9044        none      78   |
 14. | STS_51-A      11      8    1984    9078        none      67   |
     |--------------------------------------------------------------|
 15. | STS_51-C       1     24    1985    9155     3 plus       53   |
 16. | STS_51-D       4     12    1985    9233     3 plus       67   |
 17. | STS_51-B       4     29    1985    9250     3 plus       75   |
 18. | STS_51-G       6     17    1985    9299     3 plus       70   |
 19. | STS_51-F       7     29    1985    9341     1 or 2       81   |
 20. | STS_51-I       8     27    1985    9370     1 or 2       76   |
 21. | STS_51-J      10      3    1985    9407        none      79   |
 22. | STS_61-A      10     30    1985    9434     3 plus       75   |
```

```
23. |  STS_61-B      11    26   1985   9461      1 or 2      76 |
    |-----------------------------------------------------------|
24. |  STS_61-C       1    12   1986   9508      3 plus      58 |
25. |  STS_51-L       1    28   1986   9524         .        31 |
    +-----------------------------------------------------------+
```

This chapter examines three of the *shuttle.dta* variables:

distress   The number of "thermal distress incidents," in which hot gas blow-through or charring damaged joint seals of a flight's booster rockets. Burn-through of a booster joint seal precipitated the *Challenger* disaster. Many previous flights had experienced less severe damage, so the joint seals were known to be a source of possible danger.

temp   The calculated joint temperature at launch time, in degrees Fahrenheit. Temperature depends largely on weather. Rubber O-rings sealing the booster rocket joints become less flexible when cold.

date   Date, measured in days elapsed since January 1, 1960 (an arbitrary starting point). *date* is generated from the month, day, and year of launch using the **mdy** (month-day-year to elapsed time; see **help dates**) function:

```
. generate date = mdy(month, day, year)
. label variable date "Date (days since 1/1/60)"
```

Launch date matters because several changes over the course of the shuttle program might have made it riskier. Booster rocket walls were thinned to save weight and increase payloads, and joint seals were subjected to higher-pressure testing. Furthermore, the reusable shuttle hardware was aging. So we might ask, did the probability of booster joint damage (one or more distress incidents) increase with launch date?

distress is a labeled numeric variable:

```
. tabulate distress
```

Thermal distress incidents	Freq.	Percent	Cum.
none	9	39.13	39.13
1 or 2	6	26.09	65.22
3 plus	8	34.78	100.00
Total	23	100.00	

Ordinarily, **tabulate** displays the labels, but the **nolabel** option reveals that the underlying numerical codes are 0 = "none", 1 = "1 or 2", and 2 = "3 plus."

```
. tabulate distress, nolabel
```

Thermal distress incidents	Freq.	Percent	Cum.
0	9	39.13	39.13
1	6	26.09	65.22
2	8	34.78	100.00
Total	23	100.00	

We can use these codes to create a new dummy variable, *any*, coded 0 for no distress and 1 for one or more distress incidents:

```
. generate any = distress
(2 missing values generated)

. replace any = 1 if distress == 2
(8 real changes made)

. label variable any "Any thermal distress"
```

To see what this accomplished,

```
. tabulate distress any
```

```
Thermal     | Any thermal distress
distress    |
incidents   |          0          1 |     Total
------------+----------------------+----------
      none  |          9          0 |         9
    1 or 2  |          0          6 |         6
    3 plus  |          0          8 |         8
------------+----------------------+----------
     Total  |          9         14 |        23
```

Logistic regression models how a {0,1} dichotomy such as *any* depends on one or more *x* variables. The syntax of **logit** resembles that of **regress** and most other model-fitting commands, with the dependent variable listed first.

```
. logit any date, coef
```

```
Iteration 0:   log likelihood = -15.394543
Iteration 1:   log likelihood =  -13.01923
Iteration 2:   log likelihood = -12.991146
Iteration 3:   log likelihood = -12.991096
Logit estimates                              Number of obs   =        23
                                             LR chi2(1)      =      4.81
                                             Prob > chi2     =    0.0283
Log likelihood = -12.991096                  Pseudo R2       =    0.1561


------------------------------------------------------------------------
        any |     Coef.   Std. Err.      z    P>|z|   [95% Conf. Interval]
------------+-----------------------------------------------------------
       date |   .0020907   .0010703    1.95   0.051   -6.93e-06    .0041884
      _cons |  -18.13116   9.517217   -1.91   0.057   -36.78456    .5222396
------------------------------------------------------------------------
```

The **logit** iterative estimation procedure maximizes the logarithm of the likelihood function, shown at the output's top. At iteration 0, the log likelihood describes the fit of a model including only the constant. The last log likelihood describes the fit of the final model,

$$L = -18.13116 + .0020907 date \qquad [10.1]$$

where L represents the predicted logit, or log odds, of any distress incidents:

$$L = \ln[P(any = 1) / P(any = 0)] \qquad [10.2]$$

An overall χ^2 test at the upper right evaluates the null hypothesis that all coefficients in the model, except the constant, equal zero,

$$\chi^2 = -2(\ln \mathcal{L}_i - \ln \mathcal{L}_f) \qquad [10.3]$$

where $\ln \mathcal{L}_i$ is the initial or iteration 0 (model with constant only) log likelihood, and $\ln \mathcal{L}_f$ is the final iteration's log likelihood. Here,

$$\chi^2 = -2[-15.394543 - (-12.991096)]$$
$$= 4.81$$

The probability of a greater χ^2, with 1 degree of freedom (the difference in complexity between initial and final models), is low enough (.0283) to reject the null hypothesis in this example. Consequently, *date* does have a significant effect.

Less accurate, though convenient, tests are provided by the asymptotic z (standard normal) statistics displayed with **logit** results. With one predictor variable, that predictor's z statistic and the overall χ^2 statistic test equivalent hypotheses, analogous to the usual t and F statistics in simple OLS regression. Unlike their OLS counterparts, the logit z approximation and χ^2 tests sometimes disagree (they do here). The χ^2 test has more general validity.

Like Stata's other maximum-likelihood estimation procedures, **logit** displays a pseudo R^2 with its output:

$$\text{pseudo } R^2 = 1 - \ln \mathcal{L}_f / \ln \mathcal{L}_i \qquad\qquad [10.4]$$

For this example,

$$\text{pseudo } R^2 = 1 - (-12.991096) / (-15.394543)$$
$$= .1561$$

Although they provide a quick way to describe or compare the fit of different models for the same dependent variable, pseudo R^2 statistics lack the straightforward explained-variance interpretation of true R^2 in OLS regression.

After **logit**, the **predict** command (with no options) obtains predicted probabilities,

$$Phat = 1 / (1 + e^{-L}) \qquad\qquad [10.5]$$

Graphed against *date*, these probabilities follow an S-shaped logistic curve as seen in Figure 10.1.

```
. predict Phat
. label variable Phat "Predicted P(distress >= 1)"
. graph twoway connected Phat date, sort
```

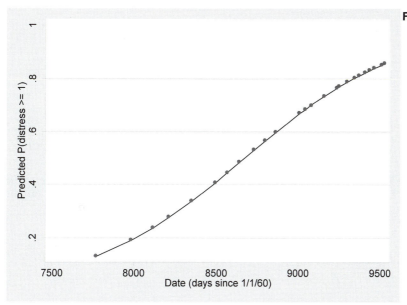

Figure 10.1

The coefficient given by **logit** (.0020907) describes *date*'s effect on the logit or log odds that any thermal distress incidents occur. Each additional day increases the predicted log odds of thermal distress by .0020907. Equivalently, we could say that each additional day multiplies predicted odds of thermal distress by $e^{.0020907} = 1.0020929$; each 100 days therefore multiplies the odds by $(e^{.0020907})^{100} = 1.23$. ($e \approx 2.71828$, the base number for natural logarithms.) Stata can make these calculations utilizing the _b[*varname*] coefficients stored after any estimation:

```
. display exp(_b[date])
1.0020929
```

```
. display exp(_b[date])^100
1.2325359
```

Or, we could simply include an **or** (odds ratio) option on the **logit** command line. An alternative way to obtain odds ratios employs the **logistic** command described in the next section. **logistic** fits exactly the same model as **logit**, but its default output table displays odds ratios rather than coefficients.

Using Logistic Regression

Here is the same regression seen earlier, but using **logistic** instead of **logit**:

. **logistic** *any date*

```
Logit estimates                                 Number of obs   =         23
                                                LR chi2(1)      =       4.81
                                                Prob > chi2     =     0.0283
Log likelihood = -12.991096                     Pseudo R2       =     0.1561
```

```
----------------------------------------------------------------------------
       any |  Odds Ratio    Std. Err.      z    P>|z|     [95% Conf. Interval]
-----------+----------------------------------------------------------------
      date |   1.002093    .0010725     1.95   0.051     .9999931    1.004197
----------------------------------------------------------------------------
```

Note the identical log likelihoods and χ^2 statistics. Instead of coefficients (b), **logistic** displays odds ratios (e^b). The numbers in the "Odds Ratio" column of the **logistic** output are amounts by which the odds favoring $y = 1$ are multiplied, with each 1-unit increase in that x variable (if other x variables' values stay the same).

After fitting a model, we can obtain a classification table and related statistics by typing

. **lstat**

```
Logistic model for any

              -------- True --------
Classified |         D              ~D  |      Total
-----------+----------------------------+-----------
     +     |        12               4  |         16
     -     |         2               5  |          7
-----------+----------------------------+-----------
   Total   |        14               9  |         23

Classified + if predicted Pr(D) >= .5
True D defined as any != 0
--------------------------------------------------
Sensitivity                     Pr( +| D)   85.71%
Specificity                     Pr( -|~D)   55.56%
Positive predictive value       Pr( D| +)   75.00%
Negative predictive value       Pr(~D| -)   71.43%
--------------------------------------------------
False + rate for true ~D        Pr( +|~D)   44.44%
False - rate for true D         Pr( -| D)   14.29%
False + rate for classified +   Pr(~D| +)   25.00%
False - rate for classified -   Pr( D| -)   28.57%
--------------------------------------------------
Correctly classified                        73.91%
--------------------------------------------------
```

By default, **lstat** employs a probability of .5 as its cutoff (although we can change this by adding a **cutoff( )** option). Symbols in the classification table have the following meanings:

D   The event of interest did occur (that is, $y = 1$) for that observation. In this example, D indicates that thermal distress occurred.

~D   The event of interest did not occur (that is, $y = 0$) for that observation. In this example, ~D corresponds to flights having no thermal distress.

+   The model's predicted probability is greater than or equal to the cutoff point. Since we used the default cutoff, + here indicates that the model predicts a .5 or higher probability of thermal distress.

–   The predicted probability is less than the cutoff. Here, – means a predicted probability of thermal distress below .5.

Thus for 12 flights, classifications are accurate in the sense that the model estimated at least a .5 probability of thermal distress, and distress did in fact occur. For 5 other flights, the model predicted less than a .5 probability, and distress did not occur. The overall "correctly classified" rate is therefore $12 + 5 = 17$ out of 23, or 73.91%. The table also gives conditional probabilities such as "sensitivity" or the percentage of observations with $P \geq .5$ given that thermal distress occurred (12 out of 14 or 85.71%).

After **logistic** or **logit**, the followup command **predict** calculates various prediction and diagnostic statistics. Discussion of the diagnostic statistics can be found in Hosmer and Lemeshow (2000).

`predict newvar`	Predicted probability that $y = 1$
`predict newvar, xb`	Linear prediction (predicted log odds that $y = 1$)
`predict newvar, stdp`	Standard error of the linear prediction
`predict newvar, dbeta`	ΔB influence statistic, analogous to Cook's D
`predict newvar, deviance`	Deviance residual for jth x pattern, d_j
`predict newvar, dx2`	Change in Pearson χ^2, written as $\Delta\chi^2$ or $\Delta\chi^2_P$
`predict newvar, ddeviance`	Change in deviance χ^2, written as ΔD or $\Delta\chi^2_D$
`predict newvar, hat`	Leverage of the jth x pattern, h_j
`predict newvar, number`	Assigns numbers to x patterns, $j = 1,2,3 \ldots J$
`predict newvar, resid`	Pearson residual for jth x pattern, r_j
`predict newvar, rstandard`	Standardized Pearson residual

Statistics obtained by the **dbeta**, **dx2**, **ddeviance**, and **hat** options do not measure the influence of individual observations, as their counterparts in ordinary regression do. Rather, these statistics measure the influence of "covariate patterns"; that is, the consequences of dropping all observations with that particular combination of x values. See Hosmer and Lemeshow (2000) for details. A later section of this chapter shows these statistics in use.

Does booster joint temperature also affect the probability of any distress incidents? We could investigate by including *temp* as a second predictor variable .

```
. logistic any date temp
```

```
Logit estimates                              Number of obs   =         23
                                             LR chi2(2)      =       8.09
                                             Prob > chi2     =     0.0175
Log likelihood = -11.350748                  Pseudo R2       =     0.2627
```

```
------------------------------------------------------------------------------
      any |  Odds Ratio   Std. Err.      z    P>|z|     [95% Conf. Interval]
----------+-------------------------------------------------------------------
     date |    1.00297    .0013675     2.17   0.030     1.000293    1.005653
     temp |   .8408309    .0987887    -1.48   0.140     .6678848    1.058561
------------------------------------------------------------------------------
```

The classification table indicates that including temperature as a predictor improved our correct classification rate to 78.26%.

```
. lstat
```

```
Logistic model for any

              -------- True --------
Classified |         D            ~D  |      Total
-----------+--------------------------+-----------
     +     |        12             3  |        15
     -     |         2             6  |         8
-----------+--------------------------+-----------
   Total   |        14             9  |        23

Classified + if predicted Pr(D) >= .5
True D defined as any != 0
--------------------------------------------------
Sensitivity                     Pr( +| D)    85.71%
Specificity                     Pr( -|~D)    66.67%
Positive predictive value       Pr( D| +)    80.00%
Negative predictive value       Pr(~D| -)    75.00%
--------------------------------------------------
False + rate for true ~D        Pr( +|~D)    33.33%
False - rate for true D         Pr( -| D)    14.29%
False + rate for classified +   Pr(~D| +)    20.00%
False - rate for classified -   Pr( D| -)    25.00%
--------------------------------------------------
Correctly classified                         78.26%
--------------------------------------------------
```

According to the fitted model, each 1-degree increase in joint temperature multiplies the odds of booster joint damage by .84 (in other words, each 1-degree warming reduces the odds of damage by about 16%). Although this effect seems strong enough to cause concern, the asymptotic z test says that it is not statistically significant ($z = -1.476$, $P = .140$). A more definitive test, however, employs the likelihood-ratio χ^2. The **lrtest** command compares nested models estimated by maximum likelihood. First, estimate a "full" model containing all variables of interest, as done above with the **logistic** *any date temp* command. Next, type an **estimates store** command, giving a name (such as *full*) to identify this first model:

```
. estimates store full
```

Now estimate a reduced model, including only a subset of the x variables from the full model. (Such reduced models are said to be "nested.") Finally, a command such as **lrtest** *full* requests a test of the nested model against the previously stored *full* model. For example (using the **quietly** prefix, because we already saw this output once),

```
. quietly logistic any date
. lrtest full
```

```
likelihood-ratio test                          LR chi2(1)  =        3.28
(Assumption: . nested in full)                 Prob > chi2 =      0.0701
```

This **lrtest** command tests the recent (presumably nested) model against the model previously saved by **estimates store**. It employs a general test statistic for nested maximum-likelihood models,

$$\chi^2 = -2(\ln \mathcal{L}_1 - \ln \mathcal{L}_0) \qquad [10.6]$$

where $\ln \mathcal{L}_0$ is the log likelihood for the first model (with all x variables), and $\ln \mathcal{L}_1$ is the log likelihood for the second model (with a subset of those x variables). Compare the resulting test statistic to a χ^2 distribution with degrees of freedom equal to the difference in complexity (number of x variables dropped) between models 0 and 1. Type **help lrtest** for more about this command, which works with any of Stata's maximum-likelihood estimation procedures (**logit**, **mlogit**, **stcox**, and many others). The overall χ^2 statistic routinely given by **logit** or **logistic** output (equation [10.3]) is a special case of [10.6].

The previous **lrtest** example performed this calculation:

$$\chi^2 = -2[-12.991096 - (-11.350748)]$$
$$= 3.28$$

with 1 degree of freedom, yielding $P = .0701$; the effect of *temp* is significant at $\alpha = .10$. Given the small sample and fatal consequences of a Type II error, $\alpha = .10$ seems a more prudent cutoff than the usual $\alpha = .05$.

Conditional Effect Plots

Conditional effect plots help in understanding what a logistic model implies about probabilities. The idea behind such plots is to draw a curve showing how the model's prediction of y changes as a function of one x variable, while holding all other x variables constant at chosen values such as their means, quartiles, or extremes. For example, we could find the predicted probability of any thermal distress incidents as a function of *temp*, holding *date* constant at its 25th percentile. The 25th percentile of *date*, found by **summarize date, detail**, is 8569 — that is, June 18, 1983.

```
. quietly logit any date temp
. generate L1 = _b[_cons] + _b[date]*8569 + _b[temp]*temp
. generate Phat1 = 1/(1 + exp(-L1))
. label variable Phat1 "P(distress >= 1 | date = 8569)"
```

L1 is the predicted logit, and *Phat1* equals the corresponding predicted probability that *distress* ≥ 1, calculated according to equation [10.5]. Similar steps find the predicted probability of any distress with *date* fixed at its 75th percentile (9341, or July 29, 1985):

```
. generate L2 = _b[_cons] + _b[date]*9341 + _b[temp]*temp

. generate Phat2 = 1/(1 + exp(-L2))

. label variable Phat2 "P(distress >= 1 | date = 9341)"
```

We can now graph the relationship between *temp* and the probability of any distress, for the two levels of *date*, as shown in Figure 10.2. Using median splines with many vertical bands (**graph twoway mspline, bands(50)**) produces smooth curves in this figure, approximating the smooth logistic functions.

```
. graph twoway mspline Phat1 temp, bands(50)
     ||  mspline Phat2 temp, bands(50)
     ||  , ytitle("Probability of thermal distress")
   ylabel(0(.2)1, grid) xlabel(, grid)
       legend(label(1 "June 1983") label(2 "July 1985")
       rows(2) position(7) ring(0))
```

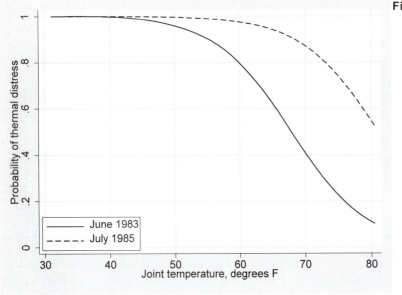

Figure 10.2

Among earlier flights (*date* = 8569, left curve), the probability of thermal distress goes from very low, at around 80° F, to near 1, below 50° F. Among later flights (*date* = 9341, right curve), however, the probability of any distress exceeds .5 even in warm weather, and climbs toward 1 on flights below 70° F. Note that *Challenger*'s launch temperature, 31° F, places it at top left in Figure 10.2. This analysis predicts almost certain booster joint damage.

Diagnostic Statistics and Plots

As mentioned earlier, the logistic regression influence and diagnostic statistics obtained by **predict** refer not to individual observations, as do the OLS regression diagnostics of Chapter 7. Rather, logistic diagnostics refer to *x* patterns. With the space shuttle data, however, each *x* pattern is unique — no two flights share the same combination of *date* and *temp* (naturally, because no two were launched the same day). Before using **predict**, we quietly refit the recent model, to be sure that model is what we think:

```
. quietly logistic any date temp

. predict Phat3
(option p assumed; Pr(any))

. label variable Phat3 "Predicted probability"

. predict dX2, dx2
(2 missing values generated)

. label variable dX2 "Change in Pearson chi-squared"

. predict dB, dbeta
(2 missing values generated)

. label variable dB "Influence"

. predict dD, ddeviance
(2 missing values generated)

. label variable dD "Change in deviance"
```

Hosmer and Lemeshow (2000) suggest plots that help in reading these diagnostics. To graph change in Pearson χ^2 versus probability of distress (Figure 10.3), type:

```
. graph twoway scatter dX2 Phat3
```

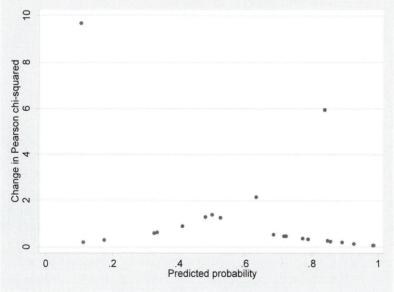

Figure 10.3

Two poorly fit x patterns, at upper right and left, stand out. We can identify these two flights (STS-2 and STS 51-A) if we include marker labels in the plot, as seen in Figure 10.4.

```
. graph twoway scatter dX2 Phat3, mlabel(flight) mlabsize(small)
```

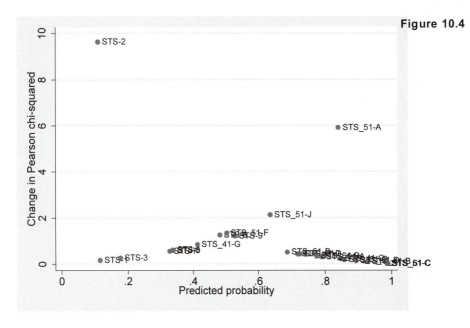

Figure 10.4

```
. list flight any date temp dX2 Phat3 if dX2 > 5
```

	flight	any	date	temp	dX2	Phat3
2.	STS-2	1	7986	70	9.630337	.1091805
4.	STS-4	.	8213	80	.	.0407113
14.	STS_51-A	0	9078	67	5.899742	.8400974
25.	STS_51-L	.	9524	31	.	.9999012

Flight STS 51-A experienced no thermal distress, despite a late launch date and cool temperature (see Figure 10.2). The model predicts a .84 probability of distress for this flight. All points along the up-to-right curve in Figure 10.4 have *any* = 0, meaning no thermal distress. Atop the up-to-left (*any* = 1) curve, flight STS-2 experienced thermal distress despite being one of the earliest flights, and launched in slightly milder weather. The model predicts only a .109 probability of distress. (Because Stata considers missing values as "high" numbers, it lists the two missing-values flights, including *Challenger*, among those with *dX2* > 5.)

Similar findings result from plotting *dD* versus predicted probability, as seen in Figure 10.5. Again, flights STS-2 (top left) and STS 51-A (top right) stand out as poorly fit. Figure 10.5 illustrates a variation on the labeled-marker scatterplot. Instead of putting the flight-number labels near the markers, as done earlier in Figure 10.4, we make the markers themselves invisible and place labels where the markers would have been in Figure 10.5.

```
. graph twoway scatter dD Phat3, msymbol(i) mlabposition(0)
    mlabel(flight) mlabsize(small)
```

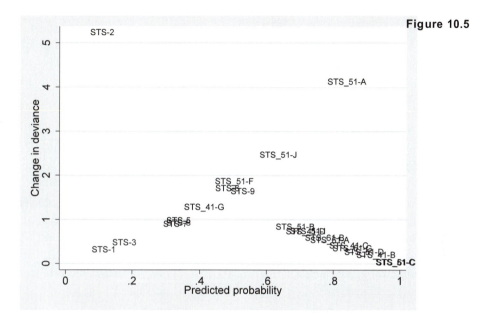

Figure 10.5

dB measures an *x* pattern's influence in logistic regression, as Cook's *D* measures an individual observation's influence in OLS. For a logistic-regression analogue to the OLS diagnostic plot in Figure 7.7, we can make the plotting symbols proportional to influence as done in Figure 10.6. Figure 10.6 reveals that the two worst-fit observations are also the most influential.

```
. graph twoway scatter dD Phat3 [aweight = dB], msymbol(oh)
```

Figure 10.6

Poorly fit and influential observations deserve special attention because they both contradict the main pattern of the data and pull model estimates in their contrary direction. Of course, simply removing such outliers allows a "better fit" with the remaining data — but this is circular reasoning. A more thoughtful reaction would be to investigate what makes the outliers unusual. Why did shuttle flight STS-2, but not STS 51-A, experience booster joint damage? Seeking an answer might lead investigators to previously overlooked variables or to otherwise respecify the model.

Logistic Regression with Ordered-Category *y*

logit and **logistic** fit only models that have two-category {0,1} *y* variables. We need other methods for models in which *y* takes on more than two categories. For example,

ologit Ordered logistic regression, where *y* is an ordinal (ordered-category) variable. The numerical values representing the categories do not matter, except that higher numbers mean "more." For example, the *y* categories might be {1 = "poor," 2 = "fair," 3 = "excellent"}.

mlogit Multinomial logistic regression, where *y* has multiple but unordered categories such as {1 = "Democrat," 2 = "Republican," 3 = "undeclared"}.

If *y* is {0,1}, **logit** (or **logistic**), **ologit**, and **mlogit** all produce essentially the same estimates.

We earlier simplified the three-category ordinal variable *distress* into a dichotomy, *any*. **logit** and **logistic** require {0,1} dependent variables. **ologit**, on the other hand, is designed for ordinal variables like *distress* that have more than two categories. The numerical codes representing these categories do not matter, so long as higher numerical values mean "more" of whatever is being measured. Recall that *distress* has categories 0 = "none," 1 = "1 or 2," and 2 = "3 plus" incidents of booster-joint distress.

Ordered logistic regression indicates that *date* and *temp* both affect *distress*, with the same signs (positive for *date*, negative for *temp*) seen in our earlier analyses:

```
. ologit distress date temp, nolog
```

```
Ordered logit estimates                          Number of obs  =         23
                                                 LR chi2(2)     =      12.32
                                                 Prob > chi2    =     0.0021
Log likelihood =  -18.79706                      Pseudo R2      =     0.2468


-------------------------------------------------------------------------------
   distress |      Coef.   Std. Err.      z     P>|z|    [95% Conf. Interval]
------------+------------------------------------------------------------------
       date |    .003286   .0012662     2.60    0.009    .0008043    .0057677
       temp |  -.1733752   .0834473    -2.08    0.038    -.336929   -.0098215
------------+------------------------------------------------------------------
      _cut1 |   16.42813   9.554813            (Ancillary parameters)
      _cut2 |   18.12227   9.722293
-------------------------------------------------------------------------------
```

Likelihood-ratio tests are more accurate than the asymptotic *z* tests shown. First, have **estimates store** preserve in memory the results from the full model (with two predictors) just estimated. Arbitrarily, we can name this model *A*.

```
. estimates store A
```

Next, fit a simpler model without *temp*, store its results as model *B*, and ask for a likelihood-ratio test of whether the fit of reduced model *B* differs significantly from that of the full model, model *A*:

```
. quietly ologit distress date
. estimates store B
. lrtest B A
```

```
likelihood-ratio test                                LR chi2(1)   =       6.12
(Assumption: B nested in A)                          Prob > chi2 =      0.0133
```

The **lrtest** output notes its assumption that model *B* is nested in model *A* — meaning that the parameters estimated in *B* are a subset of those in *A*, and that both models are estimated from the same pool of observations (which can be tricky when the data contain missing values). This likelihood-ratio test indicates that *B* 's fit is significantly poorer. Because the presence of *temp* as a predictor in model *A* is the only difference, the likelihood-ratio test thus informs us that *temp*'s contribution is significant. Similar steps find that *date* also has a significant effect.

```
. quietly ologit distress temp
. estimates store C
. lrtest C A
```

```
likelihood-ratio test                                LR chi2(1)   =      10.33
(Assumption: C nested in A)                          Prob > chi2 =      0.0013
```

The **estimates store** and **lrtest** commands provide flexible tools for comparing nested maximum-likelihood models. Type **help lrtest** and **help estimates** for details, including more advanced options.

The ordered-logit model estimates a score, *S*, as a linear function of *date* and *temp*:

$$S = .003286date - .1733752temp$$

Predicted probabilities depend on the value of *S*, plus a logistically distributed disturbance *u*, relative to the estimated cut points:

$P(distress="none") = P(S+u \le \_cut1) = P(S+u \le 16.42813)$

$P(distress="1 or 2") = P(\_cut1 < S+u \le \_cut2) = P(16.42813 < S+u \le 18.12227)$

$P(distress="3 plus") = P(\_cut2 < S+u) = P(18.12227 < S+u)$

After **ologit**, **predict** calculates predicted probabilities for each category of the dependent variable. We supply **predict** with names for these probabilities. For example: *none* could denote the probability of no distress incidents (first category of *distress*); *onetwo* the probability of 1 or 2 incidents (second category of *distress*); and *threeplus* the probability of 3 or more incidents (third and last category of *distress*):

```
. quietly ologit distress date temp
. predict none onetwo threeplus
(option p assumed; predicted probabilities)
```

This creates three new variables:

```
. describe none onetwo threeplus

              storage   display     value
variable name   type    format      label       variable label
---------------------------------------------------------------------------
none            float   %9.0g                    Pr(distress==0)
onetwo          float   %9.0g                    Pr(distress==1)
threeplus       float   %9.0g                    Pr(distress==2)
```

Predicted probabilities for *Challenger*'s last flight, the 25th in these data, are unsettling:

```
. list flight none onetwo threeplus if flight == 25

      +-------------------------------------------+
      |   flight       none      onetwo   threep~s |
      |-------------------------------------------|
  25. | STS_51-L    .0000754   .0003346    .99959 |
      +-------------------------------------------+
```

Our model, based on the analysis of 23 pre-*Challenger* shuttle flights, predicts little chance (P = .000075) of *Challenger* experiencing no booster joint damage, a scarcely greater likelihood of one or two incidents (P = .0003), but virtual certainty (P = .9996) of three or more damage incidents.

See Long (1997) or Hosmer and Lemeshow (2000) for more on ordered logistic regression and related techniques. The *Base Reference Manual* explains Stata's implementation.

Multinomial Logistic Regression

When the dependent variable's categories have no natural ordering, we resort to multinomial logistic regression, also called polytomous logistic regression. The **mlogit** command makes this straightforward. If y has only two categories, **mlogit** fits the same model as **logistic**. Otherwise, though, an **mlogit** model is more complex. This section presents an extended example interpreting **mlogit** results, using data (*NWarctic.dta*) from a survey of high school students in Alaska's Northwest Arctic borough (Hamilton and Seyfrit 1993).

```
Contains data from C:\data\NWarctic.dta
  obs:            259                    NW Arctic high school students
                                          (Hamilton & Seyfrit 1993)
  vars:             3                    20 Jul 2005 10:40
  size:         2,590  (99.9% of memory free)
---------------------------------------------------------------------------
              storage   display     value
variable name   type    format      label       variable label
---------------------------------------------------------------------------
life            byte    %8.0g       migrate     Expect to live most of life?
ties            float   %9.0g                   Social ties to community scale
kotz            byte    %8.0g       kotz        Live in Kotzebue or smaller
                                                 village?
---------------------------------------------------------------------------
```

Variable *life* indicates where students say they expect to live most of the rest of their lives: in the same region (Northwest Arctic), elsewhere in Alaska, or outside of Alaska:

```
. tabulate life, plot
```

```
Expect to |
live most |
   of life? |      Freq.
-----------+-----------+------------------------------------------------------
     same |        92 |*******************************************
 other AK |       120 |********************************************************
 leave AK |        47 |*********************
-----------+-----------+------------------------------------------------------
    Total |       259
```

Kotzebue (population near 3,000) is the Northwest Arctic's regional hub and largest city. More than a third of these students live in Kotzebue. The rest live in smaller villages of 200 to 700 people. The relatively cosmopolitan Kotzebue students less often expect to stay where they are, and lean more towards leaving the state:

. **tabulate** *life kotz*, **chi2**

```
Expect to | Live in Kotzebue or
live most |    smaller village?
   of life? |   village   Kotzebue |     Total
-----------+----------------------+----------
     same |        75         17 |        92
 other AK |        80         40 |       120
 leave AK |        11         36 |        47
-----------+----------------------+----------
    Total |       166         93 |       259
```

Pearson chi2(2) = 46.2992   Pr = 0.000

mlogit can replicate this simple analysis (although its likelihood-ratio chi-squared need not exactly equal the Pearson chi-squared found by **tabulate**):

. **mlogit** *life kotz*, **nolog base(1) rrr**

```
Multinomial logistic regression                  Number of obs   =        259
                                                 LR chi2(2)      =      46.23
                                                 Prob > chi2     =     0.0000
Log likelihood = -244.64465                      Pseudo R2       =     0.0863


------------------------------------------------------------------------------
        life |      RRR    Std. Err.      z    P>|z|     [95% Conf. Interval]
-------------+----------------------------------------------------------------
other AK     |
        kotz |  2.205882   .7304664    2.39   0.017     1.152687    4.221369
-------------+----------------------------------------------------------------
leave AK     |
        kotz |   14.4385   6.307555    6.11   0.000     6.132946    33.99188
------------------------------------------------------------------------------
(Outcome life==same is the comparison group)
```

base(1) specifies that category 1 of y (*life* = "same") is the base category for comparison. The **rrr** option instructs **mlogit** to show relative risk ratios, which resemble the odds ratios given by **logistic**.

Referring back to the **tabulate** output, we can calculate that among Kotzebue students the odds favoring "leave Alaska" over "stay in the same area" are

$$P(\text{leave AK}) / P(\text{same}) = (36/93) / (17/93)$$
$$= 2.1176471$$

Among other students the odds favoring "leave Alaska" over "same area" are

$$P(\text{leave AK}) / P(\text{same}) = (11/166) / (75/166)$$
$$= .1466667$$

Thus, the odds favoring "leave Alaska" over "same area" are 14.4385 times higher for Kotzebue students than for others:

$$2.1176471 / .1466667 = 14.4385$$

This multiplier, a ratio of two odds, equals the relative risk ratio (14.4385) displayed by **mlogit**.

In general, the relative risk ratio for category j of y, and predictor x_k, equals the amount by which predicted odds favoring $y = j$ (compared with $y = $ base) are multiplied, per 1-unit increase in x_k, other things being equal. In other words, the relative risk ratio rrr_{jk} is a multiplier such that, if all x variables except x_k stay the same,

$$\text{rrr}_{jk} \times \frac{P(y = j \mid x_k)}{P(y = \text{base} \mid x_k)} = \frac{P(y = j \mid x_k + 1)}{P(y = \text{base} \mid x_k + 1)}$$

ties is a continuous scale indicating the strength of students' social ties to family and community. We include *ties* as a second predictor:

```
. mlogit life kotz ties, nolog base(1) rrr
```

```
Multinomial logistic regression              Number of obs    =        259
                                             LR chi2(4)       =      91.96
                                             Prob > chi2      =     0.0000
Log likelihood = -221.77969                  Pseudo R2        =     0.1717

-----------------------------------------------------------------------------
       life |       RRR    Std. Err.      z    P>|z|     [95% Conf. Interval]
------------+----------------------------------------------------------------
other AK    |
       kotz |   2.214184    .7724996     2.28   0.023     1.117483    4.387193
       ties |   .4802486    .0799184    -4.41   0.000     .3465911    .6654492
------------+----------------------------------------------------------------
leave AK    |
       kotz |  14.84604     7.146824     5.60   0.000     5.778907   38.13955
       ties |   .230262     .059085     -5.72   0.000     .1392531    .38075
-----------------------------------------------------------------------------
(Outcome life==same is the comparison group)
```

Asymptotic z tests here indicate that the four relative risk ratios, describing two x variables' effects, all differ significantly from 1.0. If a y variable has J categories, then **mlogit** models the effects of each predictor (x) variable with $J - 1$ relative risk ratios or coefficients, and hence also employs $J - 1$ z tests — evaluating two or more separate null hypotheses for each predictor. Likelihood-ratio tests evaluate the overall effect of each predictor. First, store the results from the full model, here given the name *full*:

```
. estimates store full
```

Then fit a simpler model with one of the x variables omitted, and perform a likelihood-ratio test. For example, to test the effect of *ties*, we repeat the regression with *ties* omitted:

```
. quietly mlogit life kotz
```

```
. estimates store no_ties
```

```
. lrtest no_ties full
```

```
likelihood-ratio test                              LR chi2(2)  =      45.73
(Assumption: no_ties nested in full)               Prob > chi2 =     0.0000
```

The effect of *ties* is clearly significant. Next, we run a similar test on the effect of *kotz*:

```
. quietly mlogit life ties
. estimates store no_kotz
. lrtest no_kotz full
```

```
likelihood-ratio test                              LR chi2(2)  =      39.05
(Assumption: no_kotz nested in full)               Prob > chi2 =     0.0000
```

   If our data contained missing values, the three **mlogit** commands just shown might have analyzed three overlapping subsets of observations. The full model would use only observations with nonmissing *life*, *kotz*, and *ties* values; the *kotz*-only model would bring back in any observations missing just their *ties* values; and the *ties*-only model would bring back observations missing just *kotz* values. When this happens, Stata returns an error messages saying "observations differ." In such cases, the likelihood-ratio test would be invalid. Analysts must either screen observations with **if** qualifiers attached to modeling commands, such as

```
. mlogit life kotz ties, nolog base(1) rrr
. estimates store full
. quietly mlogit life kotz if ties < .
. estimates store no_ties
. lrtest no_ties full
. quietly mlogit life ties if kotz < .
. estimates store no_kotz
. lrtest no_kotz full
```

or simply drop all observations having missing values before proceeding:

```
. drop if life >= . | kotz >= . | ties >= .
```

Dataset *NWarctic.dta* has already been screened in this fashion to drop observations with missing values.

   Both *kotz* and *ties* significantly predict *life*. What else can we say from this output? To interpret specific effects, recall that *life* = "same" is the base category. The relative risk ratios tell us that:

   Odds that a student expects migration to elsewhere in Alaska rather than staying in the same area are 2.21 times greater (increase about 121%) among Kotzebue students (*kotz*=1), adjusting for social ties to community.

   Odds that a student expects to leave Alaska rather than stay in the same area are 14.85 times greater (increase about 1385%) among Kotzebue students (*kotz*=1), adjusting for social ties to community.

   Odds that a student expects migration to elsewhere in Alaska rather than staying are multiplied by .48 (decrease about 52%) with each 1-unit (since *ties* is standardized, its units equal standard deviations) increase in social ties, controlling for Kotzebue/village residence.

Odds that a student expects to leave Alaska rather than staying are multiplied by .23 (decrease about 77%) with each 1-unit increase in social ties, controlling for Kotzebue/village residence.

predict can calculate predicted probabilities from **mlogit** . The **outcome(#)** option specifies for which *y* category we want probabilities. For example, to get predicted probabilities that *life* = "leave AK" (category 3),

```
. quietly mlogit life kotz ties

. predict PleaveAK, outcome(3)
(option p assumed; predicted probability)

. label variable PleaveAK "P(life = 3 | kotz, ties)"
```

Tabulating predicted probabilities for each value of the dependent variable shows how the model fits:

```
. table life, contents(mean PleaveAK) row

--------------------------
Expect to  |
live most  |
of life?   | mean(PleaveAK)
-----------+--------------
     same  |      .0811267
 other AK  |      .1770225
 leave AK  |      .3892264
           |
    Total  |      .1814672
--------------------------
```

A minority of these students (47/259 = 18%) expect to leave Alaska. The model averages only a .39 probability of leaving Alaska even for those who actually chose this response — reflecting the fact that although our predictors have significant effects, most variation in migration plans remains unexplained.

Conditional effect plots help to visualize what a model implies regarding continuous predictors. We can draw them using estimated coefficients (not risk ratios) to calculate probabilities:

```
. mlogit life kotz ties, nolog base(1)
```

```
Multinomial logistic regression              Number of obs   =        259
                                             LR chi2(4)      =      91.96
                                             Prob > chi2     =     0.0000
Log likelihood = -221.77969                  Pseudo R2       =     0.1717

------------------------------------------------------------------------------
        life |      Coef.   Std. Err.      z    P>|z|     [95% Conf. Interval]
-------------+----------------------------------------------------------------
other AK     |
        kotz |    .794884   .3488868     2.28   0.023     .1110784    1.47869
        ties |  -.7334513   .1664104    -4.41   0.000    -1.05961    -.407293
       _cons |    .206402   .1728053     1.19   0.232    -.1322902    .5450942
-------------+----------------------------------------------------------------
leave AK     |
        kotz |   2.697733   .4813959     5.60   0.000     1.754215    3.641252
        ties |  -1.468537   .2565991    -5.72   0.000    -1.971462   -.9656124
       _cons |  -2.115025   .3758163    -5.63   0.000    -2.851611   -1.378439
------------------------------------------------------------------------------
(Outcome life==same is the comparison group)
```

The following commands calculate predicted logits, and then the probabilities needed for conditional effect plots. *L2villag* represents the predicted logit of *life* = 2 (other Alaska) for village students. *L3kotz* is the predicted logit of *life* = 3 (leave Alaska) for Kotzebue students, and so forth:

```
. generate L2villag = .206402 +.794884*0 -.7334513*ties
. generate L2kotz = .206402 +.794884*1 -.7334513*ties
. generate L3villag = -2.115025 +2.697733*0 -1.468537*ties
. generate L3kotz = -2.115025 +2.697733*1 -1.468537*ties
```

Like other Stata modeling commands, **mlogit** saves coefficient estimates as macros. For example, [2]_b[*kotz*] refers to the coefficient on *kotz* in the model's second (*life* = 2) equation. Therefore, we could have generated the same predicted logits as follows. *L2v* will be identical to *L2villag* defined earlier, *L3k* the same as *L3kotz*, and so forth:

```
. generate L2v = [2]_b[_cons] +[2]_b[kotz]*0 +[2]_b[ties]*ties
. generate L2k = [2]_b[_cons] +[2]_b[kotz]*1 +[2]_b[ties]*ties
. generate L3v = [3]_b[_cons] +[3]_b[kotz]*0 + [3]_b[ties]*ties
. generate L3k = [3]_b[_cons] +[3]_b[kotz]*1 + [3]_b[ties]*ties
```

From either set of logits, we next calculate the predicted probabilities:

```
. generate P1villag = 1/(1 +exp(L2villag) +exp(L3villag))
. label variable P1villag "same area"
. generate P2villag = exp(L2villag)/(1+exp(L2villag)+exp(L3villag))
. label variable P2villag "other Alaska"
. generate P3villag = exp(L3villag)/(1+exp(L2villag)+exp(L3villag))
. label variable P3villag "leave Alaska"
. generate P1kotz = 1/(1 +exp(L2kotz) +exp(L3kotz))
. label variable P1kotz "same area"
. generate P2kotz = exp(L2kotz)/(1 +exp(L2kotz) +exp(L3kotz))
. label variable P2kotz "other Alaska"
. generate P3kotz = exp(L3kotz)/(1 +exp(L2kotz) +exp(L3kotz))
. label variable P3kotz "leave Alaska"
```

Figures 10.7 and 10.8 show conditional effect plots for village and Kotzebue students separately.

```
. graph twoway mspline P1villag ties, bands(50)
        || mspline P2villag ties, bands(50)
        || mspline P3villag ties, bands(50)
        || , xlabel(-3(1)3) ylabel(0(.2)1) yline(0 1) xline(0)
        legend(order(2 3 1) position(12) ring(0) label(1 "same area")
        label(2 "elsewhere Alaska") label(3 "leave Alaska") cols(1))
        ytitle("Probability")
```

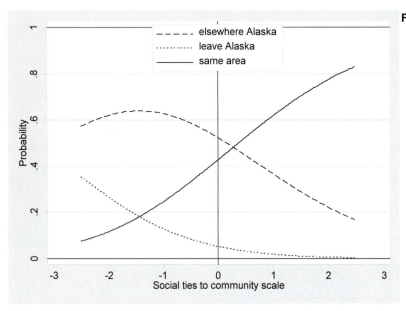

Figure 10.7

```
. graph twoway mspline P1kotz ties, bands(50)
    ||   mspline P2kotz ties, bands(50)
    ||   mspline P3kotz ties, bands(50)
    ||   , xlabel(-3(1)3) ylabel(0(.2)1) yline(0 1) xline(0)
    legend(order(3 2 1) position(12) ring(0) label(1 "same area")
    label(2 "elsewhere Alaska") label(3 "leave Alaska") cols(1))
    ytitle("Probability")
```

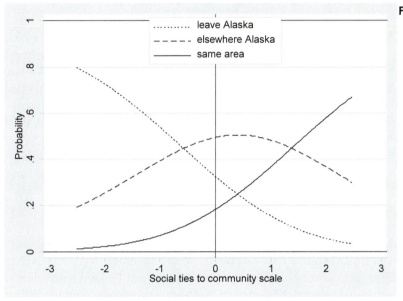

Figure 10.8

The plots indicate that among village students, social ties increase the probability of staying rather than moving elsewhere in Alaska. Relatively few village students expect to leave Alaska. In contrast, among Kotzebue students, *ties* particularly affects the probability of leaving Alaska, rather than simply moving elsewhere in the state. Only if they feel very strong social ties do Kotzebue students tend to favor staying put.

11

Survival and Event-Count Models

This chapter presents methods for analyzing event data. *Survival analysis* encompasses several related techniques that focus on times until the event of interest occurs. Although the event could be good or bad, by convention we refer to that event as a "failure." The time until failure is "survival time." Survival analysis is important in biomedical research, but it can be applied equally well to other fields from engineering to social science — for example, in modeling the time until an unemployed person gets a job, or a single person gets married. Stata offers a full range of survival analysis procedures, only a few of which are illustrated in this chapter.

We also look briefly at Poisson regression and its relatives. These methods focus not on survival times but, rather, on the rates or counts of events over a specified interval of time. Event-count methods include Poisson regression and negative binomial regression. Such models can be fit either through specialized commands, or through the broader approach of generalized linear modeling (GLM).

Consult the *Survival Analsysis and Epidemiological Tables Reference Manual* for more information about Stata's capabilities. Type **help st** to see an online overview. Selvin (1995) provides well-illustrated introductions to survival analysis and Poisson regression. I have borrowed (with permission) several of his examples. Other good introductions to survival analysis include the Stata-oriented volume by Cleves, Gould and Gutierrez (2004), a chapter in Rosner (1995), and comprehensive treatments by Hosmer and Lemeshow (1999) and Lee (1992). McCullagh and Nelder (1989) describe generalized linear models. Long (1997) has a chapter on regression models for count data (including Poisson and negative binomial), and also has some material on generalized linear models. An extensive and current treatment of generalized linear models is found in Hardin and Hilbe (2001).

Stata menu groups most relevant to this chapter include:

Statistics – Survival analysis

Graphics – Survival analysis graphs

Statistics – Count outcomes

Statistics – Generalized linear models (GLM)

Regarding epidemiological tables, not covered in this chapter, further information can be found by typing **help epitab** or exploring the menus for

Statistics – Observational/Epi. analysis.

Example Commands

Most of Stata's survival-analysis (`st*`) commands require that the data have previously been identified as survival-time by issuing an **stset** command (see following). **stset** need only be run once, and the data subsequently saved.

. `stset timevar, failure(failvar)`

Identifies single-record survival-time data. Variable *timevar* indicates the time elapsed before either a particular event (called a "failure") occurred, or the period of observation ended ("censoring"). Variable *failvar* indicates whether a failure (*failvar* = 1) or censoring (*failvar* = 0) occurred at *timevar*. The dataset contains only one record per individual. The dataset must be **stset** before any further `st*` commands will work. If we subsequently **save** the dataset, however, the **stset** definitions are saved as well. **stset** creates new variables named *_st, _d, _t*, and *_t0* that encode information necessary for subsequent `st*` commands.

. `stset timevar, failure(failvar) id(patient) enter(time start)`

Identifies multiple-record survival-time data. In this example, the variable *timevar* indicates elapsed time before failure or censoring; *failvar* indicates whether failure (1) or censoring (0) occurred at this time. *patient* is an identification number. The same individual might contribute more than one record to the data, but always has the same identification number. *start* records the time when each individual came under observation.

. `stdes`

Describes survival-time data, listing definitions set by **stset** and other characteristics of the data.

. `stsum`

Obtains summary statistics: the total time at risk, incidence rate, number of subjects, and percentiles of survival time.

. `ctset time nfail ncensor nenter, by(ethnic sex)`

Identifies count-time data. In this example, the variable *time* is a measure of time; *nfail* is the number of failures occurring at *time*. We also specified *ncensor* (number of censored observations at *time*) and *nenter* (number entering at *time*), although these can be optional. *ethnic* and *sex* are other categorical variables defining observations in these data.

. `cttost`

Converts count-time data, previously identified by a **ctset** command, into survival-time form that can be analyzed by `st*` commands.

. `sts graph`

Graphs the Kaplan–Meier survivor function. To visually compare two or more survivor functions, such as one for each value of the categorical variable *sex*, use the **by()** option,

. `sts graph, by(sex)`

To adjust, through Cox regression, for the effects of a continuous independent variable such as *age*, use the **adjustfor()** option,

. `sts graph, by(sex) adjustfor(age)`

Note: the **by()** and **adjustfor()** options work similarly with the other **sts** commands **sts list**, **sts generate**, and **sts test**.

. `sts list`
Lists the estimated Kaplan–Meier survivor (failure) function.

. `sts test sex`
Tests the equality of the Kaplan–Meier survivor function across categories of *sex*.

. `sts generate survfunc = S`
Creates a new variable arbitrarily named *survfunc*, containing the estimated Kaplan–Meier survivor function.

. `stcox x1 x2 x3`
Fits a Cox proportional hazard model, regressing time to failure on continuous or dummy variable predictors *x1–x3*.

. `stcox x1 x2 x3, strata(x4) basechazard(hazard) robust`
Fits a Cox proportional hazard model, stratified by *x4*. Stores the group-specific baseline cumulative hazard function as a new variable named *hazard*. (Baseline survivor function estimates could be obtained through a **basesur(survive)** option.) Obtains robust standard error estimates. See Chapter 9 or, for a more complete explanation of robust standard errors, consult the *User's Guide*.

. `stphplot, by(sex)`
Plots –ln(–ln(survival)) versus ln(analysis time) for each level of the categorical variable *sex*, from the previous **stcox** model. Roughly parallel curves support the Cox model assumption that the hazard ratio does not change with time. Other checks on the Cox assumptions are performed by the commands **stcoxkm** (compares Cox predicted curves with Kaplan–Meier observed survival curves) and **stphtest** (performs test based on Schoenfeld residuals). See **help stcox** for syntax and options.

. `streg x1 x2, dist(weibull)`
Fits Weibull-distribution model regression of time-to-failure on continuous or dummy variable predictors *x1* and *x2*.

. `streg x1 x2 x3 x4, dist(exponential) robust`
Fits exponential-distribution model regression of time-to-failure on continuous or dummy predictors *x1–x4*. Obtains heteroskedasticity-robust standard error estimates. In addition to Weibull and exponential, other **dist()** specifications for **streg** include lognormal, log-logistic, Gompertz, or generalized gamma distributions. Type **help streg** for more information.

. `stcurve, survival`
After **streg**, plots the survival function from this model at mean values of all the *x* variables.

. `stcurve, cumhaz at(x3=50, x4=0)`
After **streg**, plots the cumulative hazard function from this model at mean values of *x1* and *x2*, *x3* set at 50, and *x4* set at 0.

. `poisson count x1 x2 x3, irr exposure(x4)`
Performs Poisson regression of event-count variable *count* (assumed to follow a Poisson distribution) on continuous or dummy independent variables *x1–x3*. Independent-variable effects will be reported as incidence rate ratios (**irr**). The **exposure()** option identifies a variable indicating the amount of exposure, if this is not the same for all observations.

Note: A Poisson model assumes that the event probability remains constant, regardless of how many times an event occurs for each observation. If the probability does not remain constant, we should consider using **nbreg** (negative binomial regression) or **gnbreg** (generalized negative binomial regression) instead.

. glm *count x1 x2 x3*, link(log) family(poisson) lnoffset(*x4*) eform

Performs the same regression specified in the **poisson** example above, but as a generalized linear model (GLM). **glm** can fit Poisson, negative binomial, logit, and many other types of models, depending on what **link()** (link function) and **family()** (distribution family) options we employ.

Survival-Time Data

Survival-time data contain, at a minimum, one variable measuring how much time elapsed before a certain event occurred to each observation. The literature often terms this event of interest a "failure," regardless of its substantive meaning. When failure has not occurred to an observation by the time data collection ends, that observation is said to be "censored." The **stset** command sets up a dataset for survival-time analysis by identifying which variable measures time and (if necessary) which variable is a dummy indicating whether the observation failed or was censored. The dataset can also contain any number of other measurement or categorical variables, and individuals (for example, medical patients) can be represented by more than one observation.

To illustrate the use of **stset**, we will begin with an example from Selvin (1995:453) concerning 51 individuals diagnosed with HIV. The data initially reside in a raw-data file (*aids.raw*) that looks like this:

```
      1           1          1          34
      2          17          1          42
      3          37          0          47
            (rows 4–50 omitted)
     51          81          0          29
```

The first column values are case numbers (1, 2, 3, ..., 51). The second column tells how many months elapsed after the diagnosis, before that person either developed symptoms of AIDS or the study ended (1, 17, 37, ...). The third column holds a 1 if the individual developed AIDS symptoms (failure), or a 0 if no symptoms had appeared by the end of the study (censoring). The last column reports the individual's age at the time of diagnosis.

We can read the raw data into memory using **infile**, then label the variables and data and save in Stata format as file *aids1.dta*:

. infile *case time aids age* using *aids.raw*, clear
(51 observations read)

. label variable *case* "Case ID number"

. label variable *time* "Months since HIV diagnosis"

. label variable *aids* "Developed AIDS symptoms"

. label variable *age* "Age in years"

```
. label data "AIDS (Selvin 1995:453)"

. compress
case was float now byte
time was float now byte
aids was float now byte
age was float now byte

. save aids1
file c:\data\aids1.dta saved
```

The next step is to identify which variable measures time and which indicates failure/ censoring. Although not necessary with these single-record data, we can also note which variable holds individual case identification numbers. In an **stset** command, the first-named variable measures time. Subsequently, we identify with **failure()** the dummy representing whether an observation failed (1) or was censored (0). After using **stset**, we save the data again to preserve this information.

```
. stset time, failure(aids) id(case)

                id:  case
     failure event:  aids != 0 & aids < .
obs. time interval:  (time[_n-1], time]
 exit on or before:  failure

------------------------------------------------------------------------------
        51  total obs.
         0  exclusions
------------------------------------------------------------------------------
        51  obs. remaining, representing
        51  subjects
        25  failures in single failure-per-subject data
      3164  total analysis time at risk, at risk from t =          0
                              earliest observed entry t =          0
                                 last observed exit t =           97
. save, replace
file c:\data\aids1.dta saved
```

stdes yields a brief description of how our survival-time data are structured. In this simple example we have only one record per subject, so some of this information is unneeded.

```
. stdes
        failure _d:  aids
  analysis time _t:  time
              id:  case

                                  |------------- per subject -------------|
Category                 total        mean        min    median        max
---------------------------------------------------------------------------
no. of subjects             51
no. of records              51           1          1         1          1

(first) entry time                       0          0         0          0
(final) exit time                 62.03922          1        67         97

subjects with gap            0
time on gap if gap           0           .          .         .          .
time at risk              3164    62.03922          1        67         97

failures                    25    .4901961          0         0          1
---------------------------------------------------------------------------
```

The **stsum** command obtains summary statistics. We have 25 failures out of 3,164 person-months, giving an incidence rate of 25/3164 = .0079014. The percentiles of survival time derive from a Kaplan–Meier survivor function (next section). This function estimates about a 25% chance of developing AIDS within 41 months after diagnosis, and 50% within 81 months. Over the observed range of the data (up to 97 months) the probability of AIDS does not reach 75%, so there is no 75th percentile given.

```
. stsum
        failure _d:  aids
  analysis time _t:  time
              id:  case

          |                incidence     no. of   |------ Survival time -----|
          | time at risk       rate    subjects        25%       50%       75%
----------+----------------------------------------------------------------------
    total |         3164    .0079014         51         41        81         .
```

If the data happen to include a grouping or categorical variable such as *sex* (0 = male, 1 = female), we could obtain summary statistics on survival time separately for each group by a command of the following form:

```
. stsum, by(sex)
```

Later sections describe more formal methods for comparing survival times from two or more groups.

Count-Time Data

Survival-time (**st**) datasets like *aids1.dta* contain information on individual people or things, with variables indicating the time at which failure or censoring occurred for each individual. A different type of dataset called count-time (**ct**) contains aggregate data, with variables counting the number of individuals that failed or were censored at time *t*. For example, *diskdriv.dta* contains hypothetical test information on 25 disk drives. All but 5 drives failed before testing ended at 1,200 hours.

```
Contains data from C:\data\diskdriv.dta
  obs:             6                          Count-time data on disk drives
  vars:            3                          21 Jul 2005 09:34
  size:           48 (99.9% of memory free)
-------------------------------------------------------------------------------
              storage  display    value
variable name  type    format     label      variable label
-------------------------------------------------------------------------------
hours          int     %8.0g                  Hours of continuous operation
failures       byte    %8.0g                  Number of failures observed
censored       byte    %9.0g                  Number still working
-------------------------------------------------------------------------------
Sorted by:
```

. **list**

```
      +-----------------------------+
      | hours   failures   censored |
      |-----------------------------|
  1.  |   200          2          0 |
  2.  |   400          3          0 |
  3.  |   600          4          0 |
  4.  |   800          8          0 |
  5.  |  1000          3          0 |
      |-----------------------------|
  6.  |  1200          0          5 |
      +-----------------------------+
```

To set up a count-time dataset, we specify the time variable, the number-of-failures variable, and the number-censored variable, in that order. After **ctset**, the **cttost** command automatically converts our count-time data to survival-time format.

. **ctset** *hours failures censored*

```
    dataset name:  C:\data\diskdriv.dta
            time:  hours
        no. fail:  failures
        no. lost:  censored
       no. enter:  --                        (meaning all enter at time 0)
```

. **cttost**

```
(data are now st)

       failure event:  failures != 0 & failures < .
  obs. time interval:  (0, hours]
  exit on or before:  failure
              weight:  [fweight=w]


-------------------------------------------------------------------------------
      6  total obs.
      0  exclusions
-------------------------------------------------------------------------------
      6  physical obs. remaining, equal to
     25  weighted obs., representing
     20  failures in single record/single failure data
  19400  total analysis time at risk, at risk from t =          0
                              earliest observed entry t =          0
                               last observed exit t =       1200
```

```
. list

     +-----------------------------------------------+
     | hours    failures    w    _st    _d      _t    _t0 |
     |-----------------------------------------------|
  1. |  1200           0     5      1      0    1200      0 |
  2. |   200           1     2      1      1     200      0 |
  3. |   400           1     3      1      1     400      0 |
  4. |   600           1     4      1      1     600      0 |
  5. |   800           1     8      1      1     800      0 |
     |-----------------------------------------------|
  6. |  1000           1     3      1      1    1000      0 |
     +-----------------------------------------------+

. stdes

          failure _d:  failures
    analysis time _t:  hours
              weight:  [fweight=w]

                                   |------------- per subject -------------|
                    unweighted   unweighted                    unweighted
Category                 total         mean        min         median         max
----------------------------------------------------------------------------------
no. of subjects              6
no. of records               6            1          1              1           1

(first) entry time                        0          0              0           0
(final) exit time                       700        200            700        1200

subjects with gap            0
time on gap if gap           0
time at risk              4200          700        200            700        1200

failures                     5    .8333333          0              1           1
----------------------------------------------------------------------------------
```

The **cttost** command defines a set of frequency weights, w, in the resulting **st**-format dataset. **st*** commands automatically recognize and use these weights in any survival-time analysis, so the data now are viewed as containing 25 observations (25 disk drives) instead of the previous 6 (six time periods).

```
. stsum

     failure time:  hours
   failure/censor:  failures
           weight:  [fweight=w]

       |                     incidence      no. of    |---- Survival time ----|
       | time at risk           rate      subjects      25%       50%       75%
-------+------------------------------------------------------------------------
 total |        19400       .0010309            25      600       800      1000
```

Kaplan–Meier Survivor Functions

Let n_t represent the number of observations that have not failed, and are not censored, at the beginning of time period t. d_t represents the number of failures that occur to these observations during time period t. The Kaplan–Meier estimator of surviving beyond time t is the product of survival probabilities in t and the preceding periods:

$$S(t) = \prod_{j=t0}^{t} \{ (n_j - d_j) / n_j \} \qquad\qquad [11.1]$$

For example, in the AIDS data seen earlier, one of the 51 individuals developed symptoms only one month after diagnosis. No observations were censored this early, so the probability of "surviving" (meaning, not developing AIDS) beyond *time* = 1 is

$$S(1) = (51 - 1) / 51 = .9804$$

A second patient developed symptoms at *time* = 2, and a third at *time* = 9:

$$S(2) = .9804 \times (50 - 1) / 50 = .9608$$

$$S(9) = .9608 \times (49 - 1) / 49 = .9412$$

Graphing *S(t)* against *t* produces a Kaplan–Meier survivor curve, like the one seen in Figure 11.1. Stata draws such graphs automatically with the **sts graph** command. For example,

```
. use aids, clear
(AIDS (Selvin 1995:453))

. sts graph

        failure _d:  aids
  analysis time _t:  time
              id:  case
```

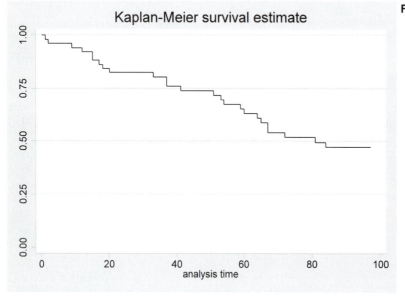

Figure 11.1

Kaplan-Meier survival estimate

For a second example of survivor functions, we turn to data in *smoking1.dta*, adapted from Rosner (1995). The observations are 234 former smokers, attempting to quit. Most did not succeed. Variable *days* records how many days elapsed between quitting and starting up again. The study lasted one year, and variable *smoking* indicates whether an individual resumed

smoking before the end of this study (*smoking* = 1, "failure") or not (*smoking* = 0, "censored").
With new data, we should begin by using **stset** to set the data up for survival-time analysis:

```
Contains data from C:\data\smoking1.dta
  obs:           234                          Smoking (Rosner 1995:607)
  vars:            8                          21 Jul 2005 09:35
  size:        3,744 (99.9% of memory free)
-------------------------------------------------------------------------------
              storage   display    value
variable name   type    format     label      variable label
-------------------------------------------------------------------------------
id              int     %9.0g                 Case ID number
days            int     %9.0g                 Days abstinent
smoking         byte    %9.0g                 Resumed smoking
age             byte    %9.0g                 Age in years
sex             byte    %9.0g      sex        Sex (female)
cigs            byte    %9.0g                 Cigarettes per day
co              int     %9.0g                 Carbon monoxide x 10
minutes         int     %9.0g                 Minutes elapsed since last cig
-------------------------------------------------------------------------------
Sorted by:
```

. **stset** *days*, **failure(***smoking***)**

```
     failure event:  smoking != 0 & smoking < .
obs. time interval:  (0, days]
 exit on or before:  failure


-------------------------------------------------------------------------------
    234  total obs.
      0  exclusions
-------------------------------------------------------------------------------
    234  obs. remaining, representing
    201  failures in single record/single failure data
  18946  total analysis time at risk, at risk from t =           0
                           earliest observed entry t =           0
                              last observed exit t =           366
```

The study involved 110 men and 124 women. Incidence rates for both sexes appear to be similar:

. **stsum, by(***sex***)**

```
      failure _d:  smoking
 analysis time _t:  days
```

sex	time at risk	incidence rate	no. of subjects	25%	Survival time 50%	75%
Male	8813	.0105526	110	4	15	68
Female	10133	.0106582	124	4	15	91
total	18946	.0106091	234	4	15	73

Figure 11.2 confirms this similarity, showing little difference between the survivor functions of men and women. That is, both sexes returned to smoking at about the same rate. The survival probabilities of nonsmokers decline very steeply during the first 30 days after quitting. For either sex, there is less than a 15% chance of surviving beyond a full year.

```
. sts graph, by(sex)
```

```
         failure _d:  smoking
   analysis time _t:  days
```

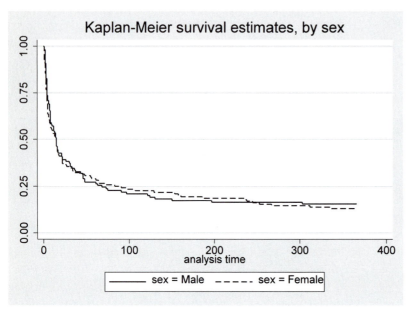

Figure 11.2

We can also formally test for the equality of survivor functions using a log-rank test. Unsurprisingly, this test finds no significant difference (P = .6772) between the smoking recidivism of men and women.

```
. sts test sex
```

```
         failure _d:  smoking
   analysis time _t:  days
```

Log-rank test for equality of survivor functions

sex		Events observed	Events expected
Male		93	95.88
Female		108	105.12
Total		201	201.00

```
            chi2(1) =      0.17
            Pr>chi2 =      0.6772
```

Cox Proportional Hazard Models

Regression methods allow us to take survival analysis further and examine the effects of multiple continuous or categorical predictors. One widely-used method known as Cox regression employs a proportional hazard model. The hazard rate for failure at time t is defined as

$$h(t) \quad = \quad \frac{\text{probability of failing between times } t \text{ and } t + \Delta t}{(\Delta t) \text{ (probability of failing after time } t)} \qquad [11.2]$$

We model this hazard rate as a function of the baseline hazard (h_0) at time t, and the effects of one or more x variables,

$$h(t) \quad = \quad h_0(t) \exp(\beta_1 x_1 + \beta_2 x_2 + \ldots + \beta_k x_k) \qquad [11.3a]$$

or, equivalently,

$$\ln[h(t)] = \quad \ln[h_0(t)] + \beta_1 x_1 + \beta_2 x_2 + \ldots + \beta_k x_k \qquad [11.3b]$$

"Baseline hazard" means the hazard for an observation with all x variables equal to 0. Cox regression estimates this hazard nonparametrically and obtains maximum-likelihood estimates of the β parameters in [11.3]. Stata's **stcox** procedure ordinarily reports hazard ratios, which are estimates of $\exp(\beta)$. These indicate proportional changes relative to the baseline hazard rate.

Does age affect the onset of AIDS symptoms? Dataset *aids.dta* contains information that helps answer this question. Note that with **stcox**, unlike most other Stata model-fitting commands, we list only the independent variable(s). The survival-analysis dependent variables, time variables, and censoring variables are understood automatically with **stset** data.

```
. use aids
(AIDS (Selvin 1995:453))


. stcox age, nolog

        failure _d:  aids
   analysis time _t:  time
             id:  case

Cox regression -- Breslow method for ties

No. of subjects =           51          Number of obs   =          51
No. of failures =           25
Time at risk    =         3164
                                         LR chi2(1)      =        5.00
Log likelihood  =    -86.576295          Prob > chi2     =      0.0254

------------------------------------------------------------------------------
        _t | Haz. Ratio   Std. Err.      z    P>|z|     [95% Conf. Interval]
-----------+------------------------------------------------------------------
       age |   1.084557    .0378623     2.33   0.020     1.01283    1.161363
------------------------------------------------------------------------------
```

We might interpret the estimated hazard ratio, 1.084557, with reference to two HIV-positive individuals whose ages are a and $a + 1$. The older person is 8.5% more likely to develop AIDS symptoms over a short period of time (that is, the ratio of their respective hazards is 1.084557). This ratio differs significantly ($P = .020$) from 1. If we wanted to state our findings for a five-year difference in age, we could raise the hazard ratio to the fifth power:

```
. display exp(_b[age])^5
1.5005865
```

Thus, the hazard of AIDS onset is about 50% higher when the second person is five years older than the first. Alternatively, we could learn the same thing (and obtain the new confidence interval) by repeating the regression after creating a new version of *age* measured in five-year units. The **nolog noshow** options below suppress display of the iteration log and the **st**-dataset description.

```
. generate age5 = age/5
. label variable age5 "age in 5-year units"
. stcox age5, nolog noshow

Cox regression -- Breslow method for ties

No. of subjects =            51          Number of obs   =           51
No. of failures =            25
Time at risk    =          3164
                                         LR chi2(1)      =         5.00
Log likelihood  =    -86.576295          Prob > chi2     =       0.0254

------------------------------------------------------------------------
         _t | Haz. Ratio   Std. Err.       z    P>|z|    [95% Conf. Interval]
------------+-----------------------------------------------------------
       age5 |   1.500587    .2619305    2.33   0.020    1.065815    2.112711
------------------------------------------------------------------------
```

Like ordinary regression, Cox models can have more than one independent variable. Dataset *heart.dta* contains survival-time data from Selvin (1995) on 35 patients with very high cholesterol levels. Variable *time* gives the number of days each patient was under observation. *coronary* indicates whether a coronary event occurred during this time (*coronary* = 1) or not (*coronary* = 0). The data also include cholesterol levels and other factors thought to affect heart disease. File *heart.dta* was previously set up for survival-time analysis by an **stset time, failure(coronary)** command, so we can go directly to **st** analysis.

```
. describe patient - ab

              storage  display    value
variable name   type   format     label      variable label
------------------------------------------------------------------------
patient         byte   %9.0g                 Patient ID number
time            int    %9.0g                 Time in days
coronary        byte   %9.0g                 Coronary event (1) or none (0)
weight          int    %9.0g                 Weight in pounds
sbp             int    %9.0g                 Systolic blood pressure
chol            int    %9.0g                 Cholesterol level
cigs            byte   %9.0g                 Cigarettes smoked per day
ab              byte   %9.0g                 Type A (1) or B (0) personality
```

```
. stdes
```

```
       failure _d:  coronary
   analysis time _t:  time
```

```
                                 |------------- per subject -------------|
Category                  total          mean           min     median          max
---------------------------------------------------------------------------------
no. of subjects              35
no. of records               35             1             1          1            1

(first) entry time                          0             0          0            0
(final) exit time                    2580.629           773       2875         3141

subjects with gap             0
time on gap if gap            0
time at risk              90322      2580.629           773       2875         3141

failures                      8       .2285714             0          0            1
---------------------------------------------------------------------------------
```

Cox regression finds that cholesterol level and cigarettes both significantly increase the hazard of a coronary event. Counterintuitively, weight appears to decrease the hazard. Systolic blood pressure and A/B personality do not have significant net effects.

```
. stcox weight sbp chol cigs ab, noshow nolog
```

```
Cox regression -- no ties

No. of subjects =            35              Number of obs   =           35
No. of failures =             8
Time at risk    =         90322
                                             LR chi2(5)      =        13.97
Log likelihood  =    -17.263231             Prob > chi2     =       0.0158
```

_t	Haz. Ratio	Std. Err.	z	P>|z|	[95% Conf. Interval]	
weight	.9349336	.0305184	-2.06	0.039	.8769919	.9967034
sbp	1.012947	.0338061	0.39	0.700	.9488087	1.081421
chol	1.032142	.0139984	2.33	0.020	1.005067	1.059947
cigs	1.203335	.1071031	2.08	0.038	1.010707	1.432676
ab	3.04969	2.985616	1.14	0.255	.4476492	20.77655

After estimating the model, **stcox** can also generate new variables holding the estimated baseline cumulative hazard and survivor functions. Since "baseline" refers to a situation with all *x* variables equal to zero, however, we first need to recenter some variables so that 0 values make sense. A patient who weighs 0 pounds, or has 0 blood pressure, does not provide a useful comparison. Guided by the minimum values actually in our data, we might shift *weight* so that 0 indicates 120 pounds, *sbp* so that 0 indicates 100, and *chol* so that 0 indicates 340:

```
. summarize patient - ab
```

Variable	Obs	Mean	Std. Dev.	Min	Max
patient	35	18	10.24695	1	35
time	35	2580.629	616.0796	773	3141
coronary	35	.2285714	.426043	0	1
weight	35	170.0857	23.55516	120	225
sbp	35	129.7143	14.28403	104	154

```
        chol |       35   369.2857   51.32284        343        645
        cigs |       35   17.14286   13.07702          0         40
          ab |       35   .5142857   .5070926          0          1
```

```
. replace weight = weight - 120
(35 real changes made)
```

```
. replace sbp = sbp - 100
(35 real changes made)
```

```
. replace chol = chol - 340
(35 real changes made)
```

```
. summarize patient - ab
```

Variable	Obs	Mean	Std. Dev.	Min	Max
patient	35	18	10.24695	1	35
time	35	2580.629	616.0796	773	3141
coronary	35	.2285714	.426043	0	1
weight	35	50.08571	23.55516	0	105
sbp	35	29.71429	14.28403	4	54
chol	35	29.28571	51.32284	3	305
cigs	35	17.14286	13.07702	0	40
ab	35	.5142857	.5070926	0	1

Zero values for all the *x* variables now make more substantive sense. To create new variables holding the baseline survivor and cumulative hazard function estimates, we repeat the regression with **basesurv()** and **basechaz()** options:

```
. stcox weight sbp chol cigs ab, noshow nolog basesurv(survivor)
     basechaz(hazard)
```

```
Cox regression -- no ties
```

```
No. of subjects =           35                 Number of obs    =           35
No. of failures =            8
Time at risk    =        90322
                                                LR chi2(5)       =        13.97
Log likelihood  =    -17.263231                 Prob > chi2      =       0.0158
```

_t	Haz. Ratio	Std. Err.	z	P>|z|	[95% Conf. Interval]	
weight	.9349336	.0305184	-2.06	0.039	.8769919	.9967034
sbp	1.012947	.0338061	0.39	0.700	.9488087	1.081421
chol	1.032142	.0139984	2.33	0.020	1.005067	1.059947
cigs	1.203335	.1071031	2.08	0.038	1.010707	1.432676
ab	3.04969	2.985616	1.14	0.255	.4476492	20.77655

Note that recentering three *x* variables had no effect on the hazard ratios, standard errors, and so forth. The command created two new variables, arbitrarily named *survivor* and *hazard*. To graph the baseline survivor function, we plot *survivor* against *time* and connect data points with in a stairstep fashion, as seen in Figure 11.3.

```
. graph twoway line survivor time, connect(stairstep) sort
```

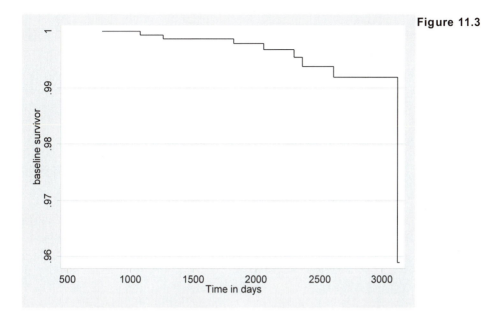

Figure 11.3

The baseline survivor function — which depicts survival probabilities for patients having "0" weight (120 pounds), "0" blood pressure (100), "0" cholesterol (340), 0 cigarettes per day, and a type B personality — declines with time. Although this decline looks precipitous at the right, notice that the probability really only falls from 1 to about .96. Given less favorable values of the predictor variables, the survival probabilities would fall much faster.

The same baseline survivor-function graph could have been obtained another way, without **stcox** . The alternative, shown in Figure 11.4, employs an **sts graph** command with **adjustfor()** option listing the predictor variables:

```
. sts graph, adjustfor(weight sbp chol cigs ab)
```

```
        failure _d:  coronary
  analysis time _t:  time
```

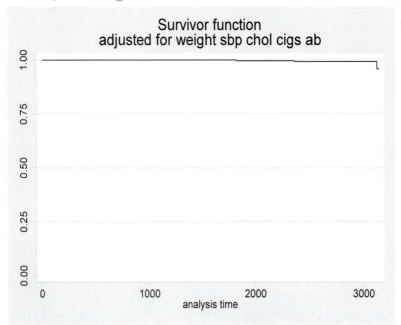

Figure 11.4

Figure 11.4, unlike Figure 11.3, follows the usual survivor-function convention of scaling the vertical axis from 0 to 1. Apart from this difference in scaling, Figures 11.3 and 11.4 depict the same curve.

Figure 11.5 graphs the estimated baseline cumulative hazard against time, using the variable (*hazard*) generated by our **stcox** command. This graph shows the baseline cumulative hazard increasing in 8 steps (because 8 patients "failed" or had coronary events), from near 0 to .033.

```
. graph twoway connected hazard time, connect(stairstep) sort
     msymbol(Oh)
```

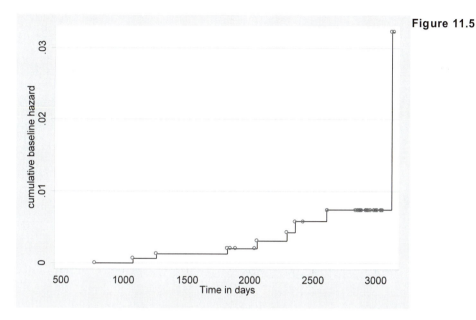

Figure 11.5

Exponential and Weibull Regression

Cox regression estimates the baseline survivor function empirically without reference to any theoretical distribution. Several alternative "parametric" approaches begin instead from assumptions that survival times do follow a known theoretical distribution. Possible distribution families include the exponential, Weibull, lognormal, log-logistic, Gompertz, or generalized gamma. Models based on any of these can be fit through the **streg** command. Such models have the same general form as Cox regression (equations [11.2] and [11.3]), but define the baseline hazard $h_0(t)$ differently. Two examples appear in this section.

If failures occur randomly, with a constant hazard, then survival times follow an exponential distribution and could be analyzed by *exponential regression*. Constant hazard means that the individuals studied do not "age," in the sense that they are no more or less likely to fail late in the period of observation than they were at its start. Over the long term, this assumption seems unjustified for machines or living organisms, but it might approximately hold if the period of observation covers a relatively small fraction of their life spans. An exponential model implies that logarithms of the survivor function, $\ln(S(t))$, are linearly related to t.

A second common parametric approach, *Weibull regression,* is based on the more general Weibull distribution. This does not require failure rates to remain constant, but allows them to increase or decrease smoothly over time. The Weibull model implies that $\ln(-\ln(S(t)))$ is a linear function of $\ln(t)$.

Graphs provide a useful diagnostic for the appropriateness of exponential or Weibull models. For example, returning to *aids.dta*, we construct a graph (Figure 11.6) of $\ln(S(t))$ versus time, after first generating Kaplan–Meier estimates of the survivor function $S(t)$. The *y*-axis labels in Figure 11.6 are given a fixed two-digit, one-decimal display format (%2.1f) and oriented horizontally, to improve their readability.

```
. use aids, clear
(AIDS (Selvin 1995:453))

. sts gen S = S

. generate logS = ln(S)

. graph twoway scatter logS time,
    ylabel(-.8(.1)0, format(%2.1f) angle(horizontal))
```

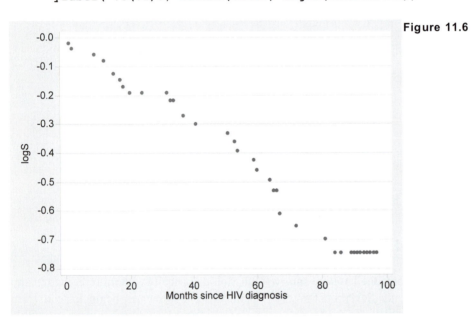

Figure 11.6

The pattern in Figure 11.6 appears somewhat linear, encouraging us to try an exponential regression:

```
.   streg age, dist(exponential) nolog noshow
```

```
Exponential regression -- log relative-hazard form

No. of subjects =         51            Number of obs   =         51
No. of failures =         25
Time at risk    =       3164
                                        LR chi2(1)      =       4.34
Log likelihood  =  -59.996976           Prob > chi2     =     0.0372

------------------------------------------------------------------------------
        _t | Haz. Ratio   Std. Err.      z    P>|z|     [95% Conf. Interval]
-----------+------------------------------------------------------------------
       age |   1.074414    .0349626     2.21   0.027     1.008028    1.145172
------------------------------------------------------------------------------
```

The hazard ratio (1.074) and standard error (.035) estimated by this exponential regression do not greatly differ from their counterparts (1.085 and .038) in our earlier Cox regression. The similarity reflects the degree of correspondence between empirical and exponential hazard functions. According to this exponential model, the hazard of an HIV-positive individual developing AIDS increases about 7.4% with each year of age.

After **streg**, the **stcurve** command draws a graph of the models' cumulative hazard, survival, or hazard functions. By default, **stcurve** draws these curves holding all *x* variables in the model at their means. We can specify other *x* values by using the **at()** option. The individuals in *aids.dta* ranged from 26 to 50 years old. We could graph the survival function at *age* = 26 by issuing a command such as

. stcurve, surviv at(*age=26*)

A more informative graph uses the **at1()** and **at2()** options to show the survival curve at two different sets of *x* values, such as the low and high extremes of *age*:

. stcurve, *survival* at1(*age=26*) at2(*age=50*) connect(direct direct)

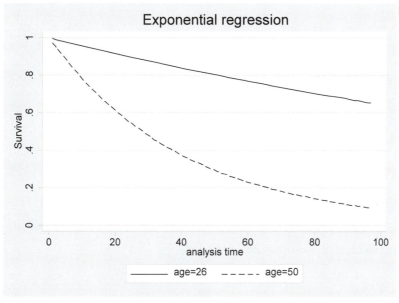

Figure 11.7

Figure 11.7 shows the predicted survival curve (for transition from HIV diagnosis to AIDS) falling more steeply among older patients. The significant *age* hazard ratio greater than 1 in our exponential regression table implied the same thing, but using **stcurve** with **at1()** and **at2()** values gives a strong visual interpretation of this effect. These options work in a similar manner with all three types of **stcurve** graphs:

stcurve, survival	Survival function.
stcurve, hazard	Hazard function.
stcurve, cumhaz	Cumulative hazard function.

Instead of the exponential distribution, **streg** can also fit survival models based on the Weibull distribution. A Weibull distribution might appear curvilinear in a plot of $\ln(S(t))$ versus *t*, but it should be linear in a plot of $\ln(-\ln(S(t)))$ versus $\ln(t)$, such as Figure 11.8. An exponential distribution, on the other hand, will appear linear in both plots and have a slope equal to 1 in the $\ln(-\ln(S(t)))$ versus $\ln(t)$ plot. In fact, the data points in Figure 11.8 are not far from a line with slope 1, suggesting that our previous exponential model is adequate.

```
. generate loglogS = ln(-ln(S))
```

```
. generate logtime = ln(time)
```

```
. graph twoway scatter loglogS logtime, ylabel(,angle(horizontal))
```

Figure 11.8

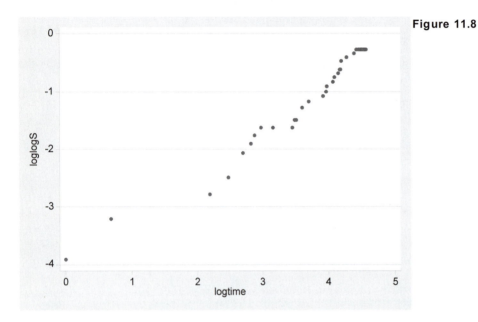

Although we do not need the additional complexity of a Weibull model with these data, results are given below for illustration.

```
. streg age, dist(weibull) noshow nolog
```

Weibull regression -- log relative-hazard form

No. of subjects =	51	Number of obs	=	51
No. of failures =	25			
Time at risk =	3164			
		LR chi2(1)	=	4.68
Log likelihood =	-59.778257	Prob > chi2	=	0.0306

_t	Haz. Ratio	Std. Err.	z	P>|z|	[95% Conf. Interval]	
age	1.079477	.0363509	2.27	0.023	1.010531	1.153127
/ln_p	.1232638	.1820858	0.68	0.498	-.2336179	.4801454
p	1.131183	.2059723			.7916643	1.616309
1/p	.8840305	.1609694			.6186934	1.263162

The Weibull regression obtains a hazard ratio estimate (1.079) intermediate between our previous Cox and exponential results. The most noticeable difference from those earlier models is the presence of three new lines at the bottom of the table. These refer to the Weibull distribution shape parameter p. A p value of 1 corresponds to an exponential model: the hazard does not change with time. $p > 1$ indicates that the hazard increases with time; $p < 1$ indicates that the hazard decreases. A 95% confidence interval for p ranges from .79 to 1.62, so we have no reason to reject an exponential ($p = 1$) model here. Different, but mathematically equivalent, parameterizations of the Weibull model focus on $\ln(p)$, p, or $1/p$, so Stata provides all three. **stcurve** draws survival, hazard, or cumulative hazard functions after **streg, dist(weibull)** just as it does after **streg, dist(exponential)** or other **streg** models.

Exponential or Weibull regression is preferable to Cox regression when survival times actually follow an exponential or Weibull distribution. When they do not, these models are misspecified and can yield misleading results. Cox regression, which makes no *a priori* assumptions about distribution shape, remains useful in a wider variety of situations.

In addition to exponential and Weibull models, **streg** can fit models based on the Gompertz, lognormal, log-logistic, or generalized gamma distributions. Type **help streg**, or consult the *Survival Analysis and Epidemiological Tables Reference Manual*, for syntax and a list of current options.

Poisson Regression

If events occur independently and with constant probability, then counts of events over a given period of time follow a Poisson distribution. Let r_j represent the incidence rate:

$$r_j = \frac{\text{count of events}}{\text{number of times event could have occurred}} \qquad [11.4]$$

The denominator in [11.4] is termed the "exposure" and is often measured in units such as person-years. We model the logarithm of incidence rate as a linear function of one or more predictor (x) variables:

$$\ln(r_t) = \beta_0 + \beta_1 x_1 + \beta_2 x_2 + \ldots + \beta_k x_k \qquad [11.5a]$$

Equivalently, the model describes logs of expected event counts:

$$\ln(\textit{expected count}) = \ln(\textit{exposure}) + \beta_0 + \beta_1 x_1 + \beta_2 x_2 + \ldots \beta_k x_k \qquad [11.5b]$$

Assuming that a Poisson process underlies the events of interest, Poisson regression finds maximum-likelihood estimates of the β parameters.

Data on radiation exposure and cancer deaths among workers at Oak Ridge National Laboratory provide an example. The 56 observations in dataset *oakridge.dta* represent 56 age/radiation-exposure categories (7 categories of age × 8 categories of radiation). For each combination, we know the number of deaths and the number of person-years of exposure.

```
Contains data from C:\data\oakridge.dta
  obs:            56                          Radiation (Selvin 1995:474)
  vars:            4                          21 Jul 2005 09:34
  size:          616 (99.9% of memory free)
-------------------------------------------------------------------------
              storage   display    value
variable name   type    format     label        variable label
-------------------------------------------------------------------------
age             byte    %9.0g      ageg         Age group
rad             byte    %9.0g                   Radiation exposure level
```

```
deaths            byte    %9.0g              Number of deaths
pyears            float   %9.0g              Person-years
------------------------------------------------------------------------
Sorted by:
```

```
. summarize

    Variable |      Obs        Mean   Std. Dev.        Min         Max
-------------+--------------------------------------------------------
         age |       56           4      2.0181          1           7
         rad |       56         4.5    2.312024          1           8
      deaths |       56    1.839286    3.178203          0          16
      pyears |       56    3807.679    10455.91         23       71382
```

```
. list in 1/6

       +----------------------------------+
       |  age   rad   deaths    pyears |
       |----------------------------------|
  1.   | < 45    1        0     29901 |
  2.   | 45-49   1        1      6251 |
  3.   | 50-54   1        4      5251 |
  4.   | 55-59   1        3      4126 |
  5.   | 60-64   1        3      2778 |
       |----------------------------------|
  6.   | 65-69   1        1      1607 |
       +----------------------------------+
```

Does the death rate increase with exposure to radiation? Poisson regression finds a statistically significant effect:

```
. poisson deaths rad, nolog exposure(pyears) irr

Poisson regression                          Number of obs   =          56
                                            LR chi2(1)      =       14.87
                                            Prob > chi2     =      0.0001
Log likelihood =  -169.7364                 Pseudo R2       =      0.0420


------------------------------------------------------------------------
      deaths |        IRR   Std. Err.       z    P>|z|     [95% Conf. Interval]
-------------+----------------------------------------------------------
         rad |   1.236469    .0603551    4.35   0.000     1.123657    1.360606
      pyears |  (exposure)
------------------------------------------------------------------------
```

For the regression above, we specified the event count (*deaths*) as the dependent variable and radiation (*rad*) as the independent variable. The Poisson "exposure" variable is *pyears*, or person-years in each category of *rad*. The **irr** option calls for incidence rate ratios rather than regression coefficients in the results table — that is, we get estimates of $\exp(\beta)$ instead of β, the default. According to this incidence rate ratio, the death rate becomes 1.236 times higher (increases by 23.6%) with each increase in radiation category. Although that ratio is statistically significant, the fit is not impressive; the pseudo R^2 (see equation [10.4]) is only .042.

To perform a goodness-of-fit test, comparing the Poisson model's predictions with the observed counts, use the follow-up command **poisgof**:

```
. poisgof
```

```
        Goodness-of-fit chi2   =   254.5475
        Prob > chi2(54)        =     0.0000
```

These goodness-of-fit test results ($\chi^2 = 254.5, P < .00005$) indicate that our model's predictions are significantly different from the actual counts — another sign that the model fits poorly.

We obtain better results when we include *age* as a second predictor. Pseudo R^2 then rises to .5966, and the goodness-of-fit test no longer leads us to reject our model.

```
. poisson deaths rad age, nolog exposure(pyears) irr
```

```
Poisson regression                              Number of obs   =         56
                                                LR chi2(2)      =     211.41
                                                Prob > chi2     =     0.0000
Log likelihood =   -71.4653                     Pseudo R2       =     0.5966

------------------------------------------------------------------------------
     deaths |        IRR   Std. Err.       z    P>|z|     [95% Conf. Interval]
------------+-----------------------------------------------------------------
        rad |   1.176673   .0593446     3.23    0.001     1.065924    1.298929
        age |   1.960034   .0997536    13.22    0.000     1.773955    2.165631
     pyears |  (exposure)
------------------------------------------------------------------------------
```

```
. poisgof
```

```
        Goodness-of-fit chi2   =   58.00534
        Prob > chi2(53)        =    0.2960
```

For simplicity, to this point we have treated *rad* and *age* as if both were continuous variables, and we expect their effects on the log death rate to be linear. In fact, however, both independent variables are measured as ordered categories. *rad* = 1, for example, means 0 radiation exposure; *rad* = 2 means 0 to 19 milliseiverts; *rad* = 3 means 20 to 39 milliseiverts; and so forth. An alternative way to include radiation exposure categories in the regression, while watching for nonlinear effects, is as a set of dummy variables. Below we use the **gen()** option of **tabulate** to create 8 dummy variables, *r1* to *r8*, representing each of the 8 values of *rad*.

```
. tabulate rad, gen(r)
```

```
  Radiation |
   exposure |
      level |      Freq.     Percent        Cum.
------------+-----------------------------------
          1 |          7       12.50       12.50
          2 |          7       12.50       25.00
          3 |          7       12.50       37.50
          4 |          7       12.50       50.00
          5 |          7       12.50       62.50
          6 |          7       12.50       75.00
          7 |          7       12.50       87.50
          8 |          7       12.50      100.00
------------+-----------------------------------
      Total |         56      100.00
```

```
. describe
```

```
Contains data from C:\data\oakridge.dta
  obs:             56                          Radiation (Selvin 1995:474)
 vars:             12                          21 Jul 2005 09:34
 size:          1,064 (99.9% of memory free)
-------------------------------------------------------------------------------
              storage   display    value
variable name   type     format    label      variable label
-------------------------------------------------------------------------------
age             byte     %9.0g      ageg       Age group
rad             byte     %9.0g                 Radiation exposure level
deaths          byte     %9.0g                 Number of deaths
pyears          float    %9.0g                 Person-years
r1              byte     %8.0g                 rad==      1.0000
r2              byte     %8.0g                 rad==      2.0000
r3              byte     %8.0g                 rad==      3.0000
r4              byte     %8.0g                 rad==      4.0000
r5              byte     %8.0g                 rad==      5.0000
r6              byte     %8.0g                 rad==      6.0000
r7              byte     %8.0g                 rad==      7.0000
r8              byte     %8.0g                 rad==      8.0000
-------------------------------------------------------------------------------
Sorted by:
```

We now include seven of these dummies (omitting one to avoid multicollinearity) as regression predictors. The additional complexity of this dummy-variable model brings little improvement in fit. It does, however, add to our interpretation. The overall effect of radiation on death rate appears to come primarily from the two highest radiation levels (*r7* and *r8*, corresponding to 100 to 119 and 120 or more milliseiverts). At these levels, the incidence rates are about four times higher.

```
. poisson deaths r2-r8 age, nolog exposure(pyears) irr

Poisson regression                              Number of obs   =         56
                                                LR chi2(8)      =     215.44
                                                Prob > chi2     =     0.0000
Log likelihood = -69.451814                     Pseudo R2       =     0.6080

-------------------------------------------------------------------------------
    deaths |       IRR   Std. Err.      z    P>|z|     [95% Conf. Interval]
-----------+-------------------------------------------------------------------
        r2 |  1.473591    .426898     1.34   0.181     .8351884    2.599975
        r3 |  1.630688   .6659257     1.20   0.231      .732428    3.630587
        r4 |  2.375967   1.088835     1.89   0.059     .9677429    5.833389
        r5 |  .7278113   .7518255    -0.31   0.758     .0961018    5.511957
        r6 |  1.168477    1.20691     0.15   0.880     .1543195    8.847472
        r7 |  4.433729   3.337738     1.98   0.048     1.013863    19.38915
        r8 |   3.89188   1.640978     3.22   0.001     1.703168    8.893267
       age |  1.961907   .1000652    13.21   0.000     1.775267    2.168169
    pyears | (exposure)
-------------------------------------------------------------------------------
```

Radiation levels 7 and 8 seem to have similar effects, so we might simplify the model by combining them. First, we test whether their coefficients are significantly different. They are not:

```
. test r7 = r8

 ( 1)  [deaths]r7 - [deaths]r8 = 0.0

         chi2(  1) =      0.03
       Prob > chi2 =    0.8676
```

Next, generate a new dummy variable *r78*, which equals 1 if either *r7* or *r8* equals 1:

. **generate r78 = (r7 | r8)**

Finally, substitute the new predictor for *r7* and *r8* in the regression:

. **poisson deaths r2-r6 r78 age, irr ex(pyears) nolog**

```
Poisson regression                              Number of obs   =        56
                                                LR chi2(7)      =    215.41
                                                Prob > chi2     =    0.0000
Log likelihood = -69.465332                     Pseudo R2       =    0.6079

------------------------------------------------------------------------------
    deaths |      IRR   Std. Err.      z    P>|z|     [95% Conf. Interval]
-----------+------------------------------------------------------------------
        r2 | 1.473602   .4269013     1.34   0.181     .8351949    2.599996
        r3 | 1.630718   .6659381     1.20   0.231     .7324415    3.630655
        r4 | 2.376065   1.08888      1.89   0.059     .9677823    5.833629
        r5 | .7278387   .7518538    -0.31   0.758     .0961055    5.512165
        r6 | 1.168507   1.206942     0.15   0.880     .1543236    8.847704
       r78 | 3.980326   1.580024     3.48   0.001     1.828214    8.665833
       age | 1.961722   .100043     13.21   0.000     1.775122    2.167937
    pyears | (exposure)
------------------------------------------------------------------------------
```

We could proceed to simplify the model further in this fashion. At each step, **test** helps to evaluate whether combining two dummy variables is justifiable.

Generalized Linear Models

Generalized linear models (GLM) have the form

$$g[E(y)] = \beta_0 + \beta_1 x_1 + \beta_2 x_2 + \ldots + \beta_k x_k, \qquad y \sim F \qquad [11.6]$$

where $g[\ ]$ is the *link function* and F the distribution family. This general formulation encompasses many specific models. For example, if $g[\ ]$ is the identity function and y follows a normal (Gaussian) distribution, we have a linear regression model:

$$E(y) = \beta_0 + \beta_1 x_1 + \beta_2 x_2 + \ldots + \beta_k x_k, \qquad y \sim \text{Normal} \qquad [11.7]$$

If $g[\ ]$ is the logit function and y follows a Bernoulli distribution, we have logit regression instead:

$$\text{logit}[E(y)] = \beta_0 + \beta_1 x_1 + \beta_2 x_2 + \ldots + \beta_k x_k, \qquad y \sim \text{Bernoulli} \qquad [11.8]$$

Because of its broad applications, GLM could have been introduced at several different points in this book. Its relevance to this chapter comes from the ability to fit event models. Poisson regression, for example, requires that $g[\ ]$ is the natural log function and that y follows a Poisson distribution:

$$\ln[E(y)] = \beta_0 + \beta_1 x_1 + \beta_2 x_2 + \ldots + \beta_k x_k, \qquad y \sim \text{Poisson} \qquad [11.9]$$

As might be expected with such a flexible method, Stata's **glm** command permits many different options. Users can specify not only the distribution family and link function, but also details of the variance estimation, fitting procedure, output, and offset. These options make **glm** a useful alternative even when applied to models for which a dedicated command (such as **regress**, **logistic**, or **poisson**) already exists.

We might represent a "generic" **glm** command as follows:

```
. glm y x1 x2 x3, family(familyname) link(linkname)
     lnoffset(exposure) eform jknife
```

where **family()** specifies the *y* distribution family, **link()** the link function, and **lnoffset()** an "exposure" variable such as that needed for Poisson regression. The **eform** option asks for regression coefficients in exponentiated form, $\exp(\beta)$ rather than β. Standard errors are estimated through jackknife (**jknife**) calculations.

Possible distribution families are

family(gaussian)	Gaussian or normal (default)
family(igaussian)	Inverse Gaussian
family(binomial)	Bernoulli binomial
family(poisson)	Poisson
family(nbinomial)	Negative binomial
family(gamma)	Gamma

We can also specify a number or variable indicating the binomial denominator *N* (number of trials), or a number indicating the negative binomial variance and deviance functions, by declaring them in the **family()** option:

```
family(binomial #)
family(binomial varname)
family(nbinomial #)
```

Possible link functions are

link(identity)	Identity (default)
link(log)	Log
link(logit)	Logit
link(probit)	Probit
link(cloglog)	Complementary log-log
link(opower #)	Odds power
link(power #)	Power
link(nbinomial)	Negative binomial
link(loglog)	Log-log
link(logc)	Log-complement

Coefficient variances or standard errors can be estimated in a variety of ways. A partial list of **glm** variance-estimating options is given below:

opg	Berndt, Hall, Hall, and Hausman "B-H-cubed" variance estimator.
oim	Observed information matrix variance estimator.
robust	Huber/White/sandwich estimator of variance.
unbiased	Unbiased sandwich estimator of variance

nwest	Heteroskedasticity and autocorrelation-consistent variance estimator.
jknife	Jackknife estimate of variance.
jknife1	One-step jackknife estimate of variance.
bstrap	Bootstrap estimate of variance. The default is 199 repetitions; specify some other number by adding the **bsrep(#)** option.

For a full list of options with some technical details, look up **glm** in the *Base Reference Manual*. A more in-depth treatment of GLM topics can be found in Hardin and Hilbe (2001).

Chapter 6 began with the simple regression of mean composite SAT scores (*csat*) on per-pupil expenditures (*expense*) of the 50 U.S. states and District of Columbia (*states.dta*):

```
. regress csat expense
```

We could fit the same model and obtain exactly the same estimates with the following command:

```
. glm csat expense, link(identity) family(gaussian)

Iteration 0:   log likelihood = -279.99869

Generalized linear models                        No. of obs      =         51
Optimization     : ML: Newton-Raphson            Residual df     =         49
                                                 Scale param     =   3577.678
Deviance         =    175306.2097                (1/df) Deviance =   3577.678
Pearson          =    175306.2097                (1/df) Pearson  =   3577.678

Variance function: V(u) = 1                      [Gaussian]
Link function    : g(u) = u                      [Identity]
Standard errors  : OIM

Log likelihood   = -279.9986936                  AIC             =   11.05877
BIC              =    175298.346

------------------------------------------------------------------------------
        csat |      Coef.   Std. Err.       z    P>|z|     [95% Conf. Interval]
-------------+----------------------------------------------------------------
     expense |  -.0222756   .0060371    -3.69   0.000    -.0341082   -.0104431
       _cons |   1060.732    32.7009    32.44   0.000     996.6399    1124.825
------------------------------------------------------------------------------
```

Because **link(identity)** and **family(gaussian)** are default options, we could actually have left them out of the previous **glm** command.

The **glm** command can do more than just duplicate our **regress** results, however. For example, we could fit the same OLS model but obtain bootstrap standard errors:

```
. glm csat expense, link(identity) family(gaussian) bstrap

Iteration 0:    log likelihood = -279.99869

Bootstrap iterations (199)
----+--- 1 ---+--- 2 ---+--- 3 ---+--- 4 ---+--- 5
.................................................. 50
.................................................. 100
.................................................. 150
..................................................

Generalized linear models                No. of obs     =         51
Optimization     : ML: Newton-Raphson    Residual df    =         49
                                         Scale param    =   4124.656
Deviance         =   175306.2097         (1/df) Deviance =  3577.678
Pearson          =   175306.2097         (1/df) Pearson  =  3577.678

Variance function: V(u) = 1              [Gaussian]
Link function    : g(u) = u              [Identity]
Standard errors  : Bootstrap

Log likelihood   =  -279.9986936         AIC            =   11.05877
BIC              =    175298.346
```

csat	Coef.	Bootstrap Std. Err.	z	P>|z|	[95% Conf. Interval]	
expense	-.0222756	.0039284	-5.67	0.000	-.0299751	-.0145762
_cons	1060.732	25.36566	41.82	0.000	1011.017	1110.448

The bootstrap standard errors reflect observed variation among coefficients estimated from 199 samples of $n = 51$ cases each, drawn by random sampling with replacement from the original $n = 51$ dataset. In this example, the bootstrap standard errors are less than the corresponding theoretical standard errors, and the resulting confidence intervals are narrower.

Similarly, we could use **glm** to repeat the first **logistic** regression of Chapter 10. In the following example, we ask for jackknife standard errors and odds ratio or exponential-form (**eform**) coefficients:

```
. glm any date, link(logit) family(bernoulli) eform jknife

Iteration 0:   log likelihood = -12.995268
Iteration 1:   log likelihood = -12.991098
Iteration 2:   log likelihood = -12.991096

Jackknife iterations (23)
----+--- 1 ---+--- 2 ---+--- 3 ---+--- 4 ---+--- 5
......................

Generalized linear models                 No. of obs      =         23
Optimization     : ML: Newton-Raphson     Residual df     =         21
                                          Scale param     =          1
Deviance      =    25.98219269            (1/df) Deviance =   1.237247
Pearson       =    22.8885488             (1/df) Pearson  =   1.089931

Variance function: V(u) = u*(1-u)         [Bernoulli]
Link function    : g(u) = ln(u/(1-u))     [Logit]
Standard errors  : Jackknife

Log likelihood   = -12.99109634           AIC             =   1.303574
BIC              =  19.71120426

------------------------------------------------------------------------
             |                Jackknife
         any |  Odds Ratio   Std. Err.     z    P>|z|    [95% Conf. Interval]
-------------+----------------------------------------------------------
        date |   1.002093    .0015486    1.35   0.176    .9990623   1.005133
------------------------------------------------------------------------
```

The final **poisson** regression of the present chapter corresponds to this **glm** model:

```
. glm deaths r2-r6 r78 age, link(log) family(poisson)
    lnoffset(pyears) eform
```

Although **glm** can replicate the models fit by many specialized commands, and adds some new capabilities, the specialized commands have their own advantages including speed and customized options. A particular attraction of **glm** is its ability to fit models for which Stata has no specialized command.

12

Principal Components, Factor, and Cluster Analysis

Principal components and factor analysis provide methods for simplification, combining many correlated variables into a smaller number of underlying dimensions. Along the way to achieving simplification, the analyst must choose from a daunting variety of options. If the data really do reflect distinct underlying dimensions, different options might nonetheless converge on similar results. In the absence of distinct underlying dimensions, however, different options often lead to divergent results. Experimenting with these options can tell us how stable a particular finding is, or how much it depends on arbitrary choices about the specific analytical technique.

Stata accomplishes principal components and factor analysis with five basic commands:

`pca`	Principal components analysis.
`factor`	Extracts factors of several different types.
`greigen`	Constructs a scree graph (plot of the eigenvalues) from the recent `pca` or `factor`.
`rotate`	Performs orthogonal (uncorrelated factors) or oblique (correlated factors) rotation, after `factor`.
`score`	Generates factor scores (composite variables) after `pca`, `factor`, or `rotate`.

The composite variables generated by `score` can subsequently be saved, listed, graphed, or analyzed like any other Stata variable.

Users who create composite variables by the older method of adding other variables together without doing factor analysis could assess their results by calculating an α reliability coefficient:

`alpha`	Cronbach's α reliability

Instead of combining variables, cluster analysis combines observations by finding non-overlapping, empirically-based typologies or groups. Cluster analysis methods are even more diverse, and less theoretical, than those of factor analysis. Stata's **cluster** command provides tools for performing cluster analysis, graphing the results, and forming new variables to identify the resulting groups.

Methods described in this chapter can be accessed through the following menus:

Statistics – Other multivariate analysis

Graphics – More statistical graphs

Statistics – Cluster analysis

Example Commands

. `pca x1-x20`

Obtains principal components of the variables *x1* through *x20*.

. `pca x1-x20, mineigen(1)`

Obtains principal components of the variables *x1* through *x20*. Retains components having eigenvalues greater than 1.

. `factor x1-x20, ml factor(5)`

Performs maximum likelihood factor analysis of the variables *x1* through *x20*. Retains only the first five factors.

. `greigen`

Graphs eigenvalues versus factor or component number from the most recent **factor** command (also known as a "scree graph").

. `rotate, varimax factors(2)`

Performs orthogonal (varimax) rotation of the first two factors from the most recent **factor** command.

. `rotate, promax factors(3)`

Performs oblique (promax) rotation of the first three factors from the most recent **factor** command.

. `score f1 f2 f3`

Generates three new factor score variables named *f1*, *f2*, and *f3*, based upon the most recent **factor** and **rotate** commands.

. `alpha x1-x10`

Calculates Cronbach's α reliability coefficient for a composite variable defined as the sum of *x1-x10*. The sense of items entering negatively is ordinarily reversed. Options can override this default, or form a composite variable by adding together either the original variables or their standardized values.

. `cluster centroidlinkage x y z w, L2 name(L2cent)`

Performs agglomerative cluster analysis with centroid linkage, using variables *x, y, z*, and *w*. Euclidean distance (**L2**) measures dissimilarity among observations. Results from this cluster analysis are saved with the name *L2cent*.

. `cluster tree, ylabel(0(.5)3) cutnumber(20) vertlabel`

Draws a cluster analysis tree graph or dendrogram showing results from the previous cluster analysis. **cutnumber(20)** specifies that the graph begins with only 20 clusters remaining, after some previous fusion of the most-similar observations. Labels are printed in a compact vertical fashion below the graph. **cluster dendrogram** does the same thing as **cluster tree**.

```
. cluster generate ctype = groups(3), name(L2cent)
```
Creates a new variable *ctype* (values of 1, 2, or 3) that classifies each observation into one of the top three groups found by the cluster analysis named *L2cent*.

Principal Components

To illustrate basic principal components and factor analysis commands, we will use a small dataset describing the nine major planets of this solar system (from Beatty et al. 1981). The data include several variables in both raw and natural logarithm form. Logarithms are employed here to reduce skew and linearize relationships among the variables.

```
Contains data from C:\data\planets.dta
  obs:            9                        Solar system data
  vars:          12                        22 Jul 2005 09:49
  size:         441 (99.9% of memory free)
-------------------------------------------------------------------------------
              storage  display    value
variable name   type   format     label      variable label
-------------------------------------------------------------------------------
planet         str7    %9s                    Planet
dsun           float   %9.0g                  Mean dist. sun, km*10^6
radius         float   %9.0g                  Equatorial radius in km
rings          byte    %8.0g      ringlbl     Has rings?
moons          byte    %8.0g                  Number of known moons
mass           float   %9.0g                  Mass in kilograms
density        float   %9.0g                  Mean density, g/cm^3
logdsun        float   %9.0g                  natural log dsun
lograd         float   %9.0g                  natural log radius
logmoons       float   %9.0g                  natural log (moons + 1)
logmass        float   %9.0g                  natural log mass
logdense       float   %9.0g                  natural log dense
-------------------------------------------------------------------------------
Sorted by:   dsun
```

To extract initial factors or principal components, use the command **factor** followed by a variable list (variables in any order) and one of the following options:

pcf	Principal components factoring
pf	Principal factoring (default)
ipf	Principal factoring with iterated communalities
ml	Maximum-likelihood factoring

Principal components are calculated through the specialized command **pca** . Type **help pca** or **help factor** to see options for these commands.

To obtain principal components factors, type

```
. factor rings logdsun - logdense, pcf
(obs=9)
```

(principal component factors; 2 factors retained)

Factor	Eigenvalue	Difference	Proportion	Cumulative
1	4.62365	3.45469	0.7706	0.7706
2	1.16896	1.05664	0.1948	0.9654
3	0.11232	0.05395	0.0187	0.9842
4	0.05837	0.02174	0.0097	0.9939
5	0.03663	0.03657	0.0061	1.0000
6	0.00006	.	0.0000	1.0000

Factor Loadings

Variable	1	2	Uniqueness
rings	0.97917	0.07720	0.03526
logdsun	0.67105	-0.71093	0.04427
lograd	0.92287	0.37357	0.00875
logmoons	0.97647	0.00028	0.04651
logmass	0.83377	0.54463	0.00821
logdense	-0.84511	0.47053	0.06439

Only the first two components have eigenvalues greater than 1, and these two components explain over 96% of the six variables' combined variance. The unimportant 3rd through 6th principal components might safely be disregarded in subsequent analysis.

Two **factor** options provide control over the number of factors extracted:

> **factors(#)** where # specifies the number of factors

> **mineigen(#)** where # specifies the minimum eigenvalue for retained factors

The principal components factoring (**pcf**) procedure automatically drops factors with eigenvalues below 1, so

```
. factor rings logdsun - logdense, pcf
```

is equivalent to

```
. factor rings logdsun - logdense, pcf mineigen(1)
```

In this example, we would also have obtained the same results by typing

```
. factor rings logdsun - logdense, pcf factors(2)
```

To see a scree graph (plot of eigenvalues versus component or factor number) after any **factor**, use the **greigen** command. A horizontal line at eigenvalue = 1 in Figure 12.1 marks the usual cutoff for retaining principal components, and again emphasizes the unimportance in this example of components 3 through 6.

```
. greigen, yline(1)
```

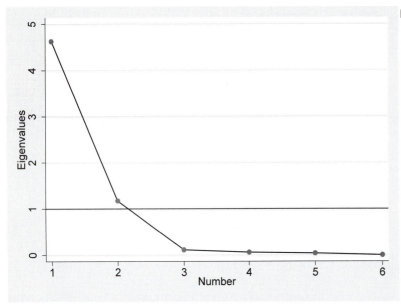

Figure 12.1

Rotation

Rotation further simplifies factor structure. After factoring, type **rotate** followed by one of these options:

varimax   Varimax orthogonal rotation, for uncorrelated factors or components (default).

promax()   Promax oblique rotation, allowing correlated factors or components. Choose a number (promax power) ≤ 4; the higher the number, the greater the degree of interfactor correlation. **promax(3)** is the default.

Two additional **rotate** options are

factors()   As it does with **factor**, this option specifies how many factors to retain.

horst   Horst modification to varimax and promax rotation.

Rotation can be performed following any factor analysis, whether it employed the **pcf**, **pf**, **ipf**, or **ml** options. In this section, we will follow through on our **pcf** example. For orthogonal (default) rotation of the first two components found in the planetary data, we type

```
. rotate
            (varimax rotation)
             Rotated Factor Loadings
     Variable |     1          2      Uniqueness
  ------------+--------------------------------
        rings |   0.52848    0.82792    0.03526
      logdsun |   0.97173    0.10707    0.04427
       lograd |   0.25804    0.96159    0.00875
     logmoons |   0.58824    0.77940    0.04651
      logmass |   0.06784    0.99357    0.00821
     logdense |  -0.88479   -0.39085    0.06439
```

This example accepts all the defaults: varimax rotation and the same number of factors retained in the last **factor**. We could have asked for the same rotation explicitly, with the following command:

```
. rotate, varimax factors(2)
```

For oblique promax rotation (allowing correlated factors) of the most recent factoring, type

```
. rotate, promax
```

```
            (promax rotation)
            Rotated Factor Loadings
  Variable |     1         2      Uniqueness
-----------+------------------------------------
     rings |   0.34664    0.76264    0.03526
   logdsun |   1.05196   -0.17270    0.04427
    lograd |   0.00599    0.99262    0.00875
  logmoons |   0.42747    0.69070    0.04651
   logmass |  -0.21543    1.08534    0.00821
  logdense |  -0.87190   -0.16922    0.06439
```

By default, this example used a promax power of 3. We could have specified the promax power and desired number of factors explicitly:

```
. rotate, promax(3) factors(2)
```

promax(4) would permit further simplification of the loading matrix, at the cost of stronger interfactor correlations and less total variance explained.

After promax rotation, *rings*, *lograd*, *logmoons*, and *logmass* load most heavily on factor 2. This appears to be a "large size/many satellites" dimension. *logdsun* and *logdense* load higher on factor 1, forming a "far out/low density" dimension. The next section shows how to create new variables representing these dimensions.

Factor Scores

Factor scores are linear composites, formed by standardizing each variable to zero mean and unit variance, and then weighting with factor score coefficients and summing for each factor. **score** performs these calculations automatically, using the most recent **rotate** or **factor** results. In the **score** command we supply names for the new variables, such as *f1* and *f2*.

```
. score f1 f2
```

```
            (based on rotated factors)
            Scoring Coefficients
  Variable |     1         2
-----------+----------------------
     rings |   0.12674    0.22099
   logdsun |   0.48769   -0.09689
    lograd |  -0.03840    0.30608
  logmoons |   0.16664    0.19543
   logmass |  -0.14338    0.34386
  logdense |  -0.39127   -0.01609
```

```
. label variable f1 "Far out/low density"

. label variable f2 "Large size/many satellites"

. list planet f1 f2
```

```
     +-------------------------------+
     | planet          f1          f2 |
     |-------------------------------|
  1. | Mercury   -1.256881   -.9172388 |
  2. |   Venus   -1.188757   -.5160229 |
  3. |   Earth   -1.035242   -.3939372 |
  4. |    Mars   -.5970106   -.6799535 |
  5. | Jupiter    .3841085    1.342658 |
     |-------------------------------|
  6. |  Saturn    .9259058    1.184475 |
  7. |  Uranus    .9347457    .7682409 |
  8. | Neptune    .8161058    .647119 |
  9. |   Pluto    1.017025   -1.43534 |
     +-------------------------------+
```

Being standardized variables, the new factor scores *f1* and *f2* have means (approximately) equal to zero and standard deviations equal to one:

```
. summarize f1 f2
```

```
   Variable |     Obs        Mean    Std. Dev.       Min        Max
 -----------+-------------------------------------------------------
        f1 |       9    9.93e-09           1   -1.256881   1.017025
        f2 |       9   -3.31e-09           1    -1.43534   1.342658
```

Thus, the factor scores are measured in units of standard deviations from their means. Mercury, for example, is about 1.26 standard deviations below average on the far out/low density (*f1*) dimension because it is actually close to the sun and high density. Mercury is .92 standard deviations below average on the large size/many satellites (*f2*) dimension because it is small and has no satellites. Saturn, in contrast, is .93 and 1.18 standard deviations above average on these two dimensions.

Promax rotation permits correlations between factor scores:

```
. correlate f1 f2
(obs=9)
```

```
             |      f1       f2
 ------------+------------------
         f1 |   1.0000
         f2 |   0.4974   1.0000
```

Scores on factor 1 have a moderate positive correlation with scores on factor 2: far out/low density planets are more likely also to be larger, with many satellites.

If we employ varimax instead of promax rotation, we get uncorrelated factor scores:

```
. quietly factor rings logdsun - logdense, pcf

. quietly rotate

. quietly score varimax1 varimax2
```

```
. correlate varimax1 varimax2
(obs=9)

             | varimax1 varimax2
-------------+------------------
    varimax1 |   1.0000
    varimax2 |   0.0000   1.0000
```

Once created by **score**, factor scores can be treated like any other Stata variable — listed, analyzed, graphed, and so forth. Graphs of principal component factors sometimes help to identify multivariate outliers or clusters of observations that stand apart from the rest. For example, Figure 12.2 reveals three distinct types of planets.

```
. graph twoway scatter f1 f2, yline(0) xline(0) mlabel(planet)
     mlabsize(medsmall) ylabel(, angle(horizontal))
     xlabel(-1.5(.5)1.5, grid)
```

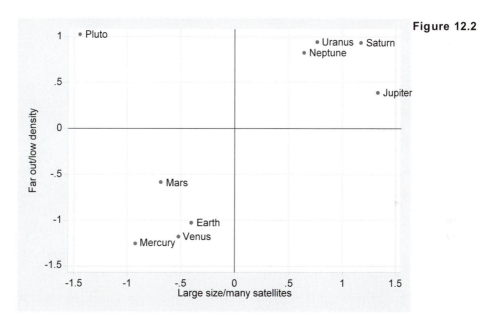

Figure 12.2

The inner, rocky planets (such as Mercury, low on "far out/low density" factor 1; low also on "large size/many satellites" factor 2) cluster together at the lower left. The outer gas giants have opposite characteristics, and cluster together at the upper right. Pluto, which physically resembles some outer-system moons, is unique among planets for being high on the "far out/low density" dimension, and at the same time low on the "large size/many satellites" dimension.

This example employed rotation. Factor scores obtained by principal components without rotation are often used to analyze large datasets in physical-science fields such as climatology and remote sensing. In these applications, principal components are called "empirical orthogonal functions." The first empirical orthogonal function, or EOF1, equals the factor score for the first unrotated principal component. EOF2 is the score for the second principal component, and so forth.

Principal Factoring

Principal factoring extracts principal components from a modified correlation matrix, in which the main diagonal consists of communality estimates instead of 1's. The **factor** options **pf** and **ipf** both perform principal factoring. They differ in how communalities are estimated:

pf   Communality estimates equal R^2 from regressing each variable on all the others.

ipf   Iterative estimation of communalities.

Whereas principal components analysis focuses on explaining the variables' variance, principal factoring explains intervariable correlations.

We apply principal factoring with iterated communalities (**ipf**) to the planetary data:

```
. factor rings logdsun - logdense, ipf
```

(obs=9)

```
          (iterated principal factors; 5 factors retained)
  Factor    Eigenvalue    Difference    Proportion    Cumulative
-------------------------------------------------------------------
    1         4.59663       3.46817        0.7903        0.7903
    2         1.12846       1.05107        0.1940        0.9843
    3         0.07739       0.06438        0.0133        0.9976
    4         0.01301       0.01176        0.0022        0.9998
    5         0.00125       0.00137        0.0002        1.0000
    6        -0.00012          .          -0.0000        1.0000
```

```
                Factor Loadings
    Variable |    1          2          3          4          5      Uniqueness
-------------+-----------------------------------------------------------------
       rings |  0.97599    0.06649    0.11374   -0.02065   -0.02234    0.02916
     logdsun |  0.65708   -0.67054    0.14114    0.04471    0.00816    0.09663
      lograd |  0.92670    0.37001   -0.04504    0.04865    0.01662   -0.00036
    logmoons |  0.96738   -0.01074    0.00781   -0.08593    0.01597    0.05636
     logmass |  0.83783    0.54576    0.00557    0.02824   -0.00714   -0.00069
    logdense | -0.84602    0.48941    0.20594   -0.00610    0.00997    0.00217
```

Only the first two factors have eigenvalues above 1. With **pcf** or **pf** factoring, we can simply disregard minor factors. Using **ipf** , however, we must decide how many factors to retain, and then repeat the analysis asking for exactly that many factors. Here we will retain two factors:

```
. factor rings logdsun - logdense, ipf factor(2)
```

(obs=9)

```
          (iterated principal factors; 2 factors retained)
  Factor    Eigenvalue    Difference    Proportion    Cumulative
-------------------------------------------------------------------
    1         4.57495       3.47412        0.8061        0.8061
    2         1.10083       1.07631        0.1940        1.0000
    3         0.02452       0.02013        0.0043        1.0043
    4         0.00439       0.00795        0.0008        1.0051
    5        -0.00356       0.02182       -0.0006        1.0045
    6        -0.02537          .          -0.0045        1.0000
```

```
                  Factor Loadings
     Variable |       1           2      Uniqueness
   -----------+----------------------------------
        rings |    0.97474     0.05374     0.04699
      logdsun |    0.65329    -0.67309     0.12016
       lograd |    0.92816     0.36047     0.00858
     logmoons |    0.96855    -0.02278     0.06139
      logmass |    0.84298     0.54616    -0.00890
     logdense |   -0.82938     0.46490     0.09599
```

After this final factor analysis, we can create composite variables by **rotate** and **score** . Rotation of the **ipf** factors produces results similar to those found earlier with **pcf** : a far out/low density dimension and a large size/many satellites dimension. When variables have a strong factor structure, as these do, the specific techniques we choose make less difference.

Maximum-Likelihood Factoring

Maximum-likelihood factoring, unlike Stata's other **factor** options, provides formal hypothesis tests that help in determining the appropriate number of factors. To obtain a single maximum-likelihood factor for the planetary data, type

. **factor** *rings logdsun - logdense*, **ml nolog factor(1)**

```
(obs=9)

              (maximum likelihood factors; 1 factor retained)
     Factor      Variance      Difference     Proportion    Cumulative
   ---------------------------------------------------------------------
        1         4.47258          .            1.0000        1.0000

   Test:  1 vs. no    factors.  Chi2(  6) =   62.02, Prob > chi2 =  0.0000
   Test:  1 vs. more factors.  Chi2(  9) =   51.73, Prob > chi2 =  0.0000

                  Factor Loadings
     Variable |       1      Uniqueness
   -----------+--------------------
        rings |    0.98726     0.02535
      logdsun |    0.59219     0.64931
       lograd |    0.93654     0.12288
     logmoons |    0.95890     0.08052
      logmass |    0.86918     0.24451
     logdense |   -0.77145     0.40487
```

The **ml** output includes two χ^2 tests:

J vs. no factors
> This tests whether the current model, with *J* factors, fits the observed correlation matrix significantly better than a no-factor model. A low probability indicates that the current model is a significant improvement over no factors.

J vs. more factors
> This tests whether the current *J*-factor model fits significantly worse than a more complicated, perfect-fit model. A low *P*-value suggests that the current model does not have enough factors.

The previous 1-factor example yields these results:

```
1 vs. no factors
```
χ^2 [6] = 62.02, P = 0.0000 (actually, meaning $P < .00005$). The 1-factor model significantly improves upon a no-factor model.

```
1 vs. more factors
```
χ^2 [9] = 51.73, P = 0.0000 ($P < .00005$). The 1-factor model is significantly worse than a perfect-fit model.

Perhaps a 2-factor model will do better:

```
. factor rings logdsun - logdense, ml nolog factor(2)
```

```
(obs=9)
```

```
                    (maximum likelihood factors; 2 factors retained)
     Factor     Variance      Difference     Proportion     Cumulative
    ------------------------------------------------------------------
       1         3.64200        1.67115         0.6489         0.6489
       2         1.97085           .            0.3511         1.0000

Test:   2 vs. no    factors.   Chi2( 12) =   134.14, Prob > chi2 =  0.0000
Test:   2 vs. more factors.   Chi2(  4) =     6.72, Prob > chi2 =  0.1513
```

```
                  Factor Loadings
      Variable |      1           2       Uniqueness
    -----------+----------------------------------
         rings |   0.86551    -0.41545     0.07829
       logdsun |   0.20920    -0.85593     0.22361
        lograd |   0.98437    -0.17528     0.00028
      logmoons |   0.81560    -0.49982     0.08497
       logmass |   0.99965     0.02639     0.00000
      logdense |  -0.46434     0.88565     0.00000
```

Now we find the following:

```
2 vs. no factors
```
χ^2 [12] = 134.14, P = 0.0000 (actually, $P < .00005$). The 2-factor model significantly improves upon a no-factor model.

```
2 vs. more factors
```
χ^2 [4] = 6.72, P = 0.1513. The 2-factor model is *not* significantly worse than a perfect-fit model.

These tests suggest that two factors provide an adequate model.

Computational routines performing maximum-likelihood factor analysis often yield "improper solutions" — unrealistic results such as negative variance or zero uniqueness. When this happens (as it did in our 2-factor **ml** example), the χ^2 tests lack formal justification. Viewed descriptively, the tests can still provide informal guidance regarding the appropriate number of factors.

Cluster Analysis — 1

Cluster analysis encompasses a variety of methods that divide observations into groups or clusters, based on their dissimilarities across a number of variables. It is most often used as an exploratory approach, for developing empirical typologies, rather than as a means of testing pre-specified hypotheses. Indeed, there exists little formal theory to guide hypothesis testing for the common clustering methods. The number of choices available at each step in the analysis is daunting, and all the more so because they can lead to many different results. This section provides no more than an entry point to begin cluster analysis. We review some basic ideas and illustrate them through a simple example. The following section considers a somewhat larger example. Stata's *Multivariate Statistics Reference Manual* introduces and defines the full range of choices available. Everitt *et al.* (2001) cover topics in more detail, including helpful comparisons among the many cluster-analysis methods.

Clustering methods fall into two broad categories, *partition* and *hierarchical*. Partition methods break the observations into a pre-set number of nonoverlapping groups. We have two ways to do this:

cluster kmeans   Kmeans cluster analysis
User specifies the number of clusters (*K*) to create. Stata then finds these through an iterative process, assigning observations to the group with the closest mean.

cluster kmedians   Kmedians cluster analysis
Similar to Kmeans, but with medians.

Partition methods tend to be computationally simpler and faster than hierarchical methods. The necessity of declaring the exact number of clusters in advance is a disadvantage for exploratory work, however.

Hierarchical methods, involve a process of smaller groups gradually fusing to form increasingly large ones. Stata takes an *agglomerative* approach in hierarchical cluster analysis: it starts out with each observation considered as its own separate "group." The closest two groups are merged, and this process continues until a specified stopping-point is reached, or all observations belong to one group. A graphical display called a *dendrogram* or *tree diagram* visualizes hierarchical clustering results. Several choices exist for the *linkage method*, which specifies what should be compared between groups that contain more than one observation:

cluster singlelinkage   Single linkage cluster analysis
Computes the dissimilarity between two groups as the dissimilarity between the closest pair of observations between the two groups. Although simple, this method has low resistance to outliers or measurement errors. Observations tend to join clusters one at a time, forming unbalanced, drawn-out groups in which members have little in common, but are linked by intermediate observations — a problem called *chaining*.

cluster completelinkage   Complete linkage cluster analysis
Uses the farthest pair of observations between the two groups. Less sensitive to outliers than single linkage, but with the opposite tendency towards clumping many observations into tight, spatially compact clusters.

cluster averagelinkage   Average linkage cluster analysis
Uses the average dissimilarity of observations between the two groups, yielding properties intermediate between single and complete linkage. Simulation studies report that this

works well for many situations and is reasonably robust (see Everitt *et al.* 2001, and sources they cite). Commonly used in archaeology.

cluster centroidlinkage   Centroid linkage cluster analysis
Centroid linkage merges the groups whose means are closest (in contrast to average linkage which looks at the average distance between elements of the two groups). This method is subject to reversals — points where a fusion takes place at a lower level of dissimilarity than an earlier fusion. Reversals signal an unstable cluster structure, are difficult to interpret, and cannot be graphed by **cluster tree**.

cluster waveragelinkage   Weighted-average linkage cluster analysis
cluster medianlinkage   Median linkage cluster analysis.
Weighted-average linkage and median linkage are variations on average linkage and centroid linkage, respectively. In both cases, the difference is in how groups of unequal size are treated when merged. In average linkage and centroid linkage, the number of elements of each group are factored into the computation, giving correspondingly larger influence to the larger group (because each observation carries the same weight). In weighted-average linkage and median linkage, the two groups are given equal weighting regardless of how many observations there are in each group. Median linkage, like centroid linkage, is subject to reversals.

cluster wardslinkage   Ward's linkage cluster analysis
Joins the two groups that result in the minimum increase in the error sum of squares. Does well with groups that are multivariate normal and of similar size, but poorly when clusters have unequal numbers of observations.

All clustering methods begin with some definition of dissimilarity (or similarity). Dissimilarity measures reflect the differentness or distance between two observations, across a specified set of variables. Generally, such measures are designed so that two identical observations have a dissimilarity of 0, and two maximally different observations have a dissimilarity of 1. Similarity measures reverse this scaling, so that identical observations have a similarity of 1. Stata's **cluster** options offer many choices of dissimilarity or similarity measures. For purposes of calculation, Stata internally transforms similarity to dissimilarity:

dissimilarity = 1 – similarity

The default dissimilarity measure is the Euclidean distance, option **L2** (or **Euclidean**). This defines the distance between observations *i* and *j* as

$$\{\textstyle\sum_k (x_{ki} - x_{kj})^2\}^{1/2}$$

where x_{ki} is the value of variable x_k for observation i, x_{kj} the value of x_k for observation j, and summation occurs over all the *x* variables considered. Other choices available for measuring the (dis)similarities between observations based on continuous variables include the squared Euclidean distance (**L2squared**),

$$\textstyle\sum_k (x_{ki} - x_{kj})^2$$

the absolute-value distance (**L1**), maximum-value distance (**Linfinity**), and correlation coefficient similarity measure (**correlation**). Choices for dissimilarities or similarities based on binary variables include simple matching (**matching**), Jaccard binary similarity coefficient (**Jaccard**), and many others. Type **help cldis** for a list and explanations.

Earlier in this chapter, a principal components analysis of variables in *planets.dta* (Figure 12.2) identified three types of planets: inner rocky planets, outer gas giants, and in a class by itself, Pluto. Cluster analysis provides an alternative approach to the question of planet "types." Because variables such as number of moons (*moons*) and mass in kilograms (*mass*) are measured in incomparable units, with hugely different variances, we should standardize in some way to avoid results dominated by the highest-variance items. A common, although not automatic, choice is standardization to zero mean and unit standard deviation. This is accomplished through the **egen** command (and using variables in log form, for the same reasons discussed earlier). **summarize** confirms that the new *z* variables have (near) zero means, and standard deviations equal to one.

```
. egen zrings = std(rings)

. egen zlogdsun = std(logdsun)

. egen zlograd = std(lograd)

. egen zlogmoon = std(logmoons)

. egen zlogmass = std(logmass)

. egen zlogdens = std(logdense)

. summ zrings - zlogdens
```

Variable	Obs	Mean	Std. Dev.	Min	Max
zrings	9	-1.99e-08	1	-.8432741	1.054093
zlogdsun	9	-1.16e-08	1	-1.393821	1.288216
zlograd	9	-3.31e-09	1	-1.3471	1.372751
zlogmoon	9	0	1	-1.207296	1.175849
zlogmass	9	-4.14e-09	1	-1.74466	1.365167
zlogdens	9	-1.32e-08	1	-1.453143	1.128901

The "three types" conclusion suggested by our principal components analysis is robust, and could have been found through cluster analysis as well. For example, we might perform a hierarchical cluster analysis with average linkage, using Euclidean distance (**L2**) as our dissimilarity measure. The option **name(*L2avg*)** gives the results from this particular analysis a name, so that we can refer to them in later commands. The results-naming feature is convenient when we need to try a number of cluster analyses and compare their outcomes.

```
. cluster averagelinkage zrings zlogdsun zlograd zlogmoon zlogmass
     zlogdens, L2 name(L2avg)
```

Nothing seems to happen, although we might notice that our dataset now contains three new variables with names based on *L2avg*. These new *L2avg** variables are not directly of interest, but can be used unobtrusively by the **cluster tree** command to draw a cluster analysis tree or dendrogram visualizing the most recent hierarchical cluster analysis results (Figure 12.3). The **label(*planet*)** option here causes planet names (values of *planet*) to appear as labels below the tree. Typing **cluster dendrogram** instead of **cluster tree** would produce the same graph.

```
. cluster tree, label(planet) ylabel(0(1)5)
```

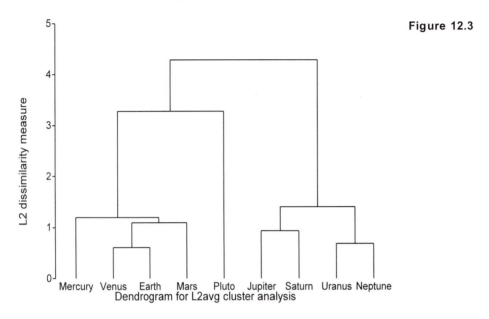

Figure 12.3

Dendrogram for L2avg cluster analysis

Dendrograms such as Figure 12.3 provide key interpretive tools for hierarchical cluster analysis. We can trace the agglomerative process from each observation its own cluster, at bottom, to all fused into one cluster, at top. Venus and Earth, and also Uranus and Neptune, are the least dissimilar or most alike pairs. They are fused first, forming the first two multi-observation clusters at a height (dissimilarity) below 1. Jupiter and Saturn, then Venus–Earth and Mars, then Venus–Earth–Mars and Mercury, and finally Jupiter–Saturn and Uranus–Neptune are fused in quick succession, all with dissimilarities around 1. At this point we have the same three groups suggested in Figure 12.2 by principal components: the inner rocky planets, the gas giants, and Pluto. The three clusters remain stable until, at much higher dissimilarity (above 3), Pluto fuses with the inner rocky planets. At a dissimilarity near 4, the final two clusters fuse.

So, how many types of planets are there? The answer, as Figure 12.3 makes clear, is "it depends." How much dissimilarity do we want to accept within each type? The long vertical lines between the three-cluster stage and the two-cluster stage in the upper part of Figure 12.3 indicate that we have three fairly distinct types. We could reduce this to two types only by fusing an observation (Pluto) that is quite dissimilar to others in its group. We could expand it to five types only by drawing distinctions between several planet groups (e.g., Mercury–Mars and Earth–Venus) that by solar-system standards are not greatly dissimilar. Thus, the dendrogram makes a case for a three-type scheme.

The **cluster generate** command creates a new variable indicating the type or group to which each observation belongs. In this example, **groups(3)** calls for three groups. The **name(L2avg)** option specifies the particular results we named *L2avg*. This option is most useful when our session included multiple cluster analyses.

```
. cluster generate plantype = groups(3), name(L2avg)

. label variable plantype "Planet type"

. list planet plantype

        +--------------------+
        |  planet   plantype |
        |--------------------|
    1.  | Mercury          1 |
    2.  |   Venus          1 |
    3.  |   Earth          1 |
    4.  |    Mars          1 |
    5.  | Jupiter          3 |
        |--------------------|
    6.  |  Saturn          3 |
    7.  |  Uranus          3 |
    8.  | Neptune          3 |
    9.  |   Pluto          2 |
        +--------------------+
```

The inner rocky planets have been coded as *plantype* = 1; the gas giants as *plantype* = 3; and Pluto, which resembles an outer-system moon more than it does other planets, is by itself as *plantype* = 2. The group designations as 1, 2, and 3 follow the left-to-right ordering of final clusters in the dendrogram (Figure 12.3). Once the data have been saved, our new typology could be used like any other categorical variable in subsequent analyses.

These planetary data have a strong pattern of natural groups, which is why such different techniques as cluster analysis and principal components point towards similar conclusions. We could have chosen other dissimilarity measures and linkage methods for this example, and still arrived at much the same place. Complex or weakly patterned data, on the other hand, often yield quite different results depending on nuances of the methods used. The clusters found by one method might not prove replicable under others, or even with slightly different analytical decisions.

Cluster Analysis — 2

Discovering a simple, robust typology to describe the nine planets was straightforward. For a more challenging example, consider the cross-national data in *nations.dta*. This dataset contains living-conditions variables that might provide a basis for classifying countries into types.

```
Contains data from C:\data\nations.dta
  obs:           109                      Data on 109 nations, ca. 1985
  vars:           15                      23 Jul 2005 18:37
  size:         4,142 (99.9% of memory free)
-------------------------------------------------------------------------------
              storage  display    value
variable name  type    format     label        variable label
-------------------------------------------------------------------------------
country        str8    %9s                      Country
pop            float   %9.0g                     1985 population in millions
birth          byte    %8.0g                     Crude birth rate/1000 people
death          byte    %8.0g                     Crude death rate/1000 people
chldmort       byte    %8.0g                     Child (1-4 yr) mortality 1985
infmort        int     %8.0g                     Infant (<1 yr) mortality 1985
life           byte    %8.0g                     Life expectancy at birth 1985
```

```
food            int      %8.0g              Per capita daily calories 1985
energy          int      %8.0g              Per cap energy consumed, kg oil
gnpcap          int      %8.0g              Per capita GNP 1985
gnpgro          float    %9.0g              Annual GNP growth % 65-85
urban           byte     %8.0g              % population urban 1985
school1         int      %8.0g              Primary enrollment % age-group
school2         int      %8.0g              Secondary enroll % age-group
school3         byte     %8.0g              Higher ed. enroll % age-group
-----------------------------------------------------------------------------
Sorted by:
```

In Chapter 8, we saw that nonlinear transformations (logs or square roots) helped to normalize distributions and linearize relationships among some of these variables. Similar arguments for nonlinear transformations could apply to cluster analysis, but to keep our example simple, we will not pursue them here. Linear transformations to standardize the variables in some fashion remain important, however. Otherwise, the variable *gnpcap*, which ranges from about $100 to $19,000 (standard deviation $4,400) would overwhelm other variables such as *life*, which ranges from 40 to 78 years (standard deviation 11 years). In the previous section, we standardized planetary data by subtracting each variable's mean, then dividing by its standard deviation, so that the resulting z-scores all had standard deviations of one. In this section we take a different approach, range standardization, which also works well for cluster analysis.

Range standardization involves dividing each variable by its range. There is no command to do this directly in Stata, but we can improvise one easily enough. The **summarize, detail** command calculates one-variable statistics, and afterwards unobtrusively stores the results in memory as macros (described in Chapter 14). A macro named $r(max)$ stores the variable's maximum, and $r(min)$ stores its minimum. Thus, to generate new variable *rpop*, defined as a range-standardized version of *pop* (population), type the commands

```
. quietly summ pop, detail
. generate rpop = pop/(r(max) - r(min))
. label variable rpop "Range-standardized population"
```

Similar commands create range-standardized versions of other living-conditions variables:

```
. quietly summ birth, detail
. generate rbirth = birth/(r(max) - r(min))
. label variable rbirth "Range-standardized bith rate"

. quietly summ infmort, detail
. generate rinf = infmort/(r(max) - r(min))
. label variable rinf "Range-standardized infant mortality"
```

and so forth, defining the 8 new variables listed below. These range-standardized variables all have ranges equal to 1.

```
. describe rpop-rschool2
```

variable name	storage type	display format	value label	variable label
rpop	float	%9.0g		Range-standardized population
rbirth	float	%9.0g		Range-standardized bith rate
rinf	float	%9.0g		Range-standardized infant mortality
rlife	float	%9.0g		Range-standardized life

```
rfood              float    %9.0g              expectancy
                                               Range-standardized food per
                                                  capita
renergy            float    %9.0g              Range-standardized energy per
                                                  capita
rgnpcap            float    %9.0g              Range-standardized GNP per
                                                  capita
rschool2           float    %9.0g              Range-standardized secondary
                                                  school %
```

. **summarize** *rpop - rschool2*

```
    Variable |       Obs        Mean    Std. Dev.        Min         Max
-------------+------------------------------------------------------------
        rpop |       109    .0374493    .1206474    .0009622    1.000962
      rbirth |       109    .7452043    .3098672    .2272727    1.227273
        rinf |       109    .4051354    .2913825     .035503    1.035503
       rlife |       109    1.621922     .291343    1.052632    2.052632
       rfood |       108    1.230213    .2644839    .7793776    1.779378
-------------+------------------------------------------------------------
     renergy |       107     .159786    .2137914    .0018464    1.001846
     rgnpcap |       109    .1666459    .2319276    .0057411    1.005741
    rschool2 |       104    .4574849    .2899882    .0196078    1.019608
```

After the variables of interest have been standardized, we can proceed with cluster analysis. As we divide more than 100 nations into "types," we have no reason to assume that each type will include a similar number of nations. Average linkage (used in our planetary example), along with some other methods, gives each observation the same weight. This tends to make larger clusters more influential as agglomeration proceeds. Weighted average and median linkage methods, on the other hand, give equal weight to each cluster regardless of how many observations it contains. Such methods consequently tend to work better for detecting clusters of unequal size. Median linkage, like centroid linkage, is subject to reversals (which will occur with these data), so the following example applies weighted average linkage. Absolute-value distance (**L1**) provides our dissimilarity measure.

. **cluster waveragelinkage** *rpop - rschool2*, **L1 name(***L1wav***)**

The full cluster analysis proves unmanageably large for a tree graph:

. **cluster tree**
```
too many leaves; consider using the cutvalue() or cutnumber() options
r(198);
```

Following the error-message advice, Figure 12.4 employs a **cutnumber(100)** option to form a dendrogram that starts with only 100 groups, after the first few fusions have taken place.

. cluster tree, ylabel(0(.5)3) cutnumber(100)

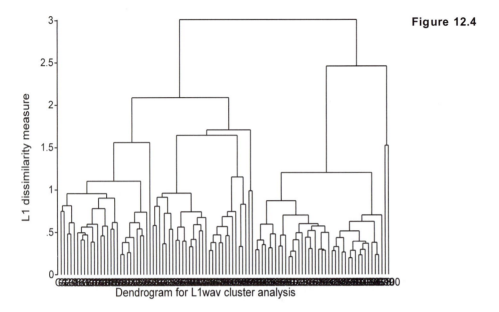

Dendrogram for L1wav cluster analysis

Figure 12.4

The bottom labels in Figure 12.4 are unreadable, but we can trace the general flow of this clustering process. Most of the fusion takes place at dissimilarities below 1. Two nations at far right are unusual; they resist fusion until about 1.5, and then form a stable two-nation group quite different from all the rest. This is one of four clusters remaining above dissimilarities of 2. The first and second of these four final clusters (reading left to right) appear heterogeneous, formed through successive fusion of a number of somewhat distinct major subgroups. The third cluster, in contrast, appears more homogeneous. It combines many nations that fused into two subgroups at dissimilarities below 1, and then fused into one group at slightly above 1.

Figure 12.5 gives another view of this analysis, this time using the **cutvalue(1)** option to show only clusters with dissimilarities above 1. The **vertlabel** option, not really needed here, calls for the bottom labels (G1, G2, etc.) to be printed vertically instead of horizontally.

```
. cluster tree, ylabel(0(.5)3) cutvalue(1) vertlabel
```

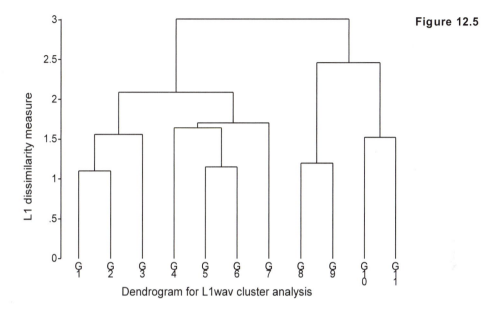

Figure 12.5

Dendrogram for L1wav cluster analysis

As Figure 12.5 shows, there are 11 groups remaining at dissimilarities above 1. For purposes of illustration, we will consider only the top four groups, which have dissimilarities above 2. **cluster generate** creates a categorical variable for the final four groups from the cluster analysis we named *L1wav*.

```
. cluster generate ctype = groups(4), name(L1wav)
. label variable ctype "Country type"
```

We could next examine which countries belong to which groups by typing

```
. by ctype:  list country
```

A more compact list of the same information appears below. This list was produced by copying and pasting data from *nations.dta* into the Data Editor to form a separate, single-purpose dataset in which the columns are country types.

```
     +-----------------------------------------+
     |  ctype1    ctype2     ctype3   ctype4 |
     |-----------------------------------------|
 1.  |  Algeria   Argentin   Banglade   China |
 2.  |   Brazil   Australi     Benin    India |
 3.  |    Burma    Austria    Bolivia          |
 4.  |    Chile    Belgium   Botswana          |
 5.  | Colombia    Canada    BurkFaso          |
     |-----------------------------------------|
 6.  | CostaRic   Denmark    Burundi           |
 7.  |   DomRep    Finland   Cameroon          |
 8.  |  Ecuador    France    CenAfrRe          |
 9.  |    Egypt    Greece    ElSalvad          |
10.  | Indonesi   HongKong   Ethiopia          |
     |-----------------------------------------|
11.  |  Jamaica   Hungary     Ghana            |
12.  |   Jordan   Ireland   Guatemal           |
13.  | Malaysia    Israel     Guinea           |
```

```
14. | Mauritiu      Italy       Haiti      |
15. |  Mexico       Japan     Honduras     |
    |-------------------------------------|
16. |  Morocco     Kuwait     IvoryCoa     |
17. |  Panama      Netherla     Kenya      |
18. | Paraguay     NewZeala    Liberia     |
19. |    Peru       Norway     Madagasc    |
20. | Philippi      Poland      Malawi     |
    |-------------------------------------|
21. | SauArabi     Portugal    Mauritan    |
22. | SriLanka     S_Korea     Mozambiq    |
23. |   Syria      Singapor     Nepal      |
24. | Thailand      Spain      Nicaragu    |
25. | Tunisia       Sweden      Niger      |
    |-------------------------------------|
26. |  Turkey      TrinToba    Nigeria     |
27. | Uruguay        U_K       Pakistan    |
28. | Venezuel      U_S_A      PapuaNG     |
29. |             UnArEmir      Rwanda     |
30. |             W_German     Senegal     |
    |-------------------------------------|
31. |             Yugoslav     SierraLe    |
32. |                          Somalia     |
33. |                           Sudan      |
34. |                          Tanzania    |
35. |                            Togo      |
    |-------------------------------------|
36. |                          YemenAR     |
37. |                          YemenPDR    |
38. |                           Zaire      |
39. |                          Zambia      |
40. |                          Zimbabwe    |
    +-------------------------------------+
```

The two-nation cluster seen at far right in Figure 12.4 turns out to be type 4, China and India. The broad, homogeneous third cluster in Figure 12.4, type 3, contains a large group of the poorest nations, mainly in Africa. The relatively diverse type 2 contains nations with higher living conditions including the U.S., Europe, and Japan. Type 1, also diverse, contains nations with intermediate conditions. Whether this or some other typology is meaningful remains a substantive question, not a statistical one, and depends on the uses for which a typology is needed. Choosing different options in the steps of our cluster analysis would have returned different results. By experimenting with a variety of reasonable choices, we could gain a sense of which findings are most stable.

Time Series Analysis

Stata's evolving time series capabilities are covered in the 350-page *Time-Series Reference Manual*. This chapter provides a brief introduction, beginning with two elementary and useful analytical tools: time plots and smoothing. We then move on to illustrate the use of correlograms, ARIMA models, and tests for stationarity and white noise. Further applications, notably periodograms and the flexible ARCH family of models, are left to the reader's explorations.

A technical and thorough treatment of time series topics is found in Hamilton (1994). Other sources include Box, Jenkins, and Reinsel (1994), Chatfield (1996), Diggle (1990), Enders (1995), Johnston and DiNardo (1997), and Shumway (1988).

Menus for time series operations come under the following headings:

Statistics – Time series

Statistics – Multivariate time series

Statistics – Cross-sectional time series

Graphics – Time series graphs

Example Commands

. `ac y, lags(8) level(95) generate(newvar)`

Graphs autocorrelations of variable *y*, with 95% confidence intervals (default), for lags 1 through 8. Stores the autocorrelations as the first 8 values of *newvar*.

. `arch D.y, arch(1/3) ar(1) ma(1)`

Fits an ARCH (autoregressive conditional heteroskedasticity) model for first differences of *y*, including first- through third-order ARCH terms, and first-order AR and MA disturbances.

. `arima y, arima(3,1,2)`

Fits a simple ARIMA(3,1,2) model. Possible options include several estimation strategies, linear constraints, and robust estimates of variance.

. `arima y, arima(3,1,2) sarima(1,0,1,12)`

Fits ARIMA model including a multiplicative seasonal component with period 12.

. `arima D.y x1 L1.x1 x2, ar(1) ma(1 12)`

Regresses first differences of *y* on *x1*, lag-1 values of *x1*, and *x2*, including AR(1), MA(1), and MA(12) disturbances.

. `corrgram y, lags(8)`
   Obtains autocorrelations, partial autocorrelations, and Q tests for lags 1 through 8.

. `dfuller y`
   Performs Dickey–Fuller unit root test for stationarity.

. `dwstat`
   After `regress`, calculates a Durbin–Watson statistic testing first-order autocorrelation.

. `egen newvar = ma(y), nomiss t(7)`
   Generates *newvar* equal to the span-7 moving average of *y*, replacing the start and end values with shorter, uncentered averages.

. `generate date = mdy(month,day,year)`
   Creates variable *date*, equal to days since January 1, 1960, from the three variables *month*, *day*, and *year*.

. `generate date = date(str_date, "mdy")`
   Creates variable *date* from the string variable *str_date*, where *str_date* contains dates in month, day, year form such as "11/19/2001", "4/18/98", or "June 12, 1948". Type `help dates` for many other date functions and options.

. `generate newvar = L3.y`
   Generates *newvar* equal to lag-3 values of *y*.

. `pac y, lags(8) yline(0) ciopts(bstyle(outline))`
   Graphs partial autocorrelations with confidence intervals and residual variance for lags 1 through 8. Draws a horizontal line at 0; shows the confidence interval as an outline, instead of a shaded area (default).

. `pergram y, generate(newvar)`
   Draws the sample periodogram (spectral density function) of variable *y* and creates *newvar* equal to the raw periodogram values.

. `prais y x1 x2`
   Performs Prais–Winsten regression of *y* on *x1* and *x2*, correcting for first-order autoregressive errors. `prais y x1 x2, corc` does Cochrane–Orcutt instead.

. `smooth 73 y, generate(newvar)`
   Generates *newvar* equal to span-7 running medians of *y*, re-smoothing by span-3 running medians. Compound smoothers such as "3RSSH" or "4253h,twice" are possible. Type `help smooth`, or `help tssmooth`, for other smoothing and filters.

. `tsset date, format(%d)`
   Defines the dataset as a time series. Time is indicated by variable *date*, which is formatted as daily. For "panel" data with parallel time series for a number of different units, such as cities, `tsset city year` identifies both panel and time variables. Most of the commands in this chapter require that the data be `tsset`.

. `tssmooth ma newvar = y, window(2 1 2)`
   Applies a moving-average filter to *y*, generating *newvar*. The `window(2 1 2)` option finds a span-5 moving average by including 2 lagged values, the current observation, and 2 leading values in the calculation of each smoothed point. Type `help tssmooth` for a list of other possible filters including weighted moving averages, exponential or double exponential, Holt–Winters, and nonlinear.

. `tssmooth nl` *newvar* `= y, smoother(4253h,twice)`
> Applies a nonlinear smoothing filter to *y*, generating *newvar*. The
> `smoother(4253h,twice)` option iteratively finds running medians of span 4, 2, 5, and
> 3, then applies Hanning, then repeats on the residuals. `tssmooth nl`, unlike other
> `tssmooth` procedures, cannot work around missing values.

. `wntestq y, lags(15)`
> Box–Pierce portmanteau *Q* test for white noise (also provided by `corrgram`).

. `xcorr x y, lags(8) xline(0)`
> Graphs cross-correlations between input (*x*) and output (*y*) variable for lags 1–8.
> `xcorr x y, table` gives a text version that includes the actual correlations (or
> include a `generate(`*newvar*`)` option to store the correlations as a variable).

Smoothing

Many time series exhibit rapid up-and-down fluctuations that make it difficult to discern underlying patterns. Smoothing such series breaks the data into two parts, one that varies gradually, and a second "rough" part containing the leftover rapid changes:

data = smooth + rough

Dataset *MILwater.dta* contains data on daily water consumption for the town of Milford, New Hampshire over seven months from January through July 1983 (Hamilton 1985b).

```
Contains data from MILwater.dta
  obs:           212                          Milford daily water use, 1/1/83
                                                - 7/31/83
  vars:            4                          27 Jul 2005 12:41
  size:        2,120 (99.9% of memory free)
-------------------------------------------------------------------------------
              storage  display    value
variable name   type   format     label      variable label
-------------------------------------------------------------------------------
month          byte    %9.0g                 Month
day            byte    %9.0g                 Date
year           int     %9.0g                 Year
water          int     %9.0g                 Water use in 1000 gallons
-------------------------------------------------------------------------------
Sorted by:
```

Before further analysis, we need to convert the month, day, and year information into a single numerical index of time. Stata's `mdy()` function does this, creating an elapsed-date variable (named *date* here) indicating the number of days since January 1, 1960.

. `generate date = mdy(month,day,year)`

. `list in 1/5`

```
     +-----------------------------------------+
     | month   day   year   water   date |
     |-----------------------------------------|
  1. |     1     1   1983     520   8401 |
  2. |     1     2   1983     600   8402 |
  3. |     1     3   1983     610   8403 |
  4. |     1     4   1983     590   8404 |
  5. |     1     5   1983     620   8405 |
     +-----------------------------------------+
```

The January 1, 1960 reference date is an arbitrary default. We can provide more understandable formatting for *date*, and also set up our data for later analyses, by using the **tsset** (**t**ime **s**eries **set**) command to identify *date* as the time index variable and to specify the **%d** (**d**aily) display option for this variable.

```
. tsset date, format(%d)
        time variable:  date, 01jan1983 to 31jul1983
. list in 1/5
```

```
      +------------------------------------+
      | month   day   year   water     date |
      |------------------------------------|
  1.  |     1     1   1983    520   01jan1983 |
  2.  |     1     2   1983    600   02jan1983 |
  3.  |     1     3   1983    610   03jan1983 |
  4.  |     1     4   1983    590   04jan1983 |
  5.  |     1     5   1983    620   05jan1983 |
      +------------------------------------+
```

Dates in the new *date* format, such as "05jan1983", are more readable than the underlying numerical values such as "8405" (days since January 1, 1960). If desired, we could use **%d** formatting to produce other formats, such as "05 Jan 1983" or "01/05/83". Stata offers a number of variable-definition, display-format, and dataset-format features that are important with time series. Many of these involve ways to input, convert, and display dates. Full descriptions of date functions are found in the *Data Management Reference Manual* and the *User's Guide*, or they can be explored within Stata by typing **help dates**.

The labeled values of *date* appear in a graph of *water* against *date*, which shows day-to-day variation, as well as an upward trend in water use as summer arrives (Figure 13.1):

```
. graph twoway line water date, ylabel(300(100)900)
```

Figure 13.1

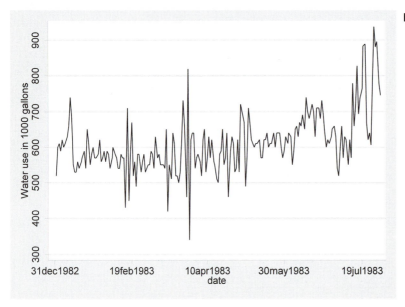

Visual inspection plays an important role in time series analysis. It often helps us to see underlying patterns in jagged series if we smooth the data by calculating a "moving average" at each point from its present, earlier, and later values. For example, a "moving average of span 3" refers to the mean of y_{t-1}, y_t, and y_{t+1}. We could use Stata's explicit subscripting to **generate** such a variable:

```
. generate water3 = (water[_n-1] + water[_n] + water[_n+1])/3
```

Or, we could apply the **ma** (moving average) function of **egen** :

```
. egen water3 = ma(water), nomiss t(3)
```

The **nomiss** option asks for shorter, uncentered moving averages in the tails; otherwise, the first and last values of *water3* would be missing. The **t(3)** option calls for moving averages of span 3. Any odd-number span ≥ 3 could be used.

For time series (**tsset**) data, powerful smoothing tools are available through the **tssmooth** commands. All but **tssmooth nl** can handle missing values.

tssmooth ma	moving-average filters, unweighted or weighted
tssmooth exponential	single exponential filters
tssmooth dexponential	double exponential filters
tssmooth hwinters	nonseasonal Holt–Winters smoothing
tssmooth shwinters	seasonal Holt–Winters smoothing
tssmooth nl	nonlinear filters

Type **help tssmooth_exponential** , **help tssmooth_hwinters** , etc. for the syntax of each command.

Figure 13.2 graphs a simple 5-day moving average of Milford water use (*water5*), together with the raw data (*water*). This **graph twoway** command overlays a line plot of smoothed *water5* values with a line plot of raw *water* values (thinner line). *X*-axis labels mark start-of-month values chosen "by hand" (8401, 8432, etc.) to make the graph more readable. Readability is also improved by formatting the labels as **%dmd** (date format, but only month followed by day). Compare Figure 13.2's labels with their default counterparts in Figure 13.1.

```
. tssmooth ma water5 = water, window(2 1 2)
The smoother applied was
    (1/5)*[x(t-2) + x(t-1) + 1*x(t) + x(t+1) + x(t+2)]; x(t)= water
```

```
. graph twoway line water5 date, clwidth(thick)
     ||   line water date, clwidth(thin) clpattern(solid)
     ||   , ylabel(300(100)900)
   xlabel(8401 8432 8460 8491 8521 8552 8582 8613,
      grid format(%dmd))
   xtitle("") ytitle(Water use in 1000 gallons)
   legend(order(2 1) position(4) ring(0) rows(2)
      label(1 "5-day average") label(2 "daily water use"))
```

Figure 13.2

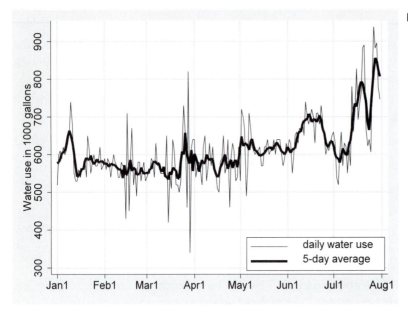

Moving averages share a drawback of other mean-based statistics: they have little resistance to outliers. Because outliers form prominent spikes in Figure 13.1, we might also try a different smoothing approach. The **tssmooth nl** command performs outlier-resistant nonlinear smoothing, employing methods and a terminology described in Velleman and Hoaglin (1981) and Velleman (1982). For example,

```
. tssmooth nl water5r = water, smoother(5)
```

creates a new variable named *water5r*, holding the values of *water* after smoothing by running medians of span 5. Compound smoothers using running medians of different spans, in combination with "hanning" (¼, ½, and ¼ -weighted moving averages of span 3) and other techniques, can be specified in Velleman's original notation. One compound smoother that seems particularly useful is called "4253h, twice." Applying this to *water*, we calculate smoothed variable *water4r*:

```
. tssmooth nl water4r = water, smoother(4253h,twice)
```

Figure 13.3 graphs new smoothed values, *water4r*. Compare Figure 13.3 with 13.2 to see how the 4253h, twice smoothing performs relative to a moving-average. Although both smoothers have similar spans, 4253h, twice does more to reduce the jagged variations.

Figure 13.3

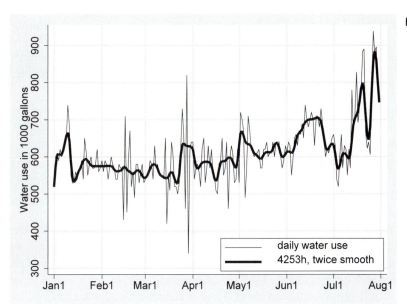

Sometimes our goal in smoothing is to look for patterns in smoothed plots. With these particular data, however, the "rough" or residuals after smoothing actually hold more interest. We can calculate the rough as the difference between data and smooth, and then graph the results in another time plot, Figure 13.4.

```
. generate rough = water - water4r

. label variable rough "Residuals from 4253h, twice"

. graph twoway line rough date,
      xlabel(8401 8432 8460 8491 8521 8552 8582 8613,
        grid format(%dmd)) xtitle("")
```

Figure 13.4

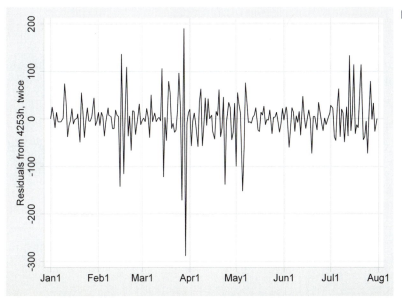

The wildest fluctuations in Figure 13.4 occur around March 27–29. Water use abruptly dropped, rose again, and then dropped even further before returning to more usual levels. On these days, local newspapers carried stories that hazardous chemical wastes had been discovered in one of the wells that supplied the town's water. Initial reports alarmed people, but they were reassured after the questionable well was taken offline.

The smoothing techniques described in this section tend to make the most sense when the observations are equally spaced in time. For time series with uneven spacing, lowess regression (see Chapter 8) provides a practical alternative.

Further Time Plot Examples

Dataset *atlantic.dta* contains time series of climate, ocean, and fisheries variables for the northern Atlantic from 1950–2000 (the original data sources include Buch 2000, and others cited in Hamilton, Brown, and Rasmussen 2003). The variables include sea temperatures on Fylla Bank off west Greenland; air temperatures in Nuuk, Greenland's capital city; two climate indexes called the North Atlantic Oscillation (NAO) and the Arctic Oscillation (AO); and catches of cod and shrimp in west Greenland waters.

```
Contains data from atlantic.dta
  obs:            51                          Greenland climate & fisheries
  vars:            8                          27 Jul 2005 12:41
  size:        1,734 (99.9% of memory free)
-------------------------------------------------------------------------------
              storage  display    value
variable name  type    format     label      variable label
-------------------------------------------------------------------------------
year            int     %ty                   Year
fylltemp        float   %9.0g                 Fylla Bank temp. at 0-40m
fyllsal         float   %9.0g                 Fylla Bank salinity at 0-40m
nuuktemp        float   %9.0g                 Nuuk air temperature
wNAO            float   %9.0g                 Winter (Dec-Mar)
                                                Lisbon-Stykkisholmur NAO
wAO             float   %9.0g                 Winter (Dec-Mar) AO index
tcod1           float   %9.0g                 Division 1 cod catch, 1000t
tshrimp1        float   %9.0g                 Division 1 shrimp catch, 1000t
-------------------------------------------------------------------------------
Sorted by:  year
```

Before analyzing these time series, we **tsset** the dataset, which tells Stata that the variable *year* contains the time-sequence information.

```
. tsset year, yearly
        time variable:  year, 1950 to 2000
```

With a **tsset** dataset, two new qualifiers become available: **tin** (times **in**) and **twithin** (times **within**). To list Fylla temperatures and NAO values for the years 1950 through 1955, type

```
. list year fylltemp wNAO if tin(1950,1955)

      +------------------------+
      | year   fylltemp   wNAO |
      |------------------------|
  1.  | 1950       2.1     1.4 |
  2.  | 1951       1.9   -1.26 |
  3.  | 1952       1.6     .83 |
```

```
4. | 1953         2.1     .18 |
5. | 1954         2.3     .13 |
   |-------------------------|
6. | 1955         1.2    -2.52 |
   +-------------------------+
```

The **twithin** qualifier works similarly, but excludes the two endpoints:

. **list** *year fylltemp wNAO* **if twithin(1950,1955)**

```
   +-------------------------+
   | year  fylltemp   wNAO |
   |-------------------------|
2. | 1951       1.9  -1.26 |
3. | 1952       1.6    .83 |
4. | 1953       2.1    .18 |
5. | 1954       2.3    .13 |
   +-------------------------+
```

We use **tssmooth nl** to define a new variable, *fyll4*, containing 4253h, twice smoothed values of *fylltemp* (data from Buch 2000).

. **tssmooth nl** *fyll4* = *fylltemp*, **smoother(4253h, twice)**

Figure 13.5 graphs raw (*fylltemp*) and smoothed (*fyll4*) Fylla Bank temperatures. Raw temperatures are shown as spike-plot deviations from the mean (1.67 °C), so this graph emphasizes both decadal cycles and annual variations.

. **graph twoway spike** *fylltemp year*, **base(1.67) yline(1.67)**
       || line *fyll4 year*, **clpattern(solid)**
       || , ytitle("Fylla Bank temperature, degrees C") ylabel(0(1)3)
       xtitle("") xtick(1955(10)1995) legend(off)

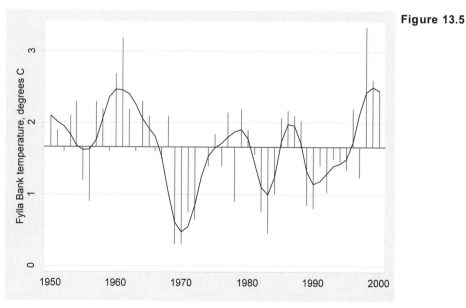

Figure 13.5

The smoothed values of Figure 13.5 exhibit irregular periods of generally warmer and cooler water. Of course, "warmer" is a relative term around Greenland; these summer sea temperatures rise no higher than 3.34 °C (37 °F).

Fylla Bank temperatures are influenced by a large-scale atmospheric pattern called the North Atlantic Oscillation, or NAO. Figure 13.6 graphs smoothed temperatures together with smoothed values of the NAO (a new variable named *wNAO4*). For this overlaid graph, temperature defines the left axis scale, **yaxis(1)**, and NAO the right, **yaxis(2)**. Further *y*-axis options specify whether they refer to axis 1 or 2. For example, a horizontal line drawn by **yline(0, axis(2))** marks the zero point of the NAO index. On both axes, numerical labels are written horizontally. The legend appears at the 5 o'clock position inside the plot space, **position(5) ring(0)**.

```
. graph twoway line fyll4 year, yaxis(1)
     ylabel(0(1)3, angle(horizontal) nogrid axis(1))
     ytitle("Fylla Bank temperature, degrees C", axis(1))
     || line wNAO4 year, yaxis(2) ytitle("Winter NAO index", axis(2))
     ylabel(-3(1)3, angle(horizontal) axis(2)) yline(0, axis(2))
     || , xtitle("") xlabel(1950(10)2000, grid) xtick(1955(5)1995)
     legend(label(1 "Fylla temperature") label(2 "NAO index") cols(1)
         position(5) ring(0))
```

Figure 13.6

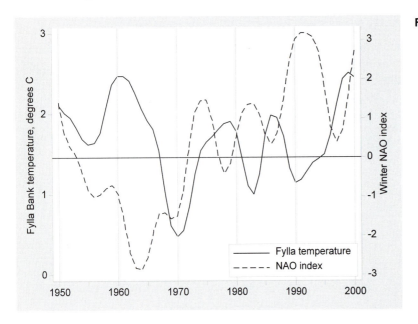

Overlaid plots provide a way to visually examine how several time series vary together. In Figure 13.6, we see evidence of a negative correlation: high-NAO periods correspond to low temperatures. The physical mechanism behind this correlation involves northerly winds that bring Arctic air and water to west Greenland during high-NAO phases. The negative temperature–NAO correlation became stronger during the later part of this time series, roughly the years 1973 to 1997. We will return to this relationship in later sections.

Lags, Leads, and Differences

Time series analysis often involves lagged variables, or values from previous times. Lags can be specified by explicit subscripting. For example, the following command creates variable *wNAO_1*, equal to the previous year's NAO value:

```
. generate wNAO_1 = wNAO[_n-1]
(1 missing value generated)
```

An alternative way to achieve the same thing, using **tsset** data, is with Stata's **L.** (lag) operator:

```
. generate wNAO_1 = L.wNAO
(1 missing values generated)
```

Lag operators are often simpler than an explicit-subscripting approach. More importantly, the lag operators also respect panel data. To generate lag 2 values, use

```
. generate wNAO_2 = L2.wNAO
(2 missing values generated)
. list year wNAO wNAO_1 wNAO_2 if tin(1950,1954)
```

```
     +-------------------------------+
     | year    wNAO   wNAO_1   wNAO_2 |
     |-------------------------------|
  1. | 1950     1.4        .        . |
  2. | 1951   -1.26      1.4        . |
  3. | 1952     .83    -1.26      1.4 |
  4. | 1953     .18      .83    -1.26 |
  5. | 1954     .13      .18      .83 |
     +-------------------------------+
```

We could have obtained this same list without generating any new variables, by instead typing

```
. list year wNAO L.wNAO L2.wNAO if tin(1950,1954)
```

The **L.** operator is one of several that simplify the analysis of **tsset** datasets. Other time series operators are **F.** (lead), **D.** (difference), and **S.** (seasonal difference). These operators can be typed in upper or lowercase — for example, **F2.wNAO** or **f2.wNAO**.

Time Series Operators

L.	Lag y_{t-1} (**L1.** means the same thing)
L2.	2-period lag y_{t-2} (similarly, **L3.**, etc. **L(1/4).** means **L1.** through **L4.**)
F.	Lead y_{t+1} (**F1.** means the same thing)
F2.	2-period lead y_{t+2} (similarly, **F3.**, etc.)
D.	Difference $y_t - y_{t-1}$ (**D1.** means the same thing)
D2.	Second difference $(y_t - y_{t-1}) - (y_{t-1} - y_{t-2})$ (similarly, **D3.**, etc.)
S.	Seasonal difference $y_t - y_{t-1}$, (which is the same as **D.**)
S2.	Second seasonal difference $(y_t - y_{t-2})$ (similarly, **S3.**, etc.)

In the case of seasonal differences, **S12.** does not mean "12th difference," but rather a first difference at lag 12. For example, if we had monthly temperatures instead of yearly, we might want to calculate **S12.*temp*** , which would be the differences between December 2000 temperature and December 1999 temperature, November 2000 temperatures and November 1999 temperature, and so forth.

Lag operators can appear directly in most analytical commands. We could regress 1973–97 *fylltemp* on *wNAO*, including as additional predictors *wNAO* values from one, two, and three years previously, without first creating any new lagged variables.

. **regress *fylltemp* wNAO L1.wNAO L2.wNAO L3.wNAO if tin(1973,1997)**

```
      Source |       SS       df       MS              Number of obs =      25
-------------+------------------------------           F(  4,     20) =    4.57
       Model |  3.1884913      4  .797122826           Prob > F      =  0.0088
    Residual |  3.48929123    20  .174464562           R-squared     =  0.4775
-------------+------------------------------           Adj R-squared =  0.3730
       Total |  6.67778254    24  .278240939           Root MSE      =  .41769


------------------------------------------------------------------------------
    fylltemp |      Coef.   Std. Err.      t    P>|t|     [95% Conf. Interval]
-------------+----------------------------------------------------------------
    wNAO     |
          -- |  -.1688424   .0412995    -4.09   0.001    -.2549917   -.0826931
          L1 |   .0043805   .0421436     0.10   0.918    -.0835294    .0922905
          L2 |  -.0472993   .050851     -0.93   0.363    -.1533725    .058774
          L3 |   .0264682   .0495416     0.53   0.599    -.0768738    .1298102
       _cons |   1.727913   .1213588    14.24   0.000     1.474763    1.981063
------------------------------------------------------------------------------
```

Equivalently, we could have typed

. **regress *fylltemp* L(0/3).wNAO if tin(1973,1997)**

The estimated model is

$$\text{predicted } fylltemp_t = 1.728 - .169wNAO_t + .004wNAO_{t-1} - .047wNAO_{t-2} + .026wNAO_{t-3}$$

Coefficients on the lagged terms are not statistically significant; it appears that current (unlagged) values of $wNAO_t$ provide the most parsimonious prediction. Indeed, if we re-estimate this model without the lagged terms, the adjusted R^2 rises from .37 to .43. Either model is very rough, however. A Durbin–Watson test for autocorrelated errors is inconclusive, but that is not reassuring given the small sample size.

. **dwstat**

```
Durbin-Watson d-statistic(  5,    25) =  1.423806
```

Autocorrelated errors, commonly encountered with time series, invalidate the usual OLS confidence intervals and tests. More suitable regression methods for time series are discussed later in this chapter.

Correlograms

Autocorrelation coefficients estimate the correlation between a variable and itself at particular lags. For example, first-order autocorrelation is the correlation between y_t and y_{t-1}. Second order refers to $\text{Cor}[y_t, y_{t-2}]$, and so forth. A correlogram graphs correlation versus lags.

Stata's **corrgram** command provides simple correlograms and related information. The maximum number of lags it shows can be limited by the data, by **matsize**, or to some arbitrary lower number that is set by specifying the **lags()** option:

```
. corrgram fylltemp, lags(9)
```

```
                                              -1      0      1 -1      0       1
     LAG      AC       PAC        Q     Prob>Q  [Autocorrelation]  [Partial Autocor]
     ------------------------------------------------------------------------------
     1      0.4038    0.4141    8.8151  0.0030      |---              |---
     2      0.1996    0.0565   11.012   0.0041      |-                |
     3      0.0788    0.0045   11.361   0.0099      |                 |
     4      0.0071   -0.0556   11.364   0.0228      |                 |
     5     -0.1623   -0.2232   12.912   0.0242     -|                -|
     6     -0.0733    0.0880   13.234   0.0395      |                 |
     7      0.0490    0.1367   13.382   0.0633      |                 |-
     8     -0.1029   -0.2510   14.047   0.0805      |                --|
     9     -0.2228   -0.2779   17.243   0.0450     -|                --|
```

Lags appear at the left side of the table, and are followed by columns for the autocorrelations (AC) and partial autocorrelations (PAC). For example, the correlation between *fylltemp*$_t$ and *fylltemp*$_{t-2}$ is .1996, and the partial autocorrelation (adjusted for lag 1) is .0565. The Q statistics (Box–Pierce portmanteau) test a series of null hypotheses that all autocorrelations up to and including each lag are zero. Because the P-values seen here are mostly below .05, we can reject the null hypothesis, and conclude that *fylltemp* shows significant autocorrelation. If none of the Q statistics had been below .05, we might conclude instead that the series was "white noise" with no significant autocorrelation.

At the right in this output are character-based plots of the autocorrelations and partial autocorrelations. Inspection of such plots plays a role in the specification of time series models. More refined graphical autocorrelation plots can be obtained through the **ac** command:

```
. ac fylltemp, lags(9)
```

The resulting correlogram, Figure 13.7, includes a shaded area marking pointwise 95% confidence intervals. Correlations outside of these intervals are individually significant.

Figure 13.7

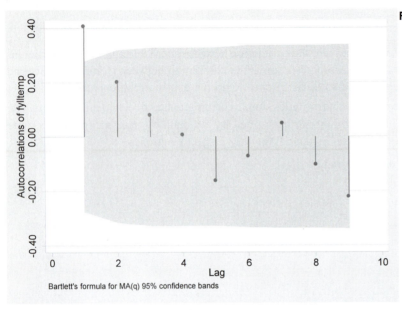

Bartlett's formula for MA(q) 95% confidence bands

A similar command, **pac**, produces the graph of partial autocorrelations seen in Figure 13.8. Approximate confidence intervals (estimating the standard error as $1/\sqrt{n}$) also appear in Figure 13.8. The default plot produced by both **ac** and **pac** has the look shown in Figure 13.7. For Figure 13.8 we chose different options, drawing a baseline at zero correlation, and indicating the confidence interval as an outline instead of a shaded area.

```
. pac fylltemp, yline(0) lags(9) ciopts(bstyle(outline))
```

Figure 13.8

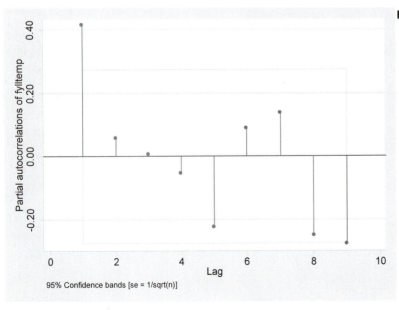

95% Confidence bands [se = 1/sqrt(n)]

Cross-correlograms help to explore relationships between two time series. Figure 13.9 shows the cross-correlogram of *wNAO* and *fylltemp* over 1973–97. The cross-correlation is substantial and negative at 0 lag, but is closer to zero at other positive or negative lags. This suggests that the relationship between the two series is "instantaneous" (in yearly data) rather than delayed or distributed over several years. Recall the nonsignificance of lagged predictors from our earlier OLS regression.

```
. xcorr wNAO fylltemp if tin(1973,1997), lags(9) xlabel(-9(1)9, grid)
```

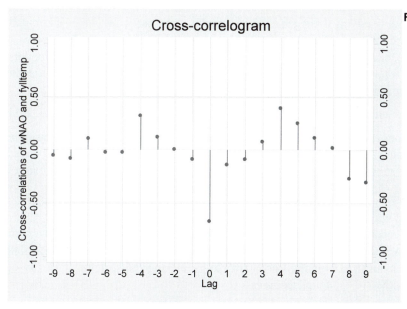

Figure 13.9

If we list our input or independent variable first in the **xcorr** command, and the output or dependent variable second — as was done for Figure 13.9 — then positive lags denote correlations between input at time t and output at time $t+1$, $t+2$, etc. Thus, we see a positive correlation of .394 between winter NAO index and Fylla temperature four years later.

The actual cross-correlation coefficients, and a text version of the cross-correlogram, can be obtained with the **table** option:

```
. xcorr wNAO fylltemp if tin(1973,1997), lags(9) table
```

```
                      -1        0        1
  LAG       CORR   [Cross-correlation]
-------------------------------------
  -9      -0.0541            |
  -8      -0.0786            |
  -7       0.1040            |
  -6      -0.0261            |
  -5      -0.0230            |
  -4       0.3185            |--
  -3       0.1212            |
  -2       0.0053            |
  -1      -0.0909            |
   0      -0.6740       -----|
   1      -0.1386          -|
   2      -0.0865            |
```

3	0.0757	|
4	0.3940	|---
5	0.2464	|-
6	0.1100	|
7	0.0183	|
8	-0.2699	--|
9	-0.3042	--|

ARIMA Models

Autoregressive integrated moving average (ARIMA) models for time series can be estimated through the **arima** command. This command encompasses simple autoregressive (AR), moving average (MA), or ARIMA models of any order. It also can estimate structural models that include one or more predictor variables and AR or MA errors. The general form of such structural models, in matrix notation, is

$$y_t = \mathbf{x}_t \beta + \mu_t \qquad [13.1]$$

where y_t is the vector of dependent-variable values at time t, $\mathbf{x}_t$ is a matrix of predictor-variable values (usually including a constant), and μ_t is a vector of disturbances. Those disturbances can be autoregressive or moving-average, of any order. For example, ARMA(1,1) disturbances are

$$\mu_t = \rho\mu_{t-1} + \theta\epsilon_{t-1} + \epsilon_t \qquad [13.2]$$

where ρ is the first-order autocorrelation parameter, θ is the first-order moving average parameter, and ϵ is a white-noise (normal i.i.d.) disturbance. **arima** fits simple models as a special case of [13.1] and [13.2], with a constant (β_0) replacing the structural term $\mathbf{x}_t\beta$. Therefore, a simple ARMA(1,1) model becomes

$$y_t = \beta_0 + \mu_t$$
$$= \beta_0 + \rho\mu_{t-1} + \theta\epsilon_{t-1} + \epsilon_t \qquad [13.3]$$

Some sources present an alternative version. In the ARMA(1,1) case, they show y_t as a function of the previous y value (y_{t-1}) and the present (ϵ_t) and lagged (ϵ_{t-1}) disturbances:

$$y_t = \alpha + \rho y_{t-1} + \theta\epsilon_{t-1} + \epsilon_t \qquad [13.4]$$

Because in the simple structural model $y_t = \beta_0 + \mu_t$, equation [13.3] (Stata's version) is equivalent to [13.4], apart from rescaling the constant $\alpha = (1-\rho)\beta_0$.

Using **arima**, an ARMA(1,1) model (equation [13.3]) can be specified in either of two ways:

```
. arima y, ar(1) ma(1)
```

or

```
. arima y, arima(1,0,1)
```

The **i** in **arima** stands for "integrated," referring to models that also involve differencing. To fit an ARIMA(2,1,1) model, use

```
. arima y, arima(2,1,1)
```

or equivalently,

```
. arima D.y, ar(1 2) ma(1)
```

Either command specifies a model in which first differences of the dependent variable $(y_t - y_{t-1})$ are a function of first differences one and two lags previous $(y_{t-1} - y_{t-2}$ and $y_{t-2} - y_{t-3})$ and also of present and previous disturbances $(\epsilon_t$ and $\epsilon_{t-1})$.

To estimate a structural model in which y_t depends on two predictor variables x (present and lagged values, x_t and x_{t-1}) and w (present values only, w_t), with ARIMA(1,0,1) errors, an appropriate command would be

```
. arima y x L.x w, arima(1,0,1)
```

Although seasonal differencing (e.g., **S12.y**) and/or seasonal lags (e.g., **L12.x**) can be included, as of this writing **arima** does not estimate multiplicative ARIMA$(p,d,q)(P,D,Q)_s$ seasonal models.

A time series y is considered "stationary" if its mean and variance do not change with time, and if the covariance between y_t and y_{t+u} depends only on the lag u, and not on the particular values of t. ARIMA modeling assumes that our series is stationary, or can be made stationary through appropriate differencing or transformation. We can check this assumption informally by inspecting time plots for trends in level or variance. Formal statistical tests for "unit roots" (a nonstationary AR(1) process in which $\rho_1 = 1$, also known as a "random walk") also help. Stata offers three unit root tests, **pperron** (Phillips–Perron), **dfuller** (augmented Dickey–Fuller), and **dfgls** (augmented Dickey–Fuller using GLS, generally a more powerful test than **dfuller**).

Applied to Fylla Bank temperatures, a **pperron** test rejects the null hypothesis of a unit root $(P < .01)$.

```
. pperron fylltemp, lag(3)
```

```
Phillips-Perron test for unit root             Number of obs   =       50
                                               Newey-West lags =        3

                         ---------- Interpolated Dickey-Fuller ---------
               Test        1% Critical       5% Critical      10% Critical
            Statistic         Value             Value             Value
------------------------------------------------------------------------
Z(rho)        -29.871        -18.900           -13.300           -10.700
Z(t)           -4.440         -3.580            -2.930            -2.600
------------------------------------------------------------------------
* MacKinnon approximate p-value for Z(t) = 0.0003
```

Similarly, a Dickey–Fuller GLS test evaluating the null hypothesis that *fylltemp* has a unit route (versus the alternative hypothesis that it is stationary with a possibly nonzero mean, but no linear time trend) rejects this null hypothesis $(P < .05)$. Both tests thus confirm the visual impression of stationarity given by Figure 13.5.

```
. dfgls fylltemp, notrend maxlag(3)
```

```
DF-GLS for fylltemp              Number of obs =      47

                DF-GLS mu        1% Critical        5% Critical        10% Critical
   [lags]    Test Statistic        Value              Value              Value
-----------------------------------------------------------------------------------
      3          -2.304            -2.620             -2.211             -1.913
      2          -2.479            -2.620             -2.238             -1.938
      1          -3.008            -2.620             -2.261             -1.959

Opt Lag (Ng-Perron seq t) = 0 [use maxlag(0)]
Min SC   = -.6735952 at lag  1 with RMSE  .6578912
Min MAIC = -.2683716 at lag  2 with RMSE  .6569351
```

For a stationary series, correlograms provide guidance about selecting a preliminary ARIMA model:

AR(p)   An autoregressive process of order p has autocorrelations that damp out gradually with increasing lag. Partial autocorrelations cut off after lag p.

MA(q)   A moving average process of order q has autocorrelations that cut off after lag q. Partial autocorrelations damp out gradually with increasing lag.

ARMA(p,q) A mixed autoregressive–moving average process has autocorrelations and partial autocorrelations that damp out gradually with increasing lag.

Correlogram spikes at seasonal lags (for example, at 12, 24, 36 in monthly data) indicate a seasonal pattern. Identification of seasonal models follows similar guidelines, but applied to autocorrelations and partial autocorrelations at seasonal lags.

Figures 13.7–13.8 weakly suggest an AR(1) process, so we will try this as a simple model for *fylltemp*.

```
. arima fylltemp, arima(1,0,0) nolog
```

```
ARIMA regression

Sample:  1950 to 2000                    Number of obs    =         51
                                         Wald chi2(1)     =       7.53
Log likelihood = -48.66274               Prob > chi2      =     0.0061
-----------------------------------------------------------------------------
             |                 OPG
   fylltemp  |    Coef.   Std. Err.      z    P>|z|     [95% Conf. Interval]
-------------+---------------------------------------------------------------
   fylltemp  |
      _cons  |   1.68923   .1513096    11.16   0.000    1.392669    1.985792
-------------+---------------------------------------------------------------
   ARMA      |
   ar        |
        L1   |  .4095759   .1492491     2.74   0.006    .1170531    .7020987
-------------+---------------------------------------------------------------
     /sigma  |   .627151   .0601859    10.42   0.000    .5091889    .7451131
-----------------------------------------------------------------------------
```

After we fit an **arima** model, its coefficients and other results are saved temporarily in Stata's usual way. For example, to see the recent model's AR(1) coefficient and s.e., type

```
. display [ARMA]_b[L1.ar]
.4095759
```

```
. display [ARMA]_se[L1.ar]
.14924909
```

The AR(1) coefficient in this example is statistically distinguishable from zero ($t = 2.74$, $p = .006$), which gives one indication of model adequacy. A second test is whether the residuals appear to be uncorrelated "white noise." We can obtain residuals (also predicted values, and other case statistics) after **arima** through **predict**:

```
. predict fyllres, resid

. corrgram fyllres, lags(15)
```

| | | | | | -1   0   1 | -1   0   1 |
LAG	AC	PAC	Q	Prob>Q	[Autocorrelation]	[Partial Autocor]
1	-0.0173	-0.0176	.0162	0.8987	|	|
2	0.0467	0.0465	.13631	0.9341	|	|
3	0.0386	0.0497	.22029	0.9742	|	|
4	0.0413	0.0496	.31851	0.9886	|	|
5	-0.1834	-0.2450	2.2955	0.8069	-|	-|
6	-0.0498	-0.0602	2.4442	0.8747	|	|
7	0.1532	0.2156	3.8852	0.7929	|-	|-
8	-0.0567	-0.0726	4.087	0.8492	|	|
9	-0.2055	-0.3232	6.8055	0.6574	-|	--|
10	-0.1156	-0.2418	7.6865	0.6594	|	-|
11	0.1397	0.2794	9.0051	0.6214	|-	|--
12	-0.0028	0.1606	9.0057	0.7024	|	|-
13	0.1091	0.0647	9.8519	0.7060	|	|
14	0.1014	-0.0547	10.603	0.7169	|	|
15	-0.0673	-0.2837	10.943	0.7566	|	--|

corrgram's Q test finds no significant autocorrelation among residuals out to lag 15. We could obtain exactly the same result by requesting a **wntestq** (white noise test Q statistic) for 15 lags.

```
. wntestq fyllres, lags(15)
```

```
Portmanteau test for white noise
---------------------------------------
 Portmanteau (Q) statistic =    10.9435
 Prob > chi2(15)           =     0.7566
```

By these criteria, our AR(1) or ARIMA(1,0,0) model appears adequate. More complicated versions, with MA or higher-order AR terms, do not offer much improvement in fit.

A similar AR(1) model fits *fylltemp* over just the years 1973–1997. During this period, however, information about the winter North Atlantic Oscillation (*wNAO*) significantly improves the predictions. For this model, we include *wNAO* as a predictor but keep an AR(1) term to account for autocorrelation of errors.

```
. arima fylltemp wNAO if tin(1973,1997), ar(1) nolog

ARIMA regression

Sample:  1973 to 1997                          Number of obs     =          25
                                               Wald chi2(2)      =       12.73
Log likelihood =  -10.3481                     Prob > chi2       =      0.0017

------------------------------------------------------------------------------
             |                 OPG
    fylltemp |      Coef.   Std. Err.      z    P>|z|     [95% Conf. Interval]
-------------+----------------------------------------------------------------
fylltemp     |
        wNAO |  -.1736227   .0531688    -3.27   0.001    -.2778317   -.0694138
       _cons |   1.703462   .1348599    12.63   0.000     1.439141    1.967782
-------------+----------------------------------------------------------------
ARMA         |
          ar |
         L1  |   .2965222    .237438     1.25   0.212    -.1688478    .7618921
-------------+----------------------------------------------------------------
      /sigma |    .36536    .0654008     5.59   0.000     .2371767    .4935432
------------------------------------------------------------------------------

. predict fyllhat
(option xb assumed; predicted values)

. label variable fyllhat "predicted temperature"

. predict fyllres2, resid

. corrgram fyllres2, lags(9)

                                         -1       0       1 -1       0        1
LAG       AC       PAC        Q   Prob>Q [Autocorrelation]   [Partial Autocor]
-------------------------------------------------------------------------------
1       0.1485    0.1529   1.1929  0.2747    |-                 |-
2      -0.1028   -0.1320   1.7762  0.4114    |                 -|
3       0.0495    0.1182   1.9143  0.5904    |                  |
4       0.0887    0.0546   2.3672  0.6686    |                  |
5      -0.1690   -0.2334   4.0447  0.5430   -|                 -|
6      -0.0234    0.0722   4.0776  0.6662    |                  |
7       0.2658    0.3062   8.4168  0.2973    |--               |--
8      -0.0726   -0.2236   8.7484  0.3640    |                 -|
9      -0.1623   -0.0999   10.444  0.3157   -|                  |
```

wNAO has a significant, negative coefficient in this model. The AR(1) coefficient now is not statistically significant. If we dropped the AR term, however, our residuals would no longer pass **corrgram**'s test for white noise. Figure 13.10 graphs the predicted values, *fyllhat*, together with the observed temperature series *fylltemp*. The model does reasonably well in fitting the main warming/cooling episodes and a few of the minor variations. To have the *y*-axis labels displayed with the same number of decimal places (0.5, 1.0, 1.5,... instead of .5, 1, 1.5,...) in this graph, we specify their format as **%2.1f** .

```
. graph twoway line fylltemp year if tin(1973, 1997)
     ||  line fyllhat year if tin(1973, 1997)
     ||  , ylabel(.5(.5)2.5, angle(horizontal) format(%2.1f))
     ytitle("Degrees C") xlabel(1975(5)1995, grid) xtitle("")
     legend(label(1 "observed temperature")
        label(2 "model prediction") position(5) ring(0) col(1))
```

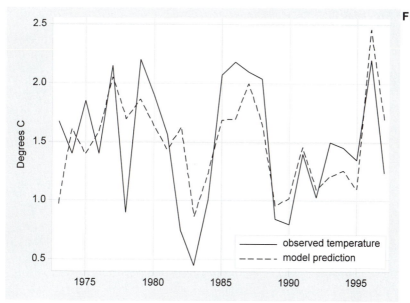

Figure 13.10

A technique called Prais–Winsten regression (**prais**), which corrects for first-order autoregressive errors, can also be illustrated with this example.

```
. prais fylltemp wNAO if tin(1973,1997), nolog
```

```
Prais-Winsten AR(1) regression -- iterated estimates

      Source |       SS       df       MS              Number of obs =      25
-------------+------------------------------           F(  1,    23) =   23.14
       Model |  3.35819258     1  3.35819258           Prob > F      =  0.0001
    Residual |  3.33743545    23  .145105889           R-squared     =  0.5016
-------------+------------------------------           Adj R-squared =  0.4799
       Total |  6.69562803    24  .278984501           Root MSE      =  .38093


-----------------------------------------------------------------------------
    fylltemp |      Coef.   Std. Err.       t    P>|t|    [95% Conf. Interval]
-------------+---------------------------------------------------------------
        wNAO |    -.17356    .037567    -4.62    0.000    -.2512733   -.0958468
       _cons |   1.703436   .1153695    14.77    0.000     1.464776    1.942096
-------------+---------------------------------------------------------------
         rho |   .2951576
-----------------------------------------------------------------------------
Durbin-Watson statistic (original)     1.344998
Durbin-Watson statistic (transformed)  1.789412
```

prais is an older method, more specialized than **arima**. Its regression-based standard errors assume that rho (ρ) is known rather than estimated. Because that assumption is untrue, the standard errors, tests, and confidence intervals given by **prais** tend to be anti-conservative, especially in small samples. **prais** provides a Durbin–Watson statistic ($d =$ 1.789). In this example, the Durbin–Watson test agrees that after fitting the model, no significant first-order autocorrelation remains.

Introduction to Programming

As mentioned in Chapters 2 and 3, we can create a simple type of program by writing any sequence of Stata commands in a text (ASCII) file. Stata's Do-file Editor (click on Window – Do-file Editor or the icon ⬧) provides a convenient way to do this. After saving the do-file, we enter Stata and type a command with the form **do *filename*** that tells Stata to read *filename.do* and execute whatever commands it contains. More sophisticated programs are possible as well, making use of Stata's built-in programming language. Many of the commands used in previous chapters actually involve programs written in Stata. These programs might have originated either from Stata Corporation or from users who wanted something beyond Stata's built-in features to accomplish a particular task.

Stata programs can access all the existing features of Stata, call other programs that call other programs in turn, and use model-fitting aids including matrix algebra and maximum likelihood estimation. Whether our purposes are broad, such as adding new statistical techniques, or narrowly specialized, such as managing a particular database, our ability to write programs in Stata greatly extends what we can do.

Substantial books (*Stata Programming Reference Manual*; *Mata Reference Manual*; *Maximum Likelihood Estimation with Stata*) have been written about Stata programming. This engaging topic is also the focus of periodic NetCourses (see www.stata.com) and a section of the *User's Guide*. The present chapter has the modest aim of introducing a few basic tools and giving examples that show how these tools can be used.

Basic Concepts and Tools

Some elementary concepts and tools, combined with the Stata capabilities described in earlier chapters, suffice to get started.

Do-files

Do-files are ASCII (text) files, created by Stata's Do-file Editor, a word processor, or any other text editor. They are typically saved with a *.do* extension. The file can contain any sequence of legitimate Stata commands. In Stata, typing the following command causes Stata to read *filename.do* and execute the commands it contains:

```
. do filename
```

Each command in *filename.do*, including the last, must end with a hard return — unless we have reset the delimiter to some other character, through a `#delimit` command. For example,

```
#delimit ;
```

This sets a semicolon as the end-of-line delimiter, so that Stata does not consider a line finished until it encounters a semicolon. Setting the semicolon as delimiter permits a single command to extend over more than one physical line. Later, we can reset "carriage return" as the usual end-of-line delimiter by typing the following command:

```
#delimit cr
```

Ado-files

Ado (automatic do) files are ASCII files containing sequences of Stata commands, much like do-files. The difference is that we need not type **do *filename*** in order to run an ado-file. Suppose we type the command

```
. clear
```

As with any command, Stata reads this and checks whether an intrinsic command by this name exists. If a **clear** command does not exist as part of the base Stata executable (and, in fact, it does not), then Stata next searches in its usual "ado" directories, trying to find a file named *clear.ado*. If Stata finds such a file (as it should), it then executes whatever commands the file contains. Ado-files have the extension *.ado*. User-written programs commonly go in a directory named C:\ado\personal, whereas the hundreds of official Stata ado-files get installed in C:\stata\ado. Type **sysdir** to see a list of the directories Stata currently uses. Type **help sysdir** or **help adopath** for advice on changing them.

The **which** command reveals whether a given command really is an intrinsic, hardcoded Stata command or one defined by an ado-file; and if it is an ado-file, where that resides. For example, **logit** is a built-in command, but the **logistic** command is defined by an ado-file named logistic.ado:

```
. which logit
built-in command:   logit
```

```
. which logistic
C:\STATA\ado\base\l\logistic.ado
*! version 3.1.9  01oct2002
```

This distinction makes no difference to most users, because **logit** and **logistic** work with similar ease and syntax when called.

Programs

Both do-files and ado-files might be viewed as types of programs, but Stata uses the word "program" in a narrower sense, to mean a sequence of commands stored in memory and executed by typing a particular program name. Do-files, ado-files, or commands typed interactively can define such programs. The definition begins with a statement that names the program. For example, to create a program named *count5*, we start with

```
program count5
```

Next should be the lines that actually define the program. Finally, we give an `end` command, followed by a hard return:

```
end
```

Once Stata has read the program-definition commands, it retains that definition of the program in memory and will run it any time we type the program's name as a command:

```
. count5
```

Programs effectively make new commands available within Stata, so most users do not need to know whether a given command comes from Stata itself or from an ado-file-defined program.

As we start to write a new program, we often create preliminary versions that are incomplete or just unsuccessful. The **program drop** command provides essential help here, allowing us to clear programs from memory so that we can define a new version For example, to clear program *count5* from memory, type

```
. program drop count5
```

To clear all programs (but not the data) from memory, type

```
. program drop _all
```

Local Macros

Macros are names (up to 31 characters) that can stand for strings, program-defined results, or user-defined values. A *local macro* exists only within the program that defines it, and cannot be referred to in another program. To create a local macro named `iterate`, standing for the number 0, type

```
local iterate = 0
```

To refer to the contents of a local macro (0 in this example), place the macro name *within left and right single quotes*. For example,

```
display `iterate'
```
0

Thus, to increase the value of `iterate` by one, we write

```
local iterate = `iterate' + 1
```

Global Macros

Global macros are similar to local macros, but once defined, they remain in memory and can be used by other programs. To refer to a global macro's contents, we *preface the macro name with a dollar sign* (instead of enclosing the name in left and right single quotes as done with local macros):

```
global distance = 73
display $distance * 2
```
146

Version

Stata's capabilities and features have changed over the years. Consequently, programs written for an older version of Stata might not run directly under the current version. The `version` command works around this problem so that old programs remain usable. Once we tell Stata for what version the program was written, Stata makes the necessary adjustments and the old program can run under a new version of Stata. For example, if we begin our program with the following statement, Stata interprets all the program's commands as it would have in Stata 6:

```
version 6
```

Comments

Stata does not attempt to execute any line that begins with an asterisk. Such lines can therefore be used to insert comments and explanation into a program, or interactively during a Stata session. For example,

```
* This entire line is a comment.
```

Alternatively, we can include a comment within an executable line. The simplest way to do so is to place the comment after a double slash, `//` (with at least one space before the double slash). For example,

```
summarize income education    // this part is the comment
```

A triple slash (also preceded by at least one space) indicates that what follows, to the end of the line, is a comment; but then the following physical line should be executed as a continuation of the first. For example,

```
summarize income education    /// this part is the comment
occupation age
```

will be executed as if we had typed

```
summarize income education occupation age
```

With or without comments, the triple slash provides an easy way to include long command lines in a program. For example, the following lines would be read as one `table` command, even though they are separated by a hard return.

```
table gender kids school if contam==1, contents(mean lived ///
   median lived count lived)
```

If our program has more than a few long commands, however, the `#delimit ;` approach (described earlier; also see **help delimit**) might be easier to write and read.

It is also possible to include comments in the middle of a command line, bracketed by `/*` and `*/` . For example,

```
summarize income /* this is the comment */ education occupation
```

If one line ends with `/*` , and the next begins with `*/` , then Stata skips over the line break and reads both lines as a single command — another line-lengthening trick sometimes found in programs.

Looping

There are a number of ways to create program loops. One simple method employs the `forvalues` command. For example, the following program counts from 1 to 5.

```
* Program that counts from one to five
program count5
   version 8.0
   forvalues i = 1/5 {
      display `i'
   }
end
```

By typing these commands, we define program `count5`. Alternatively, we could use the Do-file Editor to save the same series of commands as an ASCII file named *count5.do*. Then, typing the following causes Stata to read the file:

`. do count5`

Either way, by defining program `count5` we make this available as a new command:

`. count5`
```
1
2
3
4
5
```

The command

```
forvalues i = 1/5 {
```

assigns to local macro `i` the consecutive integers from 1 through 5. The command

```
display `i'
```

shows the contents of this macro. The name `i` is arbitrary. A slightly different notation would allow us to count from 0 to 100 by fives (0, 5, 10, ... , 100):

```
forvalues j = 0(5)100 {
```

The steps between values need not be integers. To count from 4 to 5 by increments of .01 (4.00, 4.01, 4.02, ... , 5.00), write

```
forvalues k = 4(.01)5 {
```

Any lines containing valid Stata commands, between the opening and closing curly brackets { }, will be executed repeatedly for each of the values specified. Note that nothing (on that line) follows the opening bracket, and that the closing bracket requires a line of its own.

The `foreach` command takes a different approach. Instead of specifying a set of consecutive numerical values, we give a list of items for which iteration occurs. These items could be variables, files, strings, or numerical values. Type **help foreach** to see the syntax of this command.

`forvalues` and `foreach` create loops that repeat for a pre-specified number of times. If we want looping to continue until some other condition is met, the `while` command is useful. A section of program with the following general form will repeatedly execute the commands within curly brackets, so long as *expression* evaluates to "true":

```
while expression {
    command A
    command B
    . . . .
    }
command Z
```

As in previous examples, the closing bracket } should be on its own separate line, not at the end of a command line.

When *expression* evaluates to "false," the looping stops and Stata goes on to execute *command Z*. Parallel to our previous example, here is simple program that uses a `while` loop to display onscreen the iteration numbers from 1 through 6:

```
* Program that counts from one to six
    program count6
    version 8.0
    local iterate = 1
    while `iterate' <= 6 {
        display `iterate'
    local iterate = `iterate' + 1
    }
end
```

A second example of a `while` loop appears in the *gossip.ado* program described later in this chapter. The *Programming Reference Manual* contains more about programming loops.

If . . . else

The `if` and `else` commands tell a program to do one thing if an expression is true, and something else otherwise. They are set up as follows:

```
if expression {
    command A
    command B
    . . . .
}
else {
    command Z
}
```

For example, the following program segment checks whether the content of local macro `span` is an odd number, and informs the user of the result.

```
if int(`span'/2) != (`span' - 1)/2 {
    display "span is NOT an odd number"
}
else {
    display "span IS an odd number"
}
```

Arguments

Programs define new commands. In some instances (as with the earlier example, `count5`), we intend our command to do exactly the same thing each time it is used. Often, however, we need a command that is modified by arguments such as variable names or options. There are two ways we can tell Stata how to read and understand a command line that includes arguments. The simplest of these is the `args` command.

The following do-file (*listres1.do*) defines a program that performs a two-variable regression, and then lists the observations with the largest absolute residuals.

```
* Perform simple regression and list observations with #
* largest absolute residuals.
*    listres1 Yvariable Xvariable # IDvariable
program listres1, sortpreserve
    version 8.0
    args Yvar Xvar number id
    quietly regress `Yvar' `Xvar'
    capture drop Yhat
    capture drop Resid
    capture drop Absres
    quietly predict Yhat
    quietly predict Resid, resid
    quietly gen Absres = abs(Resid)
    gsort -Absres
    drop Absres
    list `id' `Yvar' Yhat Resid in 1/`number'
end
```

The line `args Yvar Xvar number id` tells Stata that the command `listresid` should be followed by four arguments. These arguments could be numbers, variable names, or other strings separated by spaces. The first argument becomes the contents of a local macro named `Yvar`, the second a local macro named `Xvar`, and so forth. The program then uses the contents of these macros in other commands, such as the regression:

```
quietly regress `Yvar' `Xvar'
```

The program calculates absolute residuals (*Absres*), and then uses the `gsort` command (followed by a minus sign before the variable name) to sort the data in high-to-low order, with missing values last:

```
gsort -Absres
```

The option `sortpreserve` on the command line makes this program "sort-stable": it returns the data to their original order after the calculations are finished.

Dataset *nations.dta*, seen previously in Chapter 8, contains variables indicating life expectancy (*life*), per capita daily calories (*food*), and country name (*country*) for 109 countries. We can open this file, and use it to demonstrate our new program. A **do** command runs do-file *listres1.do*, thereby defining the program `listres1`:

```
. do listres1.do
```

Next, we use the newly-defined `listres1` command, followed by its four arguments. The first argument specifies the *y* variable, the second *x*, the third how many observations to list, and the fourth gives the case identifier. In this example, our command asks for a list of observations that have the five largest absolute residuals.

```
. listres1 life food 5 country
```

```
+----------------------------------------+
| country    life      Yhat       Resid |
|----------------------------------------|
1. |   Libya      60    76.6901   -16.69011 |
2. |   Bhutan     44   60.49577   -16.49577 |
3. |   Panama     72   58.13118    13.86882 |
4. |   Malawi     45   58.58232   -13.58232 |
5. | Ecuador      66   52.45305    13.54695 |
+----------------------------------------+
```

Life expectancies are lower than predicted in Libya, Bhutan, and Malawi. Conversely, life expectancies in Panama and Ecuador are higher than predicted, based on food supplies.

Syntax

The `syntax` command provides a more complicated but also more powerful way to read a command line. The following do-file named *listres2.do* is similar to our previous example, but it uses `syntax` instead of `args`:

```
* Perform simple or multiple regression and list
* observations with # largest absolute residuals.
*   listres2 yvar xvarlist [if] [in], number(#) [id(varname)]
program listres2, sortpreserve
version 8.0
syntax varlist(min=1) [if] [in], Number(integer) [Id(string)]
    marksample touse
    quietly regress `varlist' if `touse'
    capture drop Yhat
    capture drop Resid
    capture drop Absres
    quietly predict Yhat if `touse'
    quietly predict Resid if `touse', resid
    quietly gen Absres = abs(Resid)
    gsort -Absres
    drop Absres
    list `id' `1' Yhat Resid in 1/`number'
end
```

`listres2` has the same purpose as the earlier `listres1`: it performs regression, then lists observations with the largest absolute residuals. This newer version contains several improvements, however, made possible by the `syntax` command. It is not restricted to two-variable regression, as was `listres1`. `listres2` will work with any number of predictor variables, including none (in which case, predicted values equal the mean of *y*, and residuals are deviations from the mean). `listres2` permits optional `if` and `in` qualifiers. A variable identifying the observations is optional with `listres2`, instead of being required as it was with `listres1`. For example, we could regress life expectancy on *food* and *energy*, while restricting our analysis to only those countries where per capita GNP is above 500 dollars:

```
. do listres2.do

. listres2 life food energy if gnpcap > 500, n(6) i(country)

     +---------------------------------------+
     |  country   life      Yhat      Resid  |
     |---------------------------------------|
 1.  |  YemenPDR    46   61.34964  -15.34964 |
 2.  |  YemenAR     45   59.85839  -14.85839 |
 3.  |    Libya     60   73.62516  -13.62516 |
 4.  |  S_Africa    55    67.9146   -12.9146 |
 5.  |  HongKong    76   64.64022   11.35978 |
     |---------------------------------------|
 6.  |    Panama    72   61.77788   10.22212 |
     +---------------------------------------+
```

The `syntax` line in this example illustrates some general features of the command:

```
syntax varlist(min=1) [if] [in], Number(integer) [Id(string)]
```

The variable list for a `listres2` command is required to contain at least one variable name (`varlist(min=1)`). Square brackets denote optional arguments — in this example, the `if` and `in` qualifiers, and also the `id()` option. Capitalization of initial letters for the options indicates the minimum abbreviation that can be used. Because the `syntax` line in our example specified `Number(integer) Id(string)`, an actual command could be written:

```
. listres2 life food, number(6) id(country)
```

Or, equivalently,

```
. listres2 life food, n(6) i(country)
```

The contents of local macro `number` are required to be an integer, and `id` is a string (such as *country*, a variable's name).

This example also illustrates the `marksample` command, which marks the subsample (as qualified by `if` and `in`) to be used in subsequent analyses.

The syntax of `syntax` is outlined in the *Programming Manual*. Experimentation and studying other programs help in gaining fluency with this command.

Example Program: Moving Autocorrelation

The preceding sections presented basic ideas and example short programs. In this section, we apply those ideas to a slightly longer program that defines a new statistical procedure. The procedure obtains moving autocorrelations through a time series, as proposed for ocean-atmosphere data by Topliss (2001). The following do-file, *gossip.do*, defines a program that makes available a new command called `gossip`. Comments, in lines that begin with `*` or in phrases set off by `//`, explain what the program is doing. Indentation of lines has no effect on the program's execution, but makes it easier for the programmer to read.

```
capture program drop gossip     // FOR WRITING & DEBUGGING; DELETE LATER
program gossip
version 8.0
* Syntax requires user to specify two variables (Yvar and TIMEvar), and
* the span of the moving window.  Optionally, the user can ask to generate
```

```
* a new variable holding autocorrelations, to draw a graph, or both.
syntax varlist(min=1 max=2 numeric), SPan(integer) [GENerate(string) GRaph]
if int(`span'/2) != (`span' - 1)/2 {
    display as error "Span must be an odd integer"
}
else {
* The first variable in `varlist' becomes Yvar, the second TIMEvar.
    tokenize `varlist'
        local Yvar `1'
        local TIMEvar `2'
    tempvar NEWVAR
    quietly gen `NEWVAR' = .
    local miss = 0
* spanlo and spanhi are local macros holding the observation number at the
* low and high ends of a particular window.  spanmid holds the observation
* number at the center of this window.
    local spanlo = 0
    local spanhi = `span'
    local spanmid = int(`span'/2)
    while `spanlo' <=   _N -`span' {
        local spanhi = `span' + `spanlo'
        local spanlo = `spanlo' + 1
        local spanmid = `spanmid' + 1
* The next lines check whether missing values exist within the window.
* If they do exist, then no autocorrelation is calculated and we
* move on to the next window.  Users are informed that this occurred.
        quietly summ `Yvar' in `spanlo'/`spanhi'
        if r(N) != `span' {
            local miss = 1
        }
* The value of NEWVAR in observation `spanmid' is set equal to the first
* row, first column (1,1) element of the row vector of autocorrelations
* r(AC) saved by corrgram.
        else {
            quietly corrgram `Yvar' in `spanlo'/`spanhi', lag(1)
            quietly replace `NEWVAR' = el(r(AC),1,1) in `spanmid'
        }
    }
    if "`graph'" != "" {
* The following graph command illustrates the use of comments to cause
* Stata to skip over line breaks, so it reads the next two lines as if
* they were one.
        graph twoway spike `NEWVAR' `TIMEvar', yline(0) ///
            ytitle("First-order autocorrelations of `Yvar' (span `span')")
    }
    if `miss' == 1 {
        display as error "Caution:  missing values exist"
    }
    if "`generate'" != "" {
        rename `NEWVAR' `generate'
        label variable `generate' ///
            "First-order autocorrelations of `Yvar' (span `span')"
    }
}
end
```

As the comments note, gossip requires time series (tsset) data. From an existing time series variable, gossip calculates a second time series consisting of lag-1 autocorrelation coefficients within a moving window of observations — for example, a moving 9-year span. Dataset *nao.dta* contains North Atlantic climate time series that can be used for illustration:

```
Contains data from C:\data\nao.dta
  obs:          159                          North Atlantic Oscillation &
                                             mean air temperature at
                                             Stykkisholmur, Iceland
  vars:           5                          1 Aug 2005 10:50
  size:       3,498 (99.9% of memory free)
-------------------------------------------------------------------------
              storage  display    value
variable name   type   format     label    variable label
-------------------------------------------------------------------------
year            int     %ty                 Year
wNAO            float   %9.0g               Winter NAO
wNAO4           float   %9.0g               Winter NAO smoothed
temp            float   %9.0g               Mean air temperature (C)
temp4           float   %9.0g               Mean air temperature smoothed
-------------------------------------------------------------------------
Sorted by:  year
```

The variable *temp* records annual mean air temperatures at Stykkishólmur in west Iceland from 1841 to 1999. *temp4* contains smoothed values of *temp* (see Chapter 13). Figure 14.1 graphs these two time series. To visually distinguish between raw (*temp*) and smoothed (*temp4*) variables, we connect the former with very thin lines, **clwidth(vthin)**, and the latter with thick lines, **clwidth(thick)**. Type **help linewidthstyle** for a list of other line-width choices.

```
. graph twoway line temp year, clpattern(solid) clwidth(vthin)
     ||   line temp4 year, clpattern(solid) clwidth(thick)
     ||   , ytitle("Temperature, degrees C") legend(off)
```

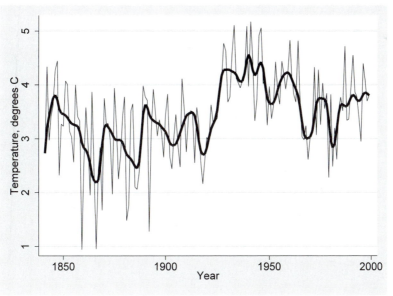

Figure 14.1

To calculate and graph a series of autocorrelations of *temp*, within a moving window of 9 years, we type the following commands. They produce the graph shown in Figure 14.2.

```
. do gossip.do

. gossip temp year, span(9) generate(autotemp) graph
```

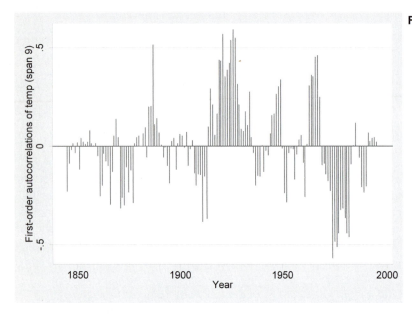

Figure 14.2

In addition to drawing Figure 14.2, gossip created a new variable named *autotemp*:

```
. describe autotemp
```

variable name	storage type	display format	value label	variable label
autotemp	float	%9.0g		First-order autocorrelations of temp (span 9)

```
. list year temp autotemp in 1/10
```

```
     +-------------------------+
     | year    temp    autotemp |
     |-------------------------|
  1. | 1841    2.73          . |
  2. | 1842    4.34          . |
  3. | 1843    2.97          . |
  4. | 1844    3.41          . |
  5. | 1845    3.62   -.2324837 |
     |-------------------------|
  6. | 1846    4.28   -.0883512 |
  7. | 1847    4.45   -.0194607 |
  8. | 1848    2.32    .0175247 |
  9. | 1849    3.27    -.03303 |
 10. | 1850    3.23    .0181154 |
     +-------------------------+
```

autotemp values are missing for the first four years (1841 to 1844). In 1845, the *autotemp* value (−.2324837) equals the lag-1 autocorrelation of *temp* over the 9-year span from 1841 to 1849. This is the same coefficient we would obtain by typing the following command:

```
. corrgram temp in 1/9, lag(1)
                                       -1      0      1 -1      0      1
   LAG     AC      PAC      Q     Prob>Q [Autocorrelation] [Partial Autocor]
-------------------------------------------------------------------------
   1      -0.2325  -0.2398  .66885  0.4135       -|                -|
```

In 1846, *autotemp* (−.0883512) equals the lag-1 autocorrelation of *temp* over the 9 years from 1842 to 1850, and so on through the data. *autotemp* values are missing for the last four years in the data (1996 to 1999), as they are for the first four.

The pronounced Arctic warming of the 1920s, visible in the temperatures of Figure 14.1, manifests in Figure 14.2 as a period of consistently positive autocorrelations. A briefer period of positive autocorrelations in the 1960s coincides with a cooling climate. Topliss (2001) suggests interpretation of such autocorrelations as indicators of changing feedbacks in ocean-atmosphere systems.

The do-file *gossip.do* was written incrementally, starting with input components such as the syntax statement and span macros, running the do-file to check how these work, and then adding other components. Not all of the trial runs produced satisfactory results. Typing the following command causes Stata to display programs line-by-line as they execute, so we can see exactly where an error occurs:

. **set trace on**

Later, we can turn this feature off by typing

. **set trace off**

gossip.do contains a first line, `capture program drop gossip`, that discards the program from memory before defining it again. This is helpful during the writing and debugging stage, when a previous version of our program might have been incomplete or incorrect. Such lines should be deleted once the program is mature, however. The next section describes further steps toward making `gossip` available as a regular Stata command.

Ado-File

Once we believe our do-file defines a program that we will want to use again, we can create an ado-file to make it available like any other Stata command. For the previous example, *gossip.do*, the change involves two steps:

1. With the Do-file Editor, delete the initial "DELETE LATER" line that had been inserted to streamline the program writing and debugging phase. We can also delete the comment lines. Doing so removes useful information, but it makes the program more compact and easier to read.

2. Save our modified file, renaming it to have an .ado extension (for example, *gossip.ado*), in a new directory. The recommended location is in C:\ado\personal; you might need to create this directory and subdirectory if they do not already exist. Other locations are possible, but review the *User's Manual* section on "Where does Stata look for ado-files?" before proceeding.

Once this is done, we can use `gossip` as a regular command within Stata. A listing of *gossip.ado* follows.

```
*!   version 2.0
*!   L. Hamilton, Statistics with Stata (2004)
program gossip
version 8.0
syntax varlist(min=1 max=2 numeric), SPan(integer) [GENerate(string) GRaph]
if int(`span'/2) != (`span' - 1)/2 {
    display as error "Span must be an odd integer"
}
else {
    tokenize `varlist'
        local Yvar `1'
        local TIMEvar `2'
    tempvar NEWVAR
    quietly gen `NEWVAR' = .
    local miss = 0
    local spanlo = 0
    local spanhi = `span'
    local spanmid = int(`span'/2)
    while `spanlo' <= _N -`span' {
        local spanhi = `span' + `spanlo'
        local spanlo = `spanlo' + 1
        local spanmid = `spanmid' + 1
        quietly summ `Yvar' in `spanlo'/`spanhi'
        if r(N) != `span' {
            local miss = 1
        }
        else {
            quietly corrgram `Yvar' in `spanlo'/`spanhi', lag(1)
            quietly replace `NEWVAR' = el(r(AC),1,1) in `spanmid'
        }
    }
    if "`graph'" != "" {
        graph twoway spike `NEWVAR' `TIMEvar', yline(0) ///
            ytitle("First-order autocorrelations of `Yvar' (span `span')")
    }
    if `miss' == 1 {
        display as error "Caution:  missing values exist"
    }
    if "`generate'" != "" {
        rename `NEWVAR' `generate'
        label variable `generate' ///
            "First-order autocorrelations of `Yvar' (span `span')"
    }
}
end
```

The program could be refined further to make it more flexible, elegant, and user-friendly. Note the inclusion of comments stating the source and "version 2.0" in the first two lines, which both begin * ! . The comment refers to version 2.0 of *gossip.ado*, not Stata (an earlier version of *gossip.ado* appeared in a previous edition of this book). The Stata version suitable for this program is specified as 8.0 by the `version` command a few lines later. Although the *! comments do not affect how the program runs, they are visible to a **which** command:

```
. which gossip
c:\ado\personal\gossip.ado
*!   version 2.0
*!   L. Hamilton, Statistics with Stata (2004)
```

Once *gossip.ado* has been saved in the C:\ado\personal directory, the command `gossip` could be used at any time. If we are following the steps in this chapter, which previously defined a preliminary version of `gossip`, then before running the new ado-file version we should drop the old definition from memory by typing

`. program drop gossip`

We are now prepared to run the final, ado-file version. To see a graph of span-15 autocorrelations of variable *wNAO* from dataset *nao.dta*, for example, we would simply open *nao.dta* and type

`. gossip wNAO year, span(15) graph`

Help File

Help files are an integral aspect of using Stata. For a user-written program such as *gossip.ado*, they become even more important because no documentation exists in the printed manuals. We can write a help file for *gossip.ado* by using Stata's Do-file Editor to create a text file named *gossip.hlp*. This help file should be saved in the same ado-file directory (for example, C:\ado\personal) as *gossip.ado*.

Any text file, saved in one of Stata's recognized ado-file directories with a name of the form *filename.hlp*, will be displayed onscreen by Stata when we type **help *filename***. For example, we might write the following in the Do-file Editor, and save it in directory C:\ado\personal as file *gossip1.hlp*. Typing **help gossip1** at any time would then cause Stata to display the text.

```
help for gossip            L. Hamilton

Moving first-order autocorrelations

gossip yvar timevar, span(#) [ generate(newvar) graph ]

Description

calculates first-order autocorrelations of time series
yvar, within a moving window of span #.  For example, if we
specify span(7) gen(new), then the first
through 3rd values of new are missing.  The 4th value of new
equals the lag-1 autocorrelation of yvar across observations 1
through 7.  The 5th value of new equals the lag-1 autocorrelation
of yvar across observations 2 through 8, and so forth.  The last
3 values of new are missing.  See Topliss (2001) for a rationale
and applications of this statistic to atmosphere-ocean data.
Statistics with Stata (2004) discusses the gossip program itself.

gossip requires tsset data.  timevar is the time
variable to be used for graphing.

Options

span(#)   specifies the width of the window for
          calculating autocorrelations.  This option is required;
          # should be an odd integer.

gen(newvar)   creates a new variable holding the
              autocorrelation coefficients.
```

```
graph      requests a spike plot of lag-1 autocorrelations vs.
        timevar.
```

Examples

```
    . gossip water month, span(13) graph
    . gossip water month, span(9) gen(autowater)
    . gossip water month, span(17) gen(autowater) graph
```

References

Hamilton, Lawrence C. 2004. Statistics with Stata. Pacific Grove,
 CA: Duxbury.

Topliss, Brenda J. 2001. "Climate variability I: A conceptual approach to
ocean-atmosphere feedback." In Abstracts for AGU Chapman Conference, The
North Atlantic Oscillation, Nov. 28 - Dec 1, 2000, Ourense, Spain.

Nicer help files containing links, text formatting, dialog boxes, and other features can be designed using Stata Markup and Control Language (SMCL). All official Stata help files, as well as log files and onscreen results, employ SMCL. The following is an SMCL version of the help file for gossip. Once this file has been saved in C:\ado\personal with the file name *gossip.hlp*, typing **help gossip** will produce a readable and official-looking display.

```
{smcl}
{* 1aug2003}{...}
{hline}
help for {hi:gossip}{right:(L. Hamilton)}
{hline}

{title:Moving first-order autocorrelations}

{p 8 12}{cmd:gossip} {it:yvar timevar} {cmd:,} {cmdab:sp:an}{cmd:(}
{it:#}{cmd:)} [ {cmdab:gen:erate}{cmd:(}{it:newvar}{cmd:)}
{cmdab:gr:aph} ]


{title:Description}

{p}{cmd:gossip} calculates first-order autocorrelations of time series
{it:yvar}, within a moving window of span {it:#}.  For example, if we
specify {cmd:span(}7{cmd:)} {cmd:gen(}{it:new}{cmd:)}, then the first
through 3rd values of {it:new} are missing.  The 4th value of {it:new}
equals the lag-1 autocorrelation of {it:yvar} across observations 1
through 7.  The 5th value of {it:new} equals the lag-1 autocorrelation
of {it:yvar} across observations 2 through 8, and so forth.  The last
3 values of {it:new} are missing.  See Topliss (2001) for a rationale
and applications of this statistic to atmosphere-ocean data.
{browse "http://www.stata.com/bookstore/sws.html":Statistics with Stata}
 (2004) discusses the {cmd:gossip} program itself.{p_end}

{p}{cmd:gossip} requires {cmd:tsset} data.  {it:timevar} is the time
variable to be used for graphing.{p_end}


{title:Options}

{p 0 4}{cmd:span(}{it:#}{cmd:)} specifies the width of the window for
calculating autocorrelations.  This option is required; {it:#} should be
 an odd integer.
```

```
{p 0 4}{cmd:gen(}{it:newvar}{cmd:)} creates a new variable holding the
autocorrelation coefficients.

{p 0 4}{cmd:graph} requests a spike plot of lag-1 autocorrelations vs.
{it:timevar}.


{title:Examples}

{p 8 12}{inp:. gossip water month, span(13) graph}{p_end}
{p 8 12}{inp:. gossip water month, span(9) gen(autowater)}{p_end}
{p 8 12}{inp:. gossip water month, span(17) gen(autowater) graph}{p_end}


{title:References}

{p 0 4}Hamilton, Lawrence C.  2004.
{browse "http://www.stata.com/bookstore/sws.html":Statistics with Stata}.
 Pacific Grove, CA:  Duxbury.{p_end}

{p 0 4}Topliss, Brenda J.  2001.  "Climate variability I:  A conceptual
approach to ocean-atmosphere feedback."  In Abstracts for AGU Chapman
Conference, The North Atlantic Oscillation, Nov. 28 - Dec 1, 2000, Ourense,
Spain. citation.{p_end}
```

The help file begins with `{smcl}`, which tells Stata to process the file as SMCL. Curly brackets `{}` enclose SMCL codes, many of which have the form `{command:text}` or `{command arguments:text}`. The following examples illustrate how these codes are interpreted.

`{hline}`	Draw a horizontal line.
`{hi:gossip}`	Highlight the text "gossip".
`{title:Moving...}`	Display the text "Moving . . ." as a title.
`{right:L Hamilton}`	Right-justify the text "L. Hamilton".
`{p 8 12}`	Format the following text as a paragraph, with the first line indented 8 columns and subsequent lines indented 12.
`{cmd:gossip}`	Display the text "gossip" as a command. That is, show "gossip" with whatever colors and font attributes are presently defined as appropriate for a command.
`{it:yvar}`	Display the text "yvar" in italics.
`{cmdab:sp:an}`	Display "span" as a command, with the letters "sp" marked as the minimum abbreviation.
`{p}`	Format the following text as a paragraph, until terminated by `{p_end}`.
`{browse "http://www.stata.com/bookstore/sws.html":Statistics...}`	Link the text "Statistics with Stata" to the web address (URL) http://www.stata.com/bookstore/sws.html. Clicking on the words "Statistics with Stata" should then launch your browser and connect it to this URL.

The *Programming Manual* supplies details about using these and many other SMCL commands.

Matrix Algebra

Matrix algebra provides essential tools for statistical modeling. Stata's matrix commands and matrix programming language (Mata) are too diverse to describe adequately here; the subject requires its own reference manual (*Mata Reference Manual*), in addition to many pages in the *Programming Reference Manual* and *User's Guide*. Consult these sources for information about the Mata language, which is new with Stata 9. The examples in this section illustrate earlier matrix commands, which also still work (hence the placement of **version 8.0** commands at the start of each program).

The built-in Stata command **regress** performs ordinary least squares (OLS) regression, among other things. But as an exercise, we could write an OLS program ourselves. *ols1.do* (following) defines a primitive regression program that does nothing except calculate and display the vector of estimated regression coefficients according to the familiar OLS equation:

$$\mathbf{b} = (\mathbf{X'X})^{-1}\mathbf{X'y}$$

```
* A very simple program, "ols1" estimates linear regression
* coefficients using ordinary least squares (OLS).
program ols1
    version 8.0
* The syntax allows only for a variable list with one or more
* numeric variables.
    syntax varlist(min=1 numeric)
* "tempname..." assigns names to temporary matrices to be used in this
* program.  When ols1 has finished, these matrices will be dropped.
    tempname crossYX crossX crossY b
* "matrix accum..." forms a cross-product matrix.  The K variables in
* varlist, and the N observations with nonmissing values on all K variables,
* comprise an N row, K column data matrix we might call yX.
* The cross product matrix crossYX equals the transpose of yX times yX.
* Written algebraically:
*          crossYX = (yX)'yX
    quietly matrix accum `crossYX' = `varlist'
* Matrix crossX extracts rows 2 through K, and columns 2 through K,
* from crossYX:
*          crossX = X'X
    matrix `crossX' = `crossYX'[2...,2...]
* Column vector crossY extracts rows 2 through K, and column 1 from crossYX:
*          crossY = X'y
    matrix `crossY' = `crossYX'[2...,1]
* The column vector b contains OLS regression coefficients, obtained by
* the classic estimating equation:
*          b = inverse(X'X)X'y
    matrix `b' = syminv(`crossX') * `crossY'
* Finally, we list the coefficient estimates, which are the contents of b.
    matrix list `b'
end
```

Comments explain each command in *ols1.do*. A comment-free version named *ols2.do* (following) gives a clearer view of the matrix commands:

```
program ols2
    version 8.0
    syntax varlist(min=1 numeric)
    tempname crossYX crossX crossY b
    quietly matrix accum `crossYX' = `varlist'
    matrix `crossX' = `crossYX'[2...,2...]
    matrix `crossY' = `crossYX'[2...,1]
    matrix `b' = syminv(`crossX') * `crossY'
    matrix list `b'
end
```

Neither *ols1.do* nor *ols2.do* make any provision for `in` or `if` qualifiers, syntax errors, or options. They also do not calculate standard errors, confidence intervals, or the other ancillary statistics we usually want with regression. To see just what they do accomplish, we will analyze a small dataset on nuclear power plants (*reactor.dta*):

```
Contains data from c:\data\reactor.dta
  obs:             5                      Reactor decommissioning costs
                                          (from Brown et al. 1986)
  vars:            6                      1 Aug 2005 10:50
  size:          130 (99.9% of memory free)
-------------------------------------------------------------------------------
              storage  display   value
variable name   type   format    label      variable label
-------------------------------------------------------------------------------
site            str14   %14s                 Reactor site
decom           byte    %8.0g                Decommissioning cost, millions
capacity        int     %8.0g                Generating capacity, megawatts
years           byte    %9.0g                Years in operation
start           int     %8.0g                Year operations started
close           int     %8.0g                Year operations closed
-------------------------------------------------------------------------------
Sorted by:  start
```

The cost of decommissioning a reactor increases with its generating capacity and with the number of years in operation, as can be seen by using **regress** :

```
. regress decom capacity years
```

Source	SS	df	MS		Number of obs =	5
Model	4666.16571	2	2333.08286		F(2, 2) =	189.42
Residual	24.6342883	2	12.3171442		Prob > F =	0.0053
					R-squared =	0.9947
					Adj R-squared =	0.9895
Total	4690.80	4	1172.70		Root MSE =	3.5096

decom	Coef.	Std. Err.	t	P>|t|	[95% Conf. Interval]	
capacity	.1758739	.0247774	7.10	0.019	.0692653	.2824825
years	3.899314	.2643087	14.75	0.005	2.762085	5.036543
_cons	-11.39963	4.330311	-2.63	0.119	-30.03146	7.23219

Our home-brewed program *ols2.do* yields exactly the same regression coefficients:

```
. do ols2.do

. ols2 decom capacity years

__000003[3,1]
                decom
capacity     .1758739
   years    3.8993139
   _cons   -11.399633
```

Although its results are correct, the minimalist `ols2` program lacks many features we would want in a useful modeling command. The following ado-file, *ols3.ado*, defines an improved program named `ols3`. This program permits `in` and `if` qualifiers, and optionally allows specification of the level for confidence intervals. It calculates and neatly displays regression coefficients in a table with their standard errors, *t* tests, and confidence intervals.

```
*! version 2.0 1aug2003
*! Matrix demonstration:  more complete OLS regression program.
program ols3, eclass
    version 8.0
    syntax varlist(min=1 numeric) [in] [if] [, Level(integer $S_level)]
    marksample touse
    tokenize "`varlist'"
    tempname crossYX crossX crossY b hat V
    quietly matrix accum `crossYX' = `varlist' if `touse'
    local nobs = r(N)
    local df = `nobs' - (rowsof(`crossYX') - 1)
    matrix `crossX' = `crossYX'[2...,2...]
    matrix `crossY' = `crossYX'[2...,1]
    matrix `b' = (syminv(`crossX') * `crossY')'
    matrix `hat' = `b' * `crossY'
    matrix `V' = syminv(`crossX') * (`crossYX'[1,1] - `hat'[1,1])/`df'
    ereturn post `b' `V', dof(`df') obs(`nobs') depname(`1') ///
        esample(`touse')
    ereturn local depvar "`1'"
    ereturn local cmd "ols3"
    if `level' < 10 | `level' > 99 {
        display as error "level( ) must be between 10 and 99 inclusive."
        exit 198
    }
    ereturn display, level(`level')
end
```

Because *ols3.ado* is an ado-file, we can simply type `ols3` as a command:

```
. ols3 decom capacity years
```

```
------------------------------------------------------------------------------
      decom |      Coef.    Std. Err.       t     P>|t|     [95% Conf. Interval]
------------+-----------------------------------------------------------------
   capacity |    .1758739    .0247774     7.10     0.019     .0692653    .2824825
      years |    3.899314    .2643087    14.75     0.005     2.762085    5.036543
      _cons |   -11.39963    4.330311    -2.63     0.119    -30.03146    7.23219
------------------------------------------------------------------------------
```

ols3.ado contains familiar elements including `syntax` and `marksample` commands, as well as `matrix` operations built upon those seen earlier in *ols1.do* and *ols2.do*. Note the use of a right single quote (') as the "matrix transpose" operator. We write the transpose of the coefficients vector (syminv(`crossX') * `crossY') as follows:

```
(syminv(`crossX') * `crossY')'
```

The `ols3` program is defined as e-class, indicating that this is a statistical model-estimation command:

```
program ols3, eclass
```

E-class programs store their results with `e()` designations. After the previous **ols3** command, these have the following contents:

```
. ereturn list

scalars:
               e(N) =   5
            e(df_r) =   2

macros:
            e(cmd) : "ols3"
         e(depvar) : "decom"

matrices:
            e(b) :  1 x 3
            e(V) :  3 x 3

functions:
         e(sample)

. display e(N)
5

. matrix list e(b)

e(b)[1,3]
        capacity        years         _cons
y1      .1758739    3.8993139    -11.399633

. matrix list e(V)

symmetric e(V)[3,3]
             capacity        years         _cons
capacity    .00061392
   years   -.00216732     .0698591
   _cons   -.01492755     -.942626    18.751591
```

The `e()` results from e-class programs remain in memory until the next e-class command. In contrast, r-class programs such as **summarize** store their results with `r()` designations, and these remain in memory only until the next e- or r-class command.

Several `ereturn` lines in *ols3.ado* save the `e()` results, then use these in the output display :

```
ereturn post `b' `V', dof(`df') obs(`nobs') depname(`1') ///
      esample(`touse')
```

The above command sets the contents of `e()` results, including the coefficient vector (b) and the variance–covariance matrix (V). This makes all the post-estimation features detailed in **help estimates** and **help postest** available. Options specify the residual degrees of freedom (df), number of observations used in estimation (nobs), dependent variable name (`1' , meaning the contents of the first macro obtained when we `tokenize varlist` ), and estimation sample marker ( `touse`).

`ereturn local depvar "`1'"`

This command sets the name of the dependent variable, macro `1` after `tokenize varlist` , to be the contents of macro `e(depvar)` .

`ereturn local cmd "ols3"`

This sets the name of the command, `ols3` , as the contents of macro `e(cmd)` .

`ereturn display, level(`level')`

The `ereturn display` command displays the coefficient table based on our previous `ereturn post` . This table follows a standard Stata format: its first two columns contain coefficient estimates (from `b`) and their standard errors (square roots of diagonal elements from `V`). Further columns are *t* statistics (first column divided by second), two-tail *t* probabilities, and confidence intervals based on the level specified in the `ols3` command line (or defaulting to 95%).

Bootstrapping

Bootstrapping refers to a process of repeatedly drawing random samples, with replacement, from the data at hand. Instead of trusting theory to describe the sampling distribution of an estimator, we approximate that distribution empirically. Drawing *k* bootstrap samples of size *n* (from an original sample also size *n)* yields *k* new estimates. The distribution of these bootstrap estimates provides an empirical basis for estimating standard errors or confidence intervals (Efron and Tibshirani 1986; for an introduction, see Stine in Fox and Long 1990). Bootstrapping seems most attractive in situations where the statistic of interest is theoretically intractable, or where the usual theory regarding that statistic rests on untenable assumptions.

Unlike Monte Carlo simulations, which fabricate their data, bootstrapping typically works from real data. For illustration, we turn to *islands.dta*, containing area and biodiversity measures for eight Pacific Island groups (from Cox and Moore 1993).

```
Contains data from c:\data\islands.dta
  obs:              8                    Pacific island biodiversity
                                           (Cox & Moore 1993)
  vars:             4                    1 Aug 2005 10:50
  size:           208 (99.9% of memory free)
-------------------------------------------------------------------------------
              storage  display   value
variable name   type   format    label      variable label
-------------------------------------------------------------------------------
island         str15   %15s                 Island group
area           float   %9.0g                Land area, km^2
birds          byte    %8.0g                Number of bird genera
plants         int     %8.0g                Number flowering plant genera
-------------------------------------------------------------------------------
Sorted by:
```

Suppose we wish to form a confidence interval for the mean number of bird genera. The usual confidence interval for a mean derives from a normality assumption. We might hesitate to make this assumption, however, given the skewed distribution that, even in this tiny sample (*n* = 8), almost leads us to reject normality:

```
. sktest birds
```

```
                    Skewness/Kurtosis tests for Normality
                                          ------- joint ------
         Variable |  Pr(Skewness)   Pr(Kurtosis)  adj chi2(2)   Prob>chi2
      -------------+---------------------------------------------------------
            birds |     0.079          0.181          4.75        0.0928
```

Bootstrapping provides a more empirical approach to forming confidence intervals. An r-class command, **summarize, detail** unobtrusively stores its results as a series of macros. Some of these macros are:

r(N)	Number of observations
r(mean)	Mean
r(skewness)	Skewness
r(min)	Minimum
r(max)	Maximum
r(p50)	50th percentile or median
r(Var)	Variance
r(sum)	Sum
r(sd)	Standard deviation

Stored results simplify the job of bootstrapping any statistic. To obtain bootstrap confidence intervals for the mean of *birds*, based on 1,000 resamplings, and save the results in new file *boot1.dta*, type the following command. The output includes a note warning about the potential problem of missing values, but that does not apply to these data.

```
. bs "summarize birds, detail"  "r(mean)", rep(1000) saving(boot1)

command:        summarize birds , detail
statistic:      _bs_1       = r(mean)

Warning:  Since summarize is not an estimation command or does not set
          e(sample), bootstrap has no way to determine which observations are
          used in calculating the statistics and so assumes that all
          observations are used.  This means no observations will be excluded
          from the resampling due to missing values or other reasons.

          If the assumption is not true, press Break, save the data, and drop
          the observations that are to be excluded.  Be sure the dataset in
          memory contains only the relevant data.

Bootstrap statistics                           Number of obs    =         8
                                               Replications     =      1000

------------------------------------------------------------------------------
Variable      |  Reps  Observed      Bias   Std. Err. [95% Conf. Interval]
--------------+---------------------------------------------------------------
       _bs_1 |  1000    47.625  -.475875   12.39088   23.30986   71.94014   (N)
             |                                         25.75      74.8125    (P)
             |                                         27         78.25      (BC)
------------------------------------------------------------------------------
Note:   N  = normal
        P  = percentile
        BC = bias-corrected
```

The **bs** command states in double quotes what analysis is to be bootstrapped ("**summ birds, detail**"). Following this comes the statistic to be bootstrapped, likewise in its own double quotes ("**r(mean)**"). More than one statistic could be listed, each separated by a space. The example above specifies two options:

rep(1000)   Calls for 1,000 repetitions, or drawing 1,000 bootstrap samples.

saving(*boot1*)   Saves the 1,000 bootstrap means in a new dataset named *boot1.dta*.

The **bs** results table shows the number of repetitions performed and the "observed" (original-sample) value of the statistic being bootstrapped — in this case, the mean *birds* value 47.625. The table also shows estimates of bias, standard error, and three types of confidence intervals. "Bias" here refers to the mean of the k bootstrap values of our statistic (for example, the mean of the 1,000 bootstrap means of *birds*), minus the observed statistic. The estimated standard error equals the standard deviation of the k bootstrap statistic values (for example, the standard deviation of the 1,000 bootstrap means of *birds*). This bootstrap standard error (12.39) is less than the conventional standard error (13.38) calculated by **ci** :

```
. ci birds

    Variable |       Obs        Mean    Std. Err.      [95% Conf. Interval]
-------------+-------------------------------------------------------------
       birds |         8      47.625    13.38034       15.98552    79.26448
```

Normal-approximation (N) confidence intervals in the **bs** table are obtained as follows:

$$\text{observed sample statistic} \pm t \times \text{bootstrap standard error}$$

where t is chosen from the theoretical t distribution with $k - 1$ degrees of freedom. Their use is recommended when the bootstrap distribution appears unbiased and approximately normal.

Percentile (P) confidence intervals simply use percentiles of the bootstrap distribution (for a 95% interval, the 2.5th and 97.5th percentiles) as lower and upper bounds. These might be appropriate when the bootstrap distribution appears unbiased but nonnormal.

The bias-corrected (BC) interval also employs percentiles of the bootstrap distribution, but chooses these percentiles following a normal-theory adjustment for the proportion of bootstrap values less than or equal to the observed statistic. When substantial bias exists (by one guideline, when bias exceeds 25% of one standard error), these intervals might be preferred.

Since we saved the bootstrap results in a file named *boot1.dta*, we can retrieve this and examine the bootstrap distribution more closely if desired. The **saving(*boot1*)** option created a dataset with 1,000 observations and a variable named *_bs_1*, holding the mean of each bootstrap sample.

```
Contains data from c:\data\boot1.dta
  obs:         1,000                          bs: summarize birds, detail
  vars:            1                          1 Aug 2005 15:10
  size:        8,000 (99.9% of memory free)
-------------------------------------------------------------------------------
              storage  display     value
variable name   type   format      label      variable label
-------------------------------------------------------------------------------
_bs_1           float  %9.0g                   r(mean)
-------------------------------------------------------------------------------
Sorted by:
```

```
. summarize

    Variable |      Obs      Mean    Std. Dev.        Min        Max
-------------+--------------------------------------------------------
       _bs_1 |     1000   47.14912    12.39088     14.625       92.5
```

Note that the standard deviation of these 1,000 bootstrap means equals the standard error (12.82) shown earlier in the **bs** results table. The mean of the 1,000 means minus the observed (original sample) mean equals the bias:

$$47.14912 - 47.625 = -.47588$$

Figure 14.3 shows the distribution of these 1,000 sample means, with the original-sample mean (47.625) marked by a vertical line. The distribution exhibits mild positive skew, but is not far from a theoretical normal curve.

```
. histogram _bs_1, norm bcolor(gs10) xaxis(1 2) xline(47.625)
     xlabel(47.635, axis(2)) xtitle("", axis(2))
```

Figure 14.3

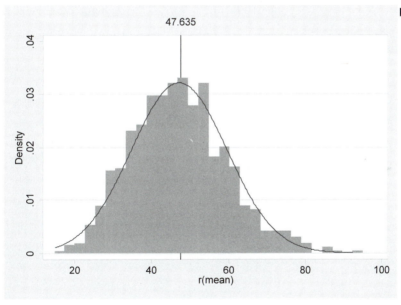

Biologists have observed that biodiversity, or the number of different kinds of plants and animals, tends to increase with island size. In *islands.dta*, we have data to test this proposition with respect to birds and flowering plants. As expected, a strong linear relationship exists between *birds* and *area*:

```
. regress birds area
```

```
      Source |       SS       df       MS              Number of obs =       8
-------------+------------------------------           F(  1,     6) =  162.96
       Model | 9669.83255      1  9669.83255           Prob > F      =  0.0000
    Residual | 356.042449      6  59.3404082           R-squared     =  0.9645
-------------+------------------------------           Adj R-squared =  0.9586
       Total | 10025.875       7  1432.26786           Root MSE      =  7.7033


-------------------------------------------------------------------------------
       birds |      Coef.   Std. Err.       t    P>|t|     [95% Conf. Interval]
-------------+-----------------------------------------------------------------
        area |   .0026512   .0002077    12.77   0.000      .002143    .0031594
       _cons |   13.97169    3.79046     3.69   0.010     4.696773   23.24662
-------------------------------------------------------------------------------
```

An e-class command, **regress** saves a set of e() results as noted earlier in this chapter. It also creates or updates a set of system variables containing the model coefficients (_b[*varname*]) and standard errors (_se[*varname*]). To bootstrap the slope and *y* intercept from the previous regression, saving the results in file *boot2.dta*, type

```
. bs "regress birds area" "_b[area] _b[_cons]", rep(1000)
     saving(boot2)
```

```
command:     regress birds area
statistics:  _bs_1      = _b[area]
             _bs_2      = _b[_cons]
```

```
Bootstrap statistics                      Number of obs    =         8
                                          Replications     =      1000


-------------------------------------------------------------------------------
Variable    |  Reps  Observed      Bias   Std. Err. [95% Conf. Interval]
------------+------------------------------------------------------------------
      _bs_1 |  1000  .0026512 -.0000737  .0003345    .0019947   .0033077   (N)
            |                                        .0019759   .0029066   (P)
            |                                          .00199   .0029246   (BC)
      _bs_2 |  1000  13.97169  .6230986  3.637705    6.833275   21.11011   (N)
            |                                        7.891942   21.74494   (P)
            |                                        6.949539   19.73012   (BC)
-------------------------------------------------------------------------------
Note:  N   = normal
       P   = percentile
       BC  = bias-corrected
```

The bootstrap distribution of coefficients on *area* is severely skewed (skewness = 4.12). Whereas the bootstrap distribution of means (Figure 14.3) appeared approximately normal, and produced bootstrap confidence intervals narrower than the theoretical confidence interval, in this regression example bootstrapping obtains larger standard errors and wider confidence intervals.

In a regression context, **bs** ordinarily performs what is called "data resampling," or resampling intact observations. An alternative procedure called "residual resampling" (resampling only the residuals) requires a bit more programming work. Two additional commands make such do-it-yourself bootstrapping easier:

bsample   Draws a sample with replacement from the existing data, replacing the data in memory.

bootstrap   Runs a user-defined program `reps()` times on bootstrap samples of size `size()`.

The *Base Reference Manual* gives examples of programs for use with **bootstrap**.

Monte Carlo Simulation

Monte Carlo simulations generate and analyze many samples of artificial data, allowing researchers to investigate the long-run behavior of their statistical techniques. The **simulate** command makes designing a simulation straightforward so that it only requires a small amount of additional programming. This section gives two examples.

To begin a simulation, we need to define a program that generates one sample of random data, analyzes it, and stores the results of interest in memory. Below we see a file defining an r-class program (one capable of storing `r()` results) named `central`. This program randomly generates 100 values of variable *x* from a standard normal distribution. It next generates 100 values of variable *w* from a "contaminated normal" distribution: N(0,1) with probability .95, and N(0,10) with probability .05. Contaminated normal distributions have often been used in robustness studies to simulate variables that contain occasional wild errors. For both variables, `central` obtains means and medians.

```
* Creates a sample containing n=100 observations of variables x and w.
* x~N(0,1)                                      x is standard normal
* w~N(0,1) with p=.95, w~N(0,10) with p=.05   w is contaminated normal
* Calculates the mean and median of x and w.
* Stored results:    r(xmean)    r(xmedian)    r(wmean)    r(wmedian)
program central, rclass
    version 8.0
    drop _all
    set obs 100
    generate x = invnorm(uniform())
    summarize x, detail
    return scalar xmean = r(mean)
    return scalar xmedian = r(p50)
    generate w = invnorm(uniform())
    replace w = 10*w if uniform() < .05
    summarize w, detail
    return scalar wmean = r(mean)
    return scalar wmedian = r(p50)
end
```

Because we defined **central** as an r-class command, like **summarize**, it can store its results in `r()` macros. **central** creates four such macros: `r(xmean)` and `r(xmedian)` for the mean and median of *x*; `r(wmean)` and `r(wmedian)` for the mean and median of *w*.

Once **central** has been defined, whether through a do-file, ado-file, or typing commands interactively, we can call this program with a **simulate** command. To create a new dataset containing means and medians of *x* and *w* from 5,000 random samples, type

```
. simulate "central"  xmean = r(xmean)   xmedian = r(xmedian)
     wmean = r(wmean)   wmedian = r(wmedian), reps(5000)

command:      central
statistics:   xmean      = r(xmean)
              xmedian    = r(xmedian)
              wmean      = r(wmean)
              wmedian    = r(wmedian)
```

This command creates new variables *xmean*, *xmedian*, *wmean*, and *wmedian*, based on the $r()$ results from each iteration of **central**.

```
. describe

Contains data
  obs:           5,000                        simulate: central
  vars:              4                        1 Aug 2005 17:50
  size:        100,000 (99.6% of memory free)
-------------------------------------------------------------------------
              storage  display    value
variable name  type    format     label     variable label
-------------------------------------------------------------------------
xmean          float   %9.0g                 r(xmean)
xmedian        float   %9.0g                 r(xmedian)
wmean          float   %9.0g                 r(wmean)
wmedian        float   %9.0g                 r(wmedian)
-------------------------------------------------------------------------
Sorted by:
```

```
. summarize

    Variable |     Obs       Mean    Std. Dev.      Min        Max
-------------+-----------------------------------------------------------
       xmean |    5000   -.0015915    .0987788  -.4112561   .3699467
     xmedian |    5000   -.0015566    .1246915  -.4647848   .4740642
       wmean |    5000   -.0004433    .2470823   -1.11406   .8774976
     wmedian |    5000    .0030762    .1303756  -.4584521   .5152998
```

The means of these means and medians, across 5,000 samples, are all close to 0 — consistent with our expectation that the sample mean and median should both provide unbiased estimates of the true population means (0) for *x* and *w*. Also as theory predicts, the mean exhibits less sample-to-sample variation than the median when applied to a normally distributed variable. The standard deviation of *xmedian* is .125, noticeably larger than the standard deviation of *xmean* (.099). When applied to the outlier-prone variable *w*, on the other hand, the opposite holds true: the standard deviation of *wmedian* is much lower than the standard deviation of *wmean* (.130 vs. .247). This Monte Carlo experiment demonstrates that the median remains a relatively stable measure of center despite wild outliers in the contaminated distribution, whereas the mean breaks down and varies much more from sample to sample. Figure 14.4 draws the comparison graphically, with box plots (and, incidentally, demonstrates how to control the shapes of box plot outlier-marker symbols).

```
. graph box xmean xmedian wmean wmedian, yline(0) legend(col(4))
    marker(1, msymbol(+)) marker(2, msymbol(Th))
    marker(3, msymbol(Oh)) marker(4, msymbol(Sh))
```

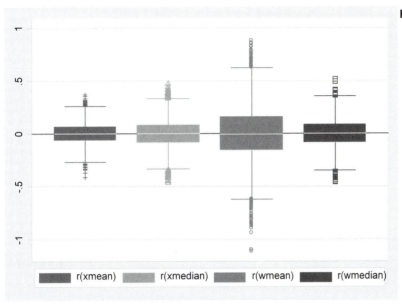

Figure 14.4

Our final example extends the inquiry to robust methods, bringing together several themes from this book. Program `regsim` generates 100 observations of *x* (standard normal) and two *y* variables. *y1* is a linear function of *x* plus standard normal errors. *y2* is also a linear function of *x*, but adding contaminated normal errors. These variables permit us to explore how various regression methods behave in the presence of normal and nonnormal errors. Four methods are employed: ordinary least squares (**regress**), robust regression (**rreg**), quantile regression (**qreg**), and quantile regression with bootstrapped standard errors (**bsqreg** , with 500 repetitions). Differences among these methods were discussed in Chapter 9. Program `regsim` applies each method to the regression of *y1* on *x* and then to the regression of *y2* on *x*. For this exercise, the program is defined by an ado-file, *regsim.ado*, saved in the C:\ado\personal directory.

```
program regsim, rclass
* Performs one iteration of a Monte Carlo simulation comparing
* OLS regression (regress) with robust (rreg) and quantile
* (qreg and bsqreg) regression.  Generates one n = 100 sample
* with x ~ N(0,1) and y variables defined by the models:
*
*    MODEL 1:       y1 = 2x + e1          e1 ~ N(0,1)
*
*    MODEL 2:       y2 = 2x + e2          e2 ~ N(0,1) with p = .95
*                                         e2 ~ N(0,10) with p = .05
*
* Bootstrap standard errors for qreg involve 500 repetitions.
*
    version 8.0
    if "`1'" == "?" {
        #delimit ;
        global S_1 "b1 b1r se1r b1q se1q se1qb
            b2 b2r se2r b2q se2q se2qb";
        #delimit cr
        exit
    }
    drop _all
    set obs 100
    generate x = invnorm(uniform())
    generate e = invnorm(uniform())
    generate y1 = 2*x + e
    reg y1 x
        return scalar B1 = _b[x]
    rreg y1 x, iterate(25)
        return scalar B1R = _b[x]
        return scalar SE1R = _se[x]
    qreg y1 x
        return scalar B1Q = _b[x]
        return scalar SE1Q = _se[x]
    bsqreg y1 x, reps(500)
        return scalar SE1QB = _se[x]
    replace e = 10 * e if uniform() < .05
    generate y2 = 2*x + e
    reg y2 x
        return scalar B2 = _b[x]
    rreg y2 x, iterate(25)
        return scalar B2R = _b[x]
        return scalar SE2R = _se[x]
    qreg y2 x
        return scalar B2Q = _b[x]
        return scalar SE2Q = _se[x]
    bsqreg y2 x, reps(500)
        return scalar SE2QB = _se[x]
end
```

The r-class program stores coefficient or standard error estimates from eight regression analyses. These results have names such as

> r(B1)   coefficient from OLS regression of *y1* on *x*
>
> r(B1R)   coefficient from robust regression of *y1* on *x*
>
> r(SE1R)   standard error of robust coefficient from model 1

and so forth. All the robust and quantile regressions involve multiple iterations: typically 5 to 10 iterations for **rreg**, about 5 for **qreg**, and several thousand for **bsqreg** with its 500 bootstrap re-estimations of about 5 iterations each, *per sample*. Thus, a single execution of *regsim* demands more than two thousand regressions. The following command calls for five repetitions.

```
. simulate "regsim"  b1 = r(B1)   b1r = r(B1R)   se1r = r(SE1R)
     b1q = r(B1Q)   se1q = r(SE1Q)   se1qb = r(SE1QB)   b2 = r(B2)
     b2r = r(B2R)   se2r = r(SE2R)   b2q = r(B2Q)   se2q = r(SE2Q)
     se2qb = r(SE2QB), reps(5)
```

You might want to run a small simulation like this as a trial to get a sense of the time required on your computer. For research purposes, however, we would need a much larger experiment. Dataset *regsim.dta* contains results from an overnight experiment involving 5,000 repetitions of `regsim` — more than 10 million regressions. The regression coefficients and standard error estimates produced by this experiment are summarized below.

```
. describe

Contains data from C:\data\regsim.dta
  obs:          5,000                       Monte Carlo estimates of b in
                                              5000 samples of n=100
  vars:            12                       2 Aug 2005 08:17
  size:       260,000 (99.0% of memory free)
-------------------------------------------------------------------------
              storage  display    value
variable name  type    format     label     variable label
-------------------------------------------------------------------------
b1            float   %9.0g                OLS b (normal errors)
b1r           float   %9.0g                Robust b (normal errors)
se1r          float   %9.0g                Robust SE[b] (normal errors)
b1q           float   %9.0g                Quantile b (normal errors)
se1q          float   %9.0g                Quantile SE[b] (normal errors)
se1qb         float   %9.0g                Quantile bootstrap SE[b]
                                             (normal errors)
b2            float   %9.0g                OLS b (contaminated errors)
b2r           float   %9.0g                Robust b (contaminated errors)
se2r          float   %9.0g                Robust SE[b] (contaminated
                                             errors)
b2q           float   %9.0g                Quantile b (contaminated errors)
se2q          float   %9.0g                Quantile SE[b] (contaminated
                                             errors)
se2qb         float   %9.0g                Quantile bootstrap SE[b]
                                             (contaminated errors)
-------------------------------------------------------------------------
Sorted by:

. summarize

    Variable |       Obs        Mean    Std. Dev.        Min         Max
-------------+-----------------------------------------------------------
          b1 |      5000    2.000828     .102018    1.631245    2.404814
         b1r |      5000    2.000989    .1052277    1.603106    2.391946
        se1r |      5000    .1041399    .0109429    .0693786    .1515421
         b1q |      5000    2.001135    .1309186    1.471802    2.536621
        se1q |      5000    .1262578    .0281738    .0532731    .2371508
-------------+-----------------------------------------------------------
       se1qb |      5000    .1362755     .032673    .0510808      .29979
          b2 |      5000    2.006001    .2484688    .9001114    3.050552
         b2r |      5000    2.000399    .1092553    1.633241    2.411423
        se2r |      5000    .1081348    .0119274    .0743103    .1560973
         b2q |      5000    2.000701     .137111    1.471802    2.536621
-------------+-----------------------------------------------------------
        se2q |      5000    .1328431    .0299644    .0542015    .2594844
       se2qb |      5000    .1436366    .0346679    .0589409    .3006417
```

Figure 14.5 draws the distributions of coefficients as box plots. To make the plot more readable we use the **legend(symxsize(2) colgap(4))** options, which set the width of symbols and the gaps between columns within the legend at less than their default size. **help legend_option** and **help relativesize** supply further information about these options.

```
. graph box b1 b1r b1q b2 b2r b2q, ytitle("Estimates of slope (b=2)")
     yline(2)
     legend(row(1) symxsize(2) colgap(4)
         label(1 "OLS 1") label(2 "robust 1") label(3 "quantile 1")
         label(4 "OLS 2") label(5 "robust 2") label(6 "quantile 2"))
```

Figure 14.5

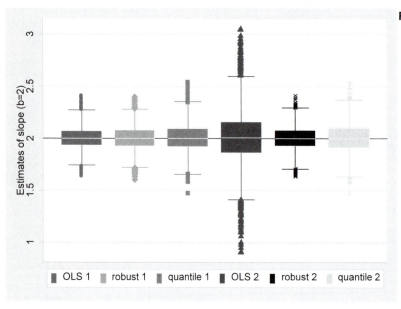

All three regression methods (OLS, robust, and quantile) produced mean coefficient estimates for both models that are not significantly different from the true value, $\beta = 2$. This can be confirmed through *t* tests such as

```
. ttest b2r = 2
```

One-sample t test

```
------------------------------------------------------------------------------
Variable |     Obs        Mean    Std. Err.    Std. Dev.   [95% Conf. Interval]
---------+--------------------------------------------------------------------
    b2r |    5000    2.000399    .0015451     .1092553     1.99737    2.003428
------------------------------------------------------------------------------
Degrees of freedom: 4999

                          Ho: mean(b2r) = 2

   Ha: mean < 2              Ha: mean != 2               Ha: mean > 2
     t =   0.2585               t =   0.2585               t =   0.2585
 P < t =   0.6020          P > |t| =   0.7960          P > t =   0.3980
```

All the regression methods thus yield unbiased estimates of β, but they differ in their sample-to-sample variation or efficiency. Applied to the normal-errors model 1, OLS proves the most efficient, as the famous Gauss–Markov theorem would lead us to expect. The observed standard deviation of OLS coefficients is .1016, compared with .1047 for robust regression and .1282 for quantile regression. Relative efficiency, expressing the OLS coefficient's observed variance as a percentage of another estimator's observed variance, provides a standard way to compare such statistics:

```
. quietly summarize b1

. global Varb1 = r(Var)

. quietly summarize b1r

. display 100*($Varb1/r(Var))
93.992612

. quietly summarize b1q

. display 100*($Varb1/r(Var))
60.722696
```

The calculations above use the r(Var) variance result from **summarize**. We first obtain the variance of the OLS estimates *b1*, and place this into global macro Varb1. Next the variances of the robust estimates *b1r*, and the quantile estimates *b1q*, are obtained and each compared with Varb1. This reveals that robust regression was about 94% as efficient as OLS when applied to the normal-errors model — close to the large-sample efficiency of 95% that this robust method theoretically should have (Hamilton 1992a). Quantile regression, in contrast, achieves a relative efficiency of only 61% with the normal-errors model.

Similar calculations for the contaminated-errors model tell a different story. OLS was the best (most efficient) estimator with normal errors, but with contaminated errors it becomes the worst:

```
. quietly summarize b2

. global Varb2 = r(Var)

. quietly summarize b2r

. display 100*($Varb2/r(Var))
517.20057

. quietly summarize b2q

. display 100*($Varb2/r(Var))
328.3971
```

Outliers in the contaminated-errors model cause OLS coefficient estimates to vary wildly from sample to sample, as can be seen in the fourth box plot of Figure 14.5. The variance of these OLS coefficients is more than five times greater than the variance of the corresponding robust coefficients, and more than three times greater than that of quantile coefficients. Put another way, both robust and quantile regression prove to be much more stable than OLS in the presence of outliers, yielding correspondingly lower standard errors and narrower confidence intervals. Robust regression outperforms quantile regression with both the normal-errors and the contaminated-errors models.

Figure 14.6 illustrates the comparison between OLS and robust regression with a scatterplot showing 5,000 pairs of regression coefficients. The OLS coefficients (vertical axis) vary much more widely around the true value, 2.0, than **rreg** coefficients (horizontal axis) do.

```
. graph twoway scatter b2 b2r, msymbol(p) ylabel(1(.5)3, grid)
     yline(2) xlabel(1(.5)3, grid) xline(2)
```

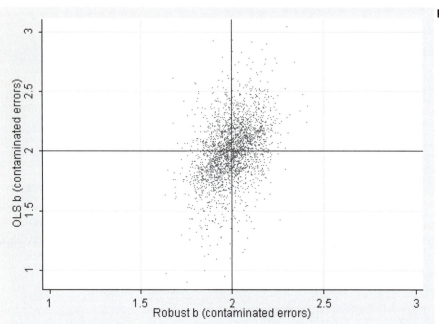

Figure 14.6

The experiment also provides information about the estimated standard errors under each method and model. Mean estimated standard errors differ from the observed standard deviations of coefficients. Discrepancies for the robust standard errors are small — less than 1%. For the theoretically-derived quantile standard errors the discrepancies appear a bit larger, between 3 and 4%. The least satisfactory estimates appear to be the bootstrapped quantile standard errors obtained by **bsqreg** . Means of the bootstrap standard errors exceed the observed standard deviation of *b1q* and *b2q* by 4 to 5%. Bootstrapping apparently over-estimated the sample-to-sample variation.

Monte Carlo simulation has become a key method in modern statistical research, and it plays a growing role in statistical teaching as well. These examples demonstrate how readily Stata supports Monte Carlo work.

References

Barron's Educational Series. 1992. *Barron's Compact Guide to Colleges*, 8th ed. New York: Barron's Educational Series.

Beatty, J. Kelly, Brian O'Leary and Andrew Chaikin (eds.). 1981. *The New Solar System*. Cambridge, MA: Sky.

Belsley, D. A., E. Kuh and R. E. Welsch. 1980. *Regression Diagnostics: Identifying Influential Data and Sources of Collinearity*. New York: John Wiley & Sons.

Box, G. E. P., G. M. Jenkins and G. C. Reinsel. 1994. *Time Series Analysis: Forecasting and Control*. 3rd ed. Englewood Cliffs, NJ: Prentice–Hall.

Brown, Lester R., William U. Chandler, Christopher Flavin, Cynthia Pollock, Sandra Postel, Linda Starke and Edward C. Wolf. 1986. *State of the World 1986*. New York: W. W. Norton.

Buch, E. 2000. *Oceanographic Investigations off West Greenland 1999*. Copenhagen: Danish Meteorological Institute.

CDC (Centers for Disease Control). 2003. Web site: http://www.cdc.gov

Chambers, John M., William S. Cleveland, Beat Kleiner and Paul A. Tukey (eds.). 1983. *Graphical Methods for Data Analysis*. Belmont, CA: Wadsworth.

Chatfield, C. 1996. *The Analysis of Time Series: An Introduction*, 5th edition. London: Chapman & Hall.

Chatterjee, S., A. S. Hadi and B. Price. 2000. *Regression Analysis by Example*, 3rd edition. New York: John Wiley & Sons.

Cleveland, William S. 1994. *The Elements of Graphing Data*. Monterey, CA: Wadsworth.

Cleves, Mario, William Gould and Roberto Gutierrez. 2004. *An Introduction to Survival Analysis Using Stata*, revised edition. College Station, TX: Stata Press.

Cook, R. Dennis and Sanford Weisberg. 1982. *Residuals and Influence in Regression*. New York: Chapman & Hall.

Cook, R. Dennis and Sanford Weisberg. 1994. *An Introduction to Regression Graphics*. New York: John Wiley & Sons.

Council on Environmental Quality. 1988. *Environmental Quality 1987–1988*. Washington, DC: Council on Environmental Quality.

Cox, C. Barry and Peter D. Moore. 1993. *Biogeography: An Ecological and Evolutionary Approach*. London: Blackwell Publishers.

Cryer, Jonathan B. and Robert B. Miller. 1994. *Statistics for Business: Data Analysis and Modeling*, 2nd edition. Belmont, CA: Duxbury Press.

Davis, Duane. 2000. *Business Research for Decision Making*, 5th edition. Belmont, CA: Duxbury Press.

DFO (Canadian Department of Fisheries and Oceans). 2003. Web site: http://www.meds-sdmm.dfo-mpo.gc.ca/alphapro/zmp/climate/IceCoverage_e.shtml

Diggle, P. J. 1990. *Time Series: A Biostatistical Introduction*. Oxford: Oxford University Press.

Efron, Bradley and R. Tibshirani. 1986. "Bootstrap methods for standard errors, confidence intervals, and other measures of statistical accuracy." *Statistical Science* 1(1):54–77.

Enders, W. 2003. *Applied Econometric Time Series*, 2nd edition. New York: John Wiley & Sons.

Everitt, Brian S., Savine Landau and Morven Leese. 2001. *Cluster Analysis*, 4th edition. London: Arnold.

Federal, Provincial, and Territorial Advisory Commission on Population Health. 1996. *Report on the Health of Canadians*. Ottawa: Health Canada Communications.

Fox, John. 1991. *Regression Diagnostics*. Newbury Park, CA: Sage Publications.

Fox, John and J. Scott Long. 1990. *Modern Methods of Data Analysis*. Beverly Hills: Sage Publications.

Frigge, Michael, David C. Hoaglin and Boris Iglewicz. 1989. "Some implementations of the boxplot." *The American Statistician* 43(1):50–54.

Gould, William, Jeffrey Pitblado and William Sribney. 2003. *Maximum Likelihood Estimation with Stata*, 2nd edition. College Station, TX: Stata Press.

Hamilton, Dave C. 2003. "The Effects of Alcohol on Perceived Attractiveness." Senior Thesis. Claremont, CA: Claremont McKenna College.

Hamilton, James D. 1994. *Time Series Analysis*. Princeton, NJ: Princeton University Press.

Hamilton, Lawrence C. 1985a. "Concern about toxic wastes: Three demographic predictors." *Sociological Perspectives* 28(4):463–486.

Hamilton, Lawrence C. 1985b. "Who cares about water pollution? Opinions in a small-town crisis." *Sociological Inquiry* 55(2):170–181.

Hamilton, Lawrence C. 1992a. *Regression with Graphics: A Second Course in Applied Statistics*. Pacific Grove, CA: Brooks/Cole.

Hamilton, Lawrence C. 1992b. "Quartiles, outliers and normality: Some Monte Carlo results." Pp. 92–95 in Joseph Hilbe (ed.) *Stata Technical Bulletin Reprints, Volume 1*. College Station, TX: Stata Press.

Hamilton, Lawrence C. 1996. *Data Analysis for Social Scientists*. Belmont, CA: Duxbury Press.

Hamilton, Lawrence C., Benjamin C. Brown and Rasmus Ole Rasmussen. 2003. "Local dimensions of climatic change: West Greenland's cod-to-shrimp transition." *Arctic* 56(3):271–282.

Hamilton, Lawrence C., Richard L. Haedrich and Cynthia M. Duncan. 2003. "Above and below the water: Social/ecological transformation in northwest Newfoundland." *Population and Environment* 25(2)101–121.

Hamilton, Lawrence C. and Carole L. Seyfrit. 1993. "Town-village contrasts in Alaskan youth aspirations." *Arctic* 46(3):255–263.

Hardin, James and Joseph Hilbe. 2001. *Generalized Linear Models and Extensions*. College Station, TX: Stata Press.

Hoaglin, David C., Frederick Mosteller and John W. Tukey (eds.). 1983. *Understanding Robust and Exploratory Data Analysis*. New York: John Wiley & Sons.

Hoaglin, David C., Frederick Mosteller and John W. Tukey (eds.). 1985. *Exploring Data Tables, Trends and Shape*. New York: John Wiley & Sons.

Hosmer, David W., Jr. and Stanley Lemeshow. 1999. *Applied Survival Analysis*. New York: John Wiley & Sons.

Hosmer, David W., Jr. and Stanley Lemeshow. 2000. *Applied Logistic Regression*, 2nd edition. New York: John Wiley & Sons.

Howell, David C. 1999. *Fundamental Statistics for the Behavioral Sciences*, 4th edition. Belmont, CA: Duxbury Press.

Howell, David C. 2002. *Statistical Methods for Psychology*, 5th edition. Belmont, CA: Duxbury Press.

Iman, Ronald L. 1994. *A Data-Based Approach to Statistics*. Belmont, CA: Duxbury Press.

Jentoft, Svein and Trond Kristoffersen. 1989. "Fishermen's co-management: The case of the Lofoten fishery." *Human Organization* 48(4):355–365.

Johnson, Anne M., Jane Wadsworth, Kaye Wellings, Sally Bradshaw and Julia Field. 1992. "Sexual lifestyles and HIV risk." *Nature* 360(3 December):410–412.

Johnston, Jack and John DiNardo. 1997. *Econometric Methods*, 4th edition. New York: McGraw-Hill.

Keller, Gerald, Brian Warrack and Henry Bartel. 2003. *Statistics for Management and Economics*, abbreviated 6th edition. Belmont, CA: Duxbury Press.

League of Conservation Voters. 1990. *The 1990 National Environmental Scorecard*. Washington, DC: League of Conservation Voters.

Lee, Elisa T. 1992. *Statistical Methods for Survival Data Analysis*, 2nd edition. New York: John Wiley & Sons.

Li, Guoying. 1985. "Robust regression." Pp. 281–343 in D. C. Hoaglin, F. Mosteller and J. W. Tukey (eds.) *Exploring Data Tables, Trends and Shape*. New York: John Wiley & Sons.

Long, J. Scott. 1997. *Regression Models for Categorical and Limited Dependent Variables*. Thousand Oaks, CA: Sage Publications.

Long, J. Scott and Jeremy Freese. 2003. *Regression Models for Categorical Outcomes Using Stata*, revised edition. College Station, TX: Stata Press.

MacKenzie, Donald. 1990. *Inventing Accuracy: A Historical Sociology of Nuclear Missile Guidance*. Cambridge, MA: MIT.

Mallows, C. L.. 1986. "Augmented partial residuals." *Technometrics* 28:313–319.

Mayewski, P. A., G. Holdsworth, M. J. Spencer, S. Whitlow, M. Twickler, M. C. Morrison, K. K. Ferland and L. D. Meeker. 1993. "Ice-core sulfate from three northern hemisphere sites: Source and temperature forcing implications." *Atmospheric Environment* 27A(17/18):2915–2919.

Mayewski, P. A., L. D. Meeker, S. Whitlow, M. S. Twickler, M. C. Morrison, P. Bloomfield, G. C. Bond, R. B. Alley, A. J. Gow, P. M. Grootes, D. A. Meese, M. Ram, K. C. Taylor and W. Wumkes. 1994. "Changes in atmospheric circulation and ocean ice cover over the North Atlantic during the last 41,000 years." *Science* 263:1747–1751.

McCullagh, D. W. Jr. and J. A. Nelder. 1989. *Generalized Linear Models*, 2nd edition. London: Chapman & Hall.

Nash, James and Lawrence Schwartz. 1987. "Computers and the writing process." *Collegiate Microcomputer* 5(1):45–48.

National Center for Education Statistics. 1992. *Digest of Education Statistics 1992*. Washington, DC: U.S. Government Printing Office.

National Center for Education Statistics. 1993. *Digest of Education Statistics 1993*. Washington, DC: U.S. Government Printing Office.

Newton, H. Joseph and Jane L. Harvill. 1997. *StatConcepts: A Visual Tour of Statistical Ideas*. Pacific Grove, CA: Duxbury Press.

Pagano, Marcello and Kim Gauvreau. 2000. *Principles of Biostatistics*, 2nd edition. Belmont, CA: Duxbury Press.

Rabe–Hesketh, Sophia and Brian Everitt. 2000. *A Handbook of Statistical Analysis Using Stata*, 2nd edition. Boca Raton, FL: Chapman & Hall.

Report of the Presidential Commission on the Space Shuttle Challenger Accident. 1986. Washington, DC.

Rosner, Bernard. 1995. *Fundamentals of Biostatistics*, 4th edition. Belmont, CA: Duxbury Press.

Selvin, Steve. 1995. *Practical Biostatistical Methods*. Belmont, CA: Duxbury Press.

Selvin, Steve. 1996. *Statistical Analysis of Epidemiologic Data*, 2nd edition. New York: Oxford University.

Seyfrit, Carole L.. 1993. *Hibernia's Generation: Social Impacts of Oil Development on Adolescents in Newfoundland*. St. John's: Institute of Social and Economic Research, Memorial University of Newfoundland.

Shumway, R. H. 1988. *Applied Statistical Time Series Analysis*. Upper Saddle River, NJ: Prentice–Hall.

Stata Corporation. 2005. *Getting Started with Stata for Macintosh*. College Station, TX: Stata Press.

Stata Corporation. 2005. *Getting Started with Stata for Unix*. College Station, TX: Stata Press.

Stata Corporation. 2005. *Getting Started with Stata for Windows*. College Station, TX: Stata Press.

Stata Corporation. 2005. *Mata Reference Manual*. College Station, TX: Stata Press.

Stata Corporation. 2005. *Stata Base Reference Manual* (3 volumes). College Station, TX: Stata Press.

Stata Corporation. 2005. *Stata Data Management Reference Manual.* College Station, TX: Stata Press.

Stata Corporation. 2005. *Stata Graphics Reference Manual.* College Station, TX: Stata Press.

Stata Corporation. 2005. *Stata Programming Reference Manual.* College Station, TX: Stata Press.

Stata Corporation. 2005. *Stata Longitudinal/Panel Data Reference Manual.* College Station, TX: Stata Press.

Stata Corporation. 2005. *Stata Multivariate Statistics Reference Manual.* College Station, TX: Stata Press.

Stata Corporation. 2005. *Stata Quick Reference and Index.* College Station, TX: Stata Press.

Stata Corporation. 2005. *Stata Survey Data Reference Manual.* College Station, TX: Stata Press.

Stata Corporation. 2005. *Stata Survival Analysis and Epidemiological Tables Reference Manual.* College Station, TX: Stata Press.

Stata Corporation. 2005. *Stata Time-Series Reference Manual.* College Station, TX: Stata Press.

Stata Corporation. 2005. *Stata User's Guide.* College Station, TX: Stata Press.

Stine, Robert and John Fox (eds.). 1997. *Statistical Computing Environments for Social Research.* Thousand Oaks, CA: Sage Publications.

Topliss, Brenda J. 2001. "Climate variability I: A conceptual approach to ocean–atmosphere feedback." In Abstracts for AGU Chapman Conference, The North Atlantic Oscillation, Nov. 28 – Dec. 1, 2000, Ourense, Spain.

Tufte, Edward R. 1997. *Visual Explanations: Images and Quantities, Evidence and Narratives.* Cheshire, CT: Graphics Press.

Tukey, John W. 1977. *Exploratory Data Analysis.* Reading, MA: Addison–Wesley.

Velleman, Paul F. 1982. "Applied Nonlinear Smoothing," pp.141–177 in Samuel Leinhardt (ed.) *Sociological Methodology 1982.* San Francisco: Jossey-Bass.

Velleman, Paul F. and David C. Hoaglin. 1981. *Applications, Basics and Computing of Exploratory Data Analysis.* Boston: Wadsworth.

Ward, Sally and Susan Ault. 1990. "AIDS knowledge, fear, and safe sex practices on campus." *Sociology and Social Research* 74(3):158–161.

Werner, Al. 1990. "Lichen growth rates for the northwest coast of Spitsbergen, Svalbard." *Arctic and Alpine Research* 22(2):129–140.

World Bank. 1987. *World Development Report 1987.* New York: Oxford University.

World Resources Institute. 1993. *The 1993 Information Please Environmental Almanac.* Boston: Houghton Mifflin.

Index

A

ac (autocorrelations), 339, 351–352

acprplot (augmented component-plus-residual plot), 197, 202–203

added-variable plot, 198, 201–202

ado-file (automatic do), 233–235, 362, 373–375

alpha (Cronbach's alpha reliability), 318–319

analysis of covariance (ANCOVA), 141–142, 153–154

analysis of variance (ANOVA)
    factorial, 142, 152–153, 156
    interaction effects, 142, 152–154, 156–157
    median, 253–255
    N-way, 152–153
    one-way, 142, 155
    predicted values, 155–158, 167
    regression model, 153–154, 249–256
    repeated-measures, 142
    robust, 249–256
    standard errors, 155–157, 167
    three-way, 142
    two-way, 142, 152–153, 156–157

anova, 142, 152–158, 167, 239

append, 13, 42–44

ARCH model (autoregressive conditional heteroskedasticity), 339

area plot, 86–87

args (arguments in program), 366–368

areg (absorb variables in regression), 179–180

ARIMA model (autoregressive integrated moving average), 339, 354–360

arithmetic operator, 26

artificial data, 14, 57–61, 241, 387–394

ASCII (text) file
    read data, 13–14, 39–42
    write data, 42
    write results (log file), 2–3, 6–7

autocode (create ordinal variables), 31, 37–38

autocorrelation, 339, 350–352, 357–358, 369–373

aweight (analytical weights), 54

axis label in graph, 66
    angle, 81–82
    format, 13, 24–25, 76, 305–306
    grid, 113–115
    suppress, 118, 129, 173

axis scale in graph, 66, 112–118

B

_b coefficients (regression), 230, 269, 273–274, 285, 356

band regression, 217–219

bar chart, 94–99, 147, 150–151

Bartlett's test for equal variances, 149–150

batch-mode program, 61

bcskew0 (transform to reduce skew), 129

beta weight (standardized regression coefficient), 160, 164–165

Bonferroni multiple-comparison test
    correlation matrix, 172–173
    one-way ANOVA, 150–151

bootstrap, 246, 315–316, 382–387, 389–394

box plot, 66, 90–91, 118–119, 147, 150–151, 389, 392

Box–Cox
   regression, 215, 226–227
   transformation, 129
Box–Pierce Q test (white noise), 341, 351, 354, 357–358
browse (Data Browser), 13
bs (bootstrap), 382–387
bsqreg (quantile regression with bootstrap), 240, 246, 389–394
by prefix, 121, 133–134

C

c chart (quality control), 105
caption in graph, 109–110
case identification number, 38–39
categorical variable, 35–39, 183–185
censored-normal regression, 264
centering to reduce multicollinearity, 212–214
chi-squared
   deviance (logistic regression), 271, 275–278
   equal variances in ANOVA, 149–150
   independence in cross-tabulation, 55, 130–133, 281
   likelihood-ratio in cross-tabulation, 130–131, 281
   likelihood-ratio in logistic regression, 267–268, 270, 272–273, 281
   probability plot, 105
   quantile plot, 105
ci (confidence interval), 124, 255
cii (immediate confidence interval), 124
classification table (logistic regression), 264, 270–272
clear (remove data from memory), 14–15, 23, 362
cluster analysis, 318–320, 329–338
coefficient of variation, 123–124
collapse, 52–53
color
   bar chart, 95–96
   pie chart, 92
   scatterplot symbols, 74
   shaded regions, 86
combine data files, 14, 42–47

combine graphs. *See* **graph combine**
comments in programs, 364, 369–370, 373–374
communality (factor analysis), 326
component-plus-residual plot, 197–198, 202–203
compress, 13, 40, 60–61
conditional effect plot, 230–232, 273–274, 284–287
confidence interval
   binomial, 124
   bootstrap, 383–384, 386
   mean, 124
   regression coefficients, 163
   regression line, 66, 85, 110–112, 160
   robust mean, 255
   Poisson, 124
constraint (linear constraints), 262
Cook and Weisberg heteroskedasticity test, 197
Cook's D, 158, 167, 197, 206–210
copy results, 4
correlation
   hypothesis test, 160, 172–173
   Kendall's tau, 131, 174–175
   matrix, 18, 59, 160, 171–174
   Pearson product-moment, 1, 18, 160, 171–173
   regression coefficient estimates, 214
   Spearman, 174
corrgram (autocorrelation), 339, 351, 357–358, 373
count-time data, 293–295
covariance
   regression coefficient estimates, 167, 173, 197, 214
   variables, 160, 173
COVRATIO, 167, 197, 206
Cox proportional hazard model, 290, 299–305
Cramer's V, 131
Cronbach's alpha, 318–319
cross-correlation, 353–354
cross-tabulation, 121, 130–136
ctset (define count-time data), 289, 293–294

cttost (convert count-time to survival-time data), 289, 294–295

cubic spline curve. *See* **graph twoway mspline**

cv (coefficient of variation), 123–124

D

Data Browser, 13
data dictionary, 41
Data Editor, 13, 15–16
data management, 12–63
database file, 41–42
date, 30, 266, 340–342
decode (numeric to string), 33–34
#delimit (end-of-line delimiter), 61, 116, 362
dendrogram, 319, 329, 331–337
describe (describe data), 3, 18
destring (string to numeric), 35
DFBETA, 158, 167, 197, 205–206, 208–210
DFITS, 167, 197, 206, 208–210
diagnostic statistics
    ANOVA, 158, 167
    logistic regression, 271, 274–278
    regression, 167, 196–214
Dickey–Fuller test, 340, 355–356
difference (time series), 349–350
display (show value onscreen), 31–32, 39, 211, 269
display format, 13, 24–25, 359
do-file, 60–61, 115–116, 361–362, 367–373
Do-File Editor, 60, 361
dot plot, 67, 95, 99–100, 150–151
drawnorm (normal variable), 13, 59
drop
    variable in memory, 22
    data in memory, 14–15, 23, 40, 56
    program in memory, 363, 373–375
dummy variable, 35–36, 176–185, 267
Durbin–Watson test, 158, 197, 350
dwstat (Durbin–Watson test), 197, 350

E

e-class, 381, 386

edit (Data Editor), 13, 15–16
effect coding for ANOVA, 250–251
efficiency of estimator, 393
egen, 33, 331, 340, 343
eigenvalue, 318–319, 321, 326
empirical orthogonal function (EOF), 325
Encapsulated Postscript (.eps) graph, 6, 116
encode (string to numeric), 13, 33–34
epidemiological tables, 288
error-bar plot, 143, 155–157
estimates store (hypothesis testing), 272–273, 278–279, 282–283
event-count model, 288, 290, 310–313
Exploratory Data Analysis (EDA), 124–126
exponential filter (time series), 343
exponential growth model, 216, 232–235
exponential regression (survival analysis), 305–307

F

factor analysis, 318–328
factor rotation, 318–319, 322–325
factor score, 318–319, 323–325
factorial ANOVA, 142, 152–153, 156
FAQs (frequently asked questions), 8
filter, 343
Fisher's exact test in cross-tabulation, 131
fixed and random effects, 162
foreach, 365
format
    axis label in graph, 76, 305–306
    input data, 40–41
    numerical display, 13, 24–25, 359
forvalues, 365
frequency table, 130–133, 138–139
frequency weights, 54–55, 66, 73–74, 120, 123, 138–140
function
    date, 30
    mathematical, 27–28
    probability, 28–30
    special, 31
    string, 31
fweight (frequency weights), 54–55, 73–74, 138–140

G

generalized linear modeling (GLM), 264, 291, 313–317

generate, 13, 23–26, 37, 39

gladder, 128

Gompertz growth model, 234–238

Goodman and Kruskal's gamma, 131

graph bar, 66–67, 94–99, 147

graph box, 66, 90–91, 118–119, 147, 389, 392

graph combine, 117–119, 147, 150–151, 222, 231–232

graph dot, 67, 95, 99–100, 150–151

graph export, 116

graph hbar, 97–98

graph hbox, 91, 150–151

graph matrix, 66, 77, 173–174

graph pie, 66, 92–94

graph twoway

    all types, 84–85

    overlays, 66, 85, 110–115, 344–345, 347–348

graph twoway area, 84, 85–86

graph twoway bar, 84

graph twoway connected, 5–6, 50–51, 66, 79–80, 83–84, 114–115, 157, 192–193

graph twoway dot, 85

graph twoway lfit, 66, 74, 85, 110, 168, 181

graph twoway lfitci, 85, 110–112, 170–171

graph twoway line, 66, 77–82, 112–115, 117, 221–222, 242, 244, 247, 344–345, 371

graph twoway lowess, 85, 88–89, 216, 219–221

graph twoway mband, 85, 216, 217–219

graph twoway mspline, 85, 182, 190, 218–219, 226, 287–288

graph twoway qfit, 85, 110, 190

graph twoway rarea, 84, 170

graph twoway rbar, 85

graph twoway rcap, 85, 89, 157

graph twoway scatter, 65–66, 72–77, 181–182, 277, 394

graph twoway spike, 84, 87–88, 347

graph use, 116

graph7, 65

gray scale, 86

greigen (graph eigenvalues), 318–319, 321–322

gsort (general sorting), 14

H

hat matrix, 167, 205–206, 210

hazard function, 290, 302, 307, 309

help, 7

help file, 7, 375–377

heteroskedasticity, 161, 197, 199–200, 223–224, 239, 256–258, 290, 315, 339

hettest (heteroskedasticity test), 197, 199–200

hierarchical linear models, 162

histogram, 65, 67–71, 385

Holt–Winters smoothing, 343

Huber/White robust standard errors, 160, 256–261

I

if qualifier, 13, 14, 19–23, 204–205, 209

if...else, 366

import data, 39–42

in qualifier, 14, 19–23, 166

incidence rate, 289–290, 293, 297, 309–310, 312

inequality, 21

infile (read ASCII data), 13–14, 40–42

infix (read fixed-format data), 41–42

influence

    logistic regression, 271, 274–278

    regression (OLS), 167, 196–198, 201, 204–208

    robust regression, 248

insert

    graph into document, 6

    table into document, 4

insheet (read spreadsheet data), 41–42

instrumental variables (2SLS), 161

interaction effect
    ANOVA, 142, 152–157, 250–253
    regression, 160, 180–185, 211–212, 259–261
interquartile range (IQR), 53, 91, 95, 103, 123–124, 126, 135
iteratively reweighted least squares (IRLS), 242
iweight (importance weights), 54

J
jackknife
    residuals, 167
    standard errors, 314–317

K
Kaplan–Meier survivor function, 289–290, 295–298
keep (keep variable or observation), 23, 173
Kendall's tau, 131, 174–175
kernel density, 65, 70, 85
Kruskal–Wallis test, 142, 151–152
kurtosis, 122–124, 126–127

L
L-estimator, 243
label data, 18
label define, 26
label values, 25–26
label variable, 16, 18
ladder of powers, 127–129
lag (time series), 349–350
lead (time series), 349–350
legend in graph, 78, 81, 112, 114–115, 157, 221, 344
letter-value display, 125–126
leverage, 158, 159, 167, 196, 198, 201–206, 210, 229, 246–248
leverage-vs.-squared-residuals plot, 198, 203–204
lfit (fit of logistic model), 264
likelihood-ratio chi-squared. *See* chi-squared

line in graph
    pattern, 81–82, 84, 115
    width, 221, 344, 371
line plot, 77–84
link function (GLM), 291, 313–317
list, 3–4, 14, 17, 19, 49, 54, 265
log, 2–3
log file, 2–3, 6–7
logarithm, 27, 127–129, 223–229
logical operator, 20
logistic growth model, 216, 233–234
logistic regression, 262–287
logistic (logistic regression), 185, 262–264, 269–278
logit (logistic regression), 267–269
looping, 365–366
lowess smoothing, 88–89, 216, 219–222
lroc (logistic ROC), 264
lrtest (likelihood-ratio test), 272–273, 278–279, 282–283
lsens (logistic sensitivity graph), 264
lstat (logistic classification table), 264, 270–272
lvr2plot (leverage-vs.-squared-residuals plot), 198, 203–204

M
M-estimator, 243
macro, 235, 334, 363, 365, 367, 370, 387
Mann–Whitney *U* test, 142, 148–149, 152
margin in graph, 110, 113, 117–118, 192–193
marker label in graph, 66, 75–76, 202, 204
marker symbol in graph, 66, 73–75, 84, 100, 183, 277
marksample, 368–369
matched-pairs test, 143, 145–146
matrix algebra, 378–382
mean, 122–124, 126,135–137, 139–140, 143–158, 387–389
median, 90–91, 122–124, 126, 135–137, 387–389
median regression. *See* quantile regression
memory, 14, 61–63
merge, 14, 44–50
missing value, 13–16, 21, 37–38

Monte Carlo, 126, 246, 387–394
moving average
    filter, 340, 343–344
    time series model, 354–360
multicollinearity, 210–214
multinomial logistic regression, 264, 278, 280–287
multiple-comparison test
    correlation matrix, 172–173
    one-way ANOVA, 150–151

N

negative exponential growth model, 233
nolabel option, 32–34
nonlinear regression, 216, 232–238
nonlinear smoothing, 340–341, 343–346
normal distribution
    artificial data, 13, 59, 241
    curve, 65
    test for, 126–129
normal probability plot. *See* quantile–normal plot
numerical variables, 16, 20, 122

O

OBDC (Open Database Connectivity), 42
odds ratio. *See* logistic regression
observation number, 38–39
omitted-variables test, 197, 199
one-sample *t* test, 143–146
one-way ANOVA, 149–152
open file, 2
order (order variables in data), 19
ordered logistic regression, 278–280
ordinal variable, 35–36
outfile (write ASCII data), 42
outlier, 126, 239–248, 344, 388–394
overlay twoway graphs, 110–115

P

p chart (quality control), 105–107
paired-difference test, 143, 145–146
panel data, 161, 191–195
partial autocorrelation, 339–340, 352
partial regression plot. *See* added-variable plot

Pearson correlation, 5, 19, 160, 171–173
percentiles, 122–124, 136
periodogram, 340
Phillips–Perron test, 355
pie chart, 66, 92–94
placement (legend in graph), 114–115
poisgof (Poisson goodness of fit test), 310–311
Poisson regression, 290–291, 309–313, 317
polynomial regression, 188–191
Portable Network Graphics (.png) graph, 6, 116
Postscript (.ps or .eps) graph, 6, 116
Prais–Winsten regression, 340, 359–360
predict (predicted values, residuals, diagnostics)
    anova, 155–158, 167
    arima, 357
    logistic, 264, 268–271, 284
    regress, 159, 165–167, 190, 196–197, 205–210, 216, 233
principal components, 318–325
print graph, 6
print results, 4
probit regression, 262–263, 314
program, 362–363
promax rotation, 319, 322–325
pweight (probability or sampling weights), 54–56
pwcorr (pairwise Pearson correlation), 160, 172–173, 174–175

Q

qladder, 128–129
quality-control graphs, 67, 105–108
quantile
    defined, 102
    quantile plot, 102–103
    quantile–normal plot, 67, 104
    quantile–quantile plot, 104–105
    regression, 239–256, 389–394
quartile, 91, 125–126
quietly, 175, 182, 188

R

r chart (quality control), 67, 106, 108

r-class, 381, 387, 390

Ramsey specification error test (RESET), 197

random data, 56–60, 241, 387–394

random number, 30, 56–59, 241

random sample, 14, 60

range (create data over range), 236

range plot, 89

range standardization, 334–335

rank, 32

rank-sum test, 142, 148–149, 152

real function, 35–36

regress (linear regression), 159–165, 239, 386, 389–394

regression

    absorb categorical variable, 179–180

    beta weight (standardized regression coefficient), 160, 164–165

    censored-normal, 264

    confidence interval, 110–112, 163, 169–171

    constant, 163

    curvilinear, 189–191, 216, 223–232

    diagnostics, 167, 196–214

    dummy variable, 176–185

    hypothesis test, 160, 175–176

    instrumental variable, 161

    line, 67, 110–112, 159–160, 168–171, 190, 242, 244, 247

    logistic, 262–287

    multinomial logistic, 264, 278, 280–287

    multiple, 164–165

    no constant, 163

    nonlinear, 232–238

    ordered logistic, 278–280

    ordinary least squares (OLS), 159–165

    Poisson, 290–291, 309–313, 317

    polynomial, 188–191

    predicted value, 165–167, 169

    probit, 262–263, 314

    residual, 165–167, 169, 205–207

    robust, 239–256, 389–394

    robust standard errors, 256–261

    stepwise, 161, 186–188

    tobit, 188, 263

transformed variables, 189–191, 216, 223–232

two-stage least squares (2SLS), 161

weighted least squares (WLS), 161, 245

relational operator, 20

relative risk ratio, 264, 281–284

rename, 16, 17

replace, 16, 25–26, 33

RESET (Ramsey test), 197

reshape, 49–52

residual, 159–160, 167, 200–208

residual-vs.-fitted (predicted values) plot, 160, 169, 188–191, 198, 200

retrieve graph, 116

robust

    anova, 249–255

    mean, 255

    regression, 239–256

    standard errors and variance, 256–261

ROC curve (receiver operating characteristic), 264

rotation (factor analysis), 318–319, 322–325

rough, 345

rreg (robust regression), 239–256, 389–394

rvfplot (residual-vs.-fitted plot), 160, 188–191, 198, 200

rvpplot (residual-vs.-predictor plot

S

sample (draw random sample), 14, 60

sampling weights, 55–56

sandwich estimator of variance, 160, 256–261

SAS data files, 42

save (save dataset), 14, 16, 23

save graph, 6

saveold (save dataset in previous Stata format), 14

scatterplot. *Also see* **graph twoway scatter**

    axis labels, 66, 72

    basic, 66–67

    marker labels, 67, 74–75, 202–204

marker symbols, 72–73, 119, 182–183
matrix, 66, 77, 173–174
weighting, 66, 74–75, 207–208
with regression line, 66, 110–112, 159–160, 181–182
Scheffé multiple-comparison test, 150–151
score (factor scores), 318–319, 323–325
scree graph (eigenvalues), 318–319, 321–322
search, 8–9
seasonal difference (time series), 349–350
serrbar (standard-error bar plot), 143, 155–157
set memory, 14, 62–63
shading
color, 86
intensity, 91
Shapiro–Francia test, 127
Shapiro–Wilk test, 127
shewart, 106
Šidák multiple-comparison test, 150, 172–173
sign test, 144–145
signed-rank test, 143 146
skewness, 122–124, 126–127
sktest (skewness–kurtosis test), 126–127, 383
slope dummy variable, 180
SMCL (Stata Markup and Control Language), 376–377
smoothing, 340–341, 343–346
sort, 14, 19, 21–22, 166
Spearman correlation, 174–175
spectral density, 340
spike plot, 84, 87–88, 347
spreadsheet data, 41–42
SPSS data files, 42
standard deviation, 122–124, 126, 135

standard error
ANOVA, 155–157
bootstrap. *See* bootstrap
mean, 124
regression prediction, 167, 169–171
robust (Huber/White), 160, 256–261
standardized regression coefficient, 160, 164–165
standardized variable, 32, 331
Stat/Transfer, 42
Stata Journal, 10–11
Statalist online forum, 10
stationary time series, 340, 355–356
stcox (Cox hazard model), 290, 299–303
stcurve (survival analysis graphs), 290, 307
stdes (describe survival-time data), 289, 292–293
stem-and-leaf display, 124–125
stepwise regression, 161, 186–188
stphplot, 290
streg (survival-analysis regression), 290, 305–309
string to numeric, 32–35
string variable, 17, 40–41
sts generate (generate survivor function), 290
sts graph (graph survivor function), 289, 296, 298
sts list (list survivor function), 290
sts test (test survivor function), 290, 298
stset (define survival-time data), 289, 291–292, 297
stsum (summarize survival-time data), 289, 293, 297
studentized residual, 167, 205, 207
subscript, 39–40, 343
summarize (summary statistics), 2, 17, 20, 31–32, 90–91, 120–124, 383
sunflower plot, 74–75
survey sampling weights, 55–56, 161, 263
survival analysis, 288–309
svy: regress (survey data regression), 161
svyset (survey data definition), 56
sw (stepwise model fitting), 186–188
symmetry plot, 100, 102

syntax (programming), 368–369

T

t test
    correlation coefficient, 160, 172–173
    means, 143–149
    robust means, 255
    unequal variance, 148
table, 121, 134–136, 152
tabstat, 120, 123–124
tabulate, 4, 15, 36–37, 56, 121, 130–133, 136
technical support, 9
test (hypothesis test for model), 160, 175–176, 312
text in graph, 109–110, 113, 222
time plot, 77–84, 343–348
time series, 339–360
tin (times in), 346–347, 350, 359
title in graph, 109–110, 112–113
tobit regression, 188, 263
transfer data, 42
transform variable, 126–129, 189–190, 216
transpose data, 47–49
tree diagram, 319, 329, 331–337
tsset (define time series data), 340, 342, 346
tssmooth (time series smoothing), 340–341, 343–346
ttest, 143–149, 392
Tukey, John, 124
twithin (times within), 346–347
two-sample test, 146–149
two-stage least squares (2SLS), 161

U

unequal variance in *t* test, 143, 148–149
uniform (random number generator), 30, 56–58, 241
unit root, 355–356
use, 2–3, 15

V

variance, 122–124, 135, 214
variance inflation factor, 197, 211–212
varimax rotation, 319, 322–325

version, 364
W
web site, 9
Weibull regression (survival analysis), 305, 307–399
weighted least squares (WLS), 161, 245
weights, 55–57, 74–75, 122–124, 138–140, 161
Welsch's distance, 167, 206–210
which, 374
while, 365–366
white noise, 341, 351, 354, 357–358
Wilcoxon rank-sum test, 142, 148–149, 152
Wilcoxon signed-rank test, 143, 146
Windows metafile (.wmf or .emf) graph, 6, 116
wntest (Box–Pierce white noise Q test), 341
word processor
    insert Stata graph into, 6
    insert Stata table into, 4

X

x axis in graph. *See* axis label in graph, axis scale in graph
x-bar chart (quality control), 106–108
xcorr (cross-correlation), 353–354
xi (expanded interaction terms), 160, 183–185
xpose (transpose data), 48–49
xtmixed (multilevel mixed-effect models), 162
xtreg (panel data regression), 161, 191–195

Y

y axis in graph. *See* axis label in graph, axis scale in graph

Z

z score (standardized variable), 32, 331